THE
EVOLUTION
UNDERGROUND

THE
EVOLUTION
UNDERGROUND

**BURROWS, BUNKERS, AND THE MARVELOUS
SUBTERRANEAN WORLD BENEATH OUR FEET**

ANTHONY J. MARTIN

PEGASUS BOOKS
NEW YORK LONDON

THE EVOLUTION UNDERGROUND

Pegasus Books Ltd.
148 W 37th Street, 13th Floor
New York, NY 10018

First Pegasus Books cloth edition February 2017

Interior design by Maria Fernandez

Library of Congress Cataloging-in-Publication Data is available.

ISBN: 978-1-68177-312-4

10 9 8 7 6 5 4 3 2 1

Printed in the United States of America
Distributed by W. W. Norton & Company

To my mother, Veronica, who loved books.

"To us, 'up' is a 'good' direction. Not so, or not necessarily so, to an ant. 'Up' is where the food comes from, to be sure; but 'down' is where security, peace, and home are to be found. 'Up' is the scorching sun; freezing night; no shelter in the beloved tunnels; exile; death."

—Ursula K. Le Guin (1974),
*"The Author of the Acacia Seeds" and
Other Extracts from the Journal of the
Association of Therolinguistics*

Contents

CHAPTER 1

The Wondrous World
of Burrows

Into the Dragon's Lair

The alligator den had a big surprise for us. Its occupant was hidden inside a dark space down an inclined tunnel, its entrance denoted by a meter-wide, half-moon-shaped hole in the middle of a pine forest. The alligator's presence was verified only by a rumbling growl, followed by an openmouthed hiss. The burrow chamber added resonance to these sounds, turning an already spooky situation into a downright portentous one. This sonic combo, intended as a warning, worked quite well in that respect, persuading all of us to issue a collective "Whoa!," take a few steps back, and assess our situation.

It was yet another moment in my teaching career when I wondered how many other professors must concern themselves with apex predators showing up in their classrooms. Nonetheless, on the plus side, if any of my students had been bored with the course material, they were now very much engaged, perhaps even wondering if this alligator-related incident would be covered on the next exam.

At that moment, my undergraduate students, a faculty colleague, and I were deep in the interior of St. Catherines Island on the Georgia coast and on our sixth day of a March 2013 spring-break field course to the Georgia barrier islands. St. Catherines is an undeveloped island used mostly for scientific research and the fifth island we had visited thus far on our trip. My colleague was geographer Michael Page, who had joined us the previous day; he had been on St. Catherines with me once before to map alligator dens in July 2012. During that time, we documented dozens of dens next to water bodies, many of which hosted alligators. In some instances, we affirmed the identity and purpose of these big holes by witnessing alligators swimming or otherwise dashing into them. With other dens, we spotted tracks and tail-drag traces crisscrossing their entrances, effectively telling us not to get any closer.

This den, though, had no such fresh warning traces outside of it, meaning the contentious alligator inside had been there for a while. When the growl-hiss greeting broadcast from the den, Michael was standing above and behind the den, whereas I was almost directly in front. We had already seen about ten alligator dens that morning, all of them empty. This lulled us into a false sense of security, a confirmation bias that affected our better judgment when approaching this one. My prejudice was further bolstered by a memory of this very same burrow, which had had absolutely no sign of an alligator in it when Michael and I had examined it the previous summer. During that visit, we photographed and measured each den we encountered, as well as recorded their locations with a global positioning system (GPS) unit. But we remembered this specific den because it had the largest entrance of any we had seen, at more than a meter (3.3 feet) wide and 40 centimeters (16 inches) tall. It was big enough that I could have crawled into it, had I been so stupid.

Size aside, what really made this den memorable was its location, which was in the middle of the woods. As everyone should know, alligators normally live in water. Yet no lakes, ponds, or streams were within sight, and the forest floor around the den was carpeted with dry pine needles. Still, this den and several others nearby were located on the bank of what used to be a human-made canal. Thus Michael and I quite reasonably surmised the canal had been submerged sometime in the past—perhaps decades ago—which encouraged alligators to move into the neighborhood and dig dens. Later, drought and other changes in local hydrology must have altered the water supply in this area. So just as in any area where humans lack the basic means for survival—like nearby coffee shops offering pumpkin-spice lattes and free wi-fi—the alligators moved somewhere else.

In this instance, I had just begun explaining to our students how this was yet another example of an abandoned den made by previous generations of now-dead alligators. This meant it only served as the trace of a former alligator in what used to be an aquatic environment that later turned into a terrestrial environment. A fine hypothesis it was, but one so rudely proven wrong by the live, two-meter-long, body-armored, and bad-tempered saurian residing in its so-called abandoned home.

To my students' credit, they had started us on the path to falsifying the notion that this big hole was gator-less. Once spotted, I greeted it like an old friend, enthusiastically striding toward its opening before delivering my little lecture to the assembled group. A few students stood back, impressed by the size of the hole and staring into its underground darkness, a seemingly bottomless pit of mystery. The whirring of zoom lenses and digitally rendered shutter sounds behind me told me they were taking plenty of pictures. I was pleased that they found this burrow as interesting as I did.

Suddenly, I was jarred out of my educational reverie when one of students said, "I see teeth in there."

"Teeth?" I asked.

"Yeah," she said, and others nodded agreement. She was looking into the den, while two others looked anxiously back and forth between their camera view-screens and the den, testing what they either observed or imagined.

"What kind of teeth?" I asked. Like a typical paleontologist, I was thinking of a disembodied skull or jaw, instead of a breathing animal bearing (or baring) those teeth.

"I don't know. Could it be a snake?"

"Sure, that's possible." I had seen alligator dens with snakes in them before. Also, unlike certain fictional archaeologists, I like snakes and relished the thought that one might be in the burrow. "But you probably wouldn't be seeing its teeth," I said, as I became more confused about this unexpected shift in the lesson plan for my students. Puzzled, I stepped closer to the entrance, which is when I received an admonition from their "classmate" who had somehow (but understandably) made it past the registrar without paying tuition.

I looked up at Michael. The disbelief probably still registered on my face, but my expression also must have wordlessly asked him, "What do we do now?"

With his GPS unit in one hand, Michael smiled, and with barely suppressed glee at the absurdity of our predicament he said, "Guess we have to mark that one as occupied."

Dens: The Swiss Army Knives of Alligator Survival

This alligator incident marked the beginning of an idea for me that had far wider implications than field-trip hijinks and close encounters with potentially dangerous foes. This idea stems from knowing how alligators descended from a lineage of crocodilians and their kin that were alive more than 200 million years ago (abbreviated *mya*), when dinosaurs were still stomping, fighting, nesting, eating, mating, peeing, pooping, and otherwise leaving their mark on the world. Yet when a meteorite smacked into the earth about 66 *mya*, this disaster and other problems caused a devastating worldwide crisis for life everywhere, whether in the oceans or on land. As a result, all of the dinosaurs that did not have the good sense to be birds died, leaving only their bones and traces. Meanwhile, alligators and other crocodilians carried on, as did a number of turtles, lizards, snakes, fishes, amphibians, insects, earthworms, mammals, and other animals we now accept as normal parts of our modern world. What did they have in their genetic or behavioral repertoire that could have helped them survive, but not dinosaurs?

Let's think about birds first. As everyone with a five-year-old child knows by now, not all dinosaur lineages went extinct, as some evolved into modern birds. The first birds descended from theropod dinosaurs about 160 *mya*; most theropods were two-legged carnivores, such as cinema stars *Velociraptor* and *Tyrannosaurus*. So far, paleontologists have discovered about forty species of feathered theropods, enough that we can now confidently assert that most (if not all) theropod dinosaurs from the Jurassic and Cretaceous Periods (about 160–66 *mya*) were feathered. (This also means the *Jurassic Park* films, including *Jurassic World*, should have been rated R, because all of the raptors and other theropods portrayed in them were naked.) Anyway, feathered and flighted avian dinosaurs somehow survived a mass extinction that took out all of their relatives 66 *mya*.

Interestingly, on that very same island of St. Catherines and others off the Georgia coast, my students and I had witnessed interactions between birds and crocodilians that made us feel like we were back in the Cretaceous. Some island interiors held ponds with small islands, where tall wading birds—such as storks, herons, and egrets—built their nests on tree branches, well above land and water surfaces. In addition to the parents, their nests were protected by what seemed like unlikely allies: alligators. Because alligators were swimming in the ponds and staying nearby in dens, they served as convincing deterrents to raccoons or any other mammals that thought they could raid a bird's nest and enjoy scrambled eggs for breakfast. This deal, however, was a Faustian bargain. As a Mafia-like payment, if a hatchling fell out of the nest and onto an island or into a pond, this hapless baby bird became an easy meal for any alligator lucky enough to be in the right place at the right time. Yet this brutal compensation is a much better deal for parent birds than having an entire egg clutch consumed by ruthless raccoons. Hence these birds and alligators may have coevolved their respective behaviors, with mutual arrangements struck by their ancestors millions of years ago.

So now let's focus on the alligators, and specifically those on St. Catherines Island. At the time I visited there with my students in 2013, the alligators had been enduring a drought for the previous few years, part of a more severe overall pattern caused by less rainfall on the island during the past several decades. This meant the normal habitats for alligators—freshwater ponds and other wetlands—had shrunk, leaving them with fewer places to stay and make a living by killing fish and other animals. One might expect such low water supplies and dire conditions would have left alligator skeletons strewn throughout a desiccated landscape. Nonetheless, they were still very much present, active, and striking fear in more than just fish, continuing to survive by spending more time in dens. Alligators likely dug these big burrows along the edges of ponds, canals, and other wetlands during times of plentiful water; the dens then remained once the wetlands vanished and were succeeded by grasslands and forests. Yet alligators could still move back into these dens and use those that intersected the groundwater table below the surface.

Thus these underground "wetlands" served the purpose of keeping alligator skins moist, while conferring many other benefits. For instance, given that these dens held fresh water on an island where such supplies had become more precious, they also provided a tempting source of water for other animals, such as mammals and birds. Their thirst then neatly delivered the alligators' groceries to them. All the alligators had to do was wait just within burrow entrances and snatch whatever looked large enough to eat. My students and I found evidence of this ambush strategy on that same field trip: two dens that had fresh carcasses outside of them. One den had just the remains of a vulture, its bones and feathers stuck in the entrance, whereas another had the remains of a raccoon about a meter away from its opening. A meter farther from the raccoon, though, was another dead vulture; the still-red bloodiness of both bodies suggested they had been killed in quick succession. So it was easy to think how the raccoon, once dispatched and only partially eaten, would have attracted the attention of vultures, supplying the den-dwelling alligator with a two-course meal. Similarly, older and long-abandoned dens in parts of the island had bone collections adorning their fronts, usually consisting of a jumbled mix of deer and raccoon parts. A few of these bones even held round, conical holes showing exactly where alligator teeth had punched into them. All of this trace evidence told us the alligators could switch from aquatic to terrestrial predation if necessary, like a shark deciding it was going to turn into a lion. This surprising behavioral transformation and adaptability in alligators was made possible through their dens, which during times of environmental change became all-purpose hunting lodges.

In addition to keeping their occupants wet and enabling them to ambush prey, dens served another important purpose, which was protection. For example, drought conditions on St. Catherines and other Georgia islands had increased the fuel load of dry pine needles and dead wood in island interiors, which bolstered the likelihood of lightning-caused fires racing through forests and grasslands alike. Sure enough, one such fire in the summer of 2012 scorched part of St. Catherines, with the blaze blackening a marshy area on the east side of the island. This same place had enough alligator dens in it that the island manager, Royce Hayes, had nicknamed it the "Nerve Center," as in, you get really nervous when surrounded by so many alligator dens. The day after the fire had run its course, Royce and his wife, Christa, went there to survey its effects on the marsh, including the local fauna. There they were amazed to find fresh alligator tracks on top of a wildfire ash layer, made by alligators that apparently stayed safe and secure in their dens during the fire, and then emerged for a little walkabout.

If this use of alligator dens doesn't impress as a form of protection, then think of alligator babies. That's right: cute little alligator babies, which easily fit on the palm of an average adult human hand when newly hatched. Only later do they grow up to become monsters—much like how human children eventually turn into teenagers. Despite being so adorable, nearly everything bigger than a baby alligator—including other alligators—regards it as an appetizer. Hence these little tykes need defending, which is partially provided by their overprotective mothers, but also by dens. Alligator mothers stay with their offspring for as long as two years after they hatch, and if dens are nearby, they will use these not only as places with plenty of fresh water (which baby alligators need), but also for hiding the kids from trouble.

I have seen (or caused) the latter behavior many times on St. Catherines and other Georgia islands. My walking near a den or a small pond with baby alligators sets off their alarm calls, which consist of a series of high-pitched grunts: Imagine choking Kermit the Frog, only multiplied by a dozen. These noises send a clear signal that you could die, because a big momma gator is close by and now knows her babies are in danger. Once the babies sound the alarm, the mother either crawls or swims into the den headfirst, leading the way for her wee ones. Still grunting, they align and scramble together toward and into the den to be with mum. By then, she will have turned around in a large chamber very close to the burrow entrance, ready to defend her offspring against anything that might try to bring them harm, human or otherwise.

I have often wondered whether this reaction in alligators, triggered by an upright biped like myself, is an innate response to wading birds, such as the previously mentioned herons, egrets, and storks. These birds—perhaps avenging alligator-caused deaths of their chicks—are also known to hunt baby alligators. Still, not one is willing to approach a den with a large adult alligator in it, and instead will grudgingly respect its family values. In several instances, I have seen the mother's massive head just behind the den entrance, almost daring you to get closer and test her evolutionary legacy.

Dens protect alligators of all ages in another way, which is from cold winters and hot summers. As most people know, alligators are "cold-blooded," or if you want to impress your friends with your scientific vocabulary, you can say they are ectothermic. This means they cannot regulate their own body temperatures and instead have to rely on their surrounding environment to keep themselves within a range that allows for life to go on. For alligators, the ideal is about 27–32°C (80–90°F); any higher or lower than this range, they have problems. Surprisingly, though, alligators can live farther away from the equator than any modern crocodilians. (Alligators and crocodiles belong to the same evolutionarily related group, or clade, named *Crocodylia* or *Crocodilia*, with spelling depending on who you ask.) In my experience, when you play word-association games with people and say "alligator," people will respond with "Florida." But in North America, these big reptiles can live as far north as North Carolina, and how they accomplish this trick is by using dens. These burrows bestow a Goldilocks effect by averaging the temperatures of cold winters and hot summers, making it just right all year. On the Georgia coast, where summer temperatures can easily exceed 32°C (90°F) and water temperatures approach those of hot tubs, alligators duck into dens to cool down. Conversely, I have also seen large alligators out sunning themselves on near-freezing days in December, implying that a den was close by and kept them warm enough to get out for a little solar therapy. Similarly, cave enthusiasts (spelunkers) understand the mollifying effects of being underground quite well, enjoying what feels like cool and warm cave interiors during summer and winter (respectively), when the cave is actually the same temperature all year.

All of this brings us back to the unexpected burrow occupant my class and I encountered, while neatly answering the perfectly reasonable scientific inquiry: "What the heck was a large adult alligator doing in the middle of a forest?" Remember how I said we were visiting in March? The timing of our trip suggests this big critter had likely entered the den sometime during the winter, when temperatures dipped low enough and long enough that it needed to stay sufficiently warm to survive. We were there at the cusp of spring on the Georgia coast, when outdoor temperatures were edging closer to the alligator heaven of 27–32°C (80–90°F) instead of the crystalline cold of winter. Yet the weather in early March, with average lows around 10°C (50°F), was still not quite warm enough to coax this one out of its temporary refuge. Case in point: Photographs of my students from that day show them bundled up, some with hoods covering their heads. What was the year-round average temperature in this part of Georgia? More like 20°C (67°F), meaning if you lived underground all year, there would be no need to set a thermostat, as it would stay that way all of the time. While the weather outside was dipping below freezing, this big alligator and many of its compatriots had probably overwintered in dens that remained close to 70°F all winter. What cavers and other underground enthusiasts have learned through experience, alligators figured out through natural selection.

Given the multifaceted uses of dens, it is now easy to see how a simple statement can be made about the role dens have played in the evolutionary history of alligators: no dens, no alligators. This bold statement is backed up by a quick look at alligators' living close relatives, such as the Chinese alligator (*Alligator sinensis*), which dig extensive tunnels in riverbanks to make dens, as well as other crocodilians that burrow to survive. In fact, more than half of all crocodilian species (14 out of 23) dig and live in burrows during times of environmental stress, such as droughts. Then consider how many salamanders, frogs, toads, turtles, lizards, snakes, and other ectothermic animals live at far higher latitudes than alligators. Nearly all of these animals accomplish this feat by spending winters underground or otherwise protected. Even self-heating endotherms—namely, birds and mammals—decrease their chances of freezing or sweltering by seeking shelter below ground surfaces. In short, these animals can't move up unless they get down.

The Evolution Underground

These insights we gain from studying alligators' dens suggest that at least some of the ancestors of modern-day alligators and crocodiles, and perhaps their bird companions, likely used burrows to get past the environmental hazards of the past. For an example of burrowing birds, just think of those charming, family-oriented, unstoppable krill-eating marchers, penguins. All penguin species live in the Southern Hemisphere and all polar bears live in the Northern Hemisphere, meaning that the only place you would ever see a polar bear eating a penguin is in a badly managed zoo. Yet despite the stereotype of penguins living only in Antarctica and huddling together for warmth there, most species actually live in a wide variety of environments. Moreover, greater than half of all penguin species make and live in burrows, which they use for—you guessed it—raising young, protecting themselves and their chicks from predators, and avoiding the harsh conditions of their outside environments. (Incidentally, the oldest known fossil penguins date from about 62 *mya*, just after the extinction of their non-avian dinosaur cousins. Coincidence? Maybe, but it is good food for future thought.) So alligator dens are by no means a unique instance of burrows allowing their makers to survive long enough to pass on genes to the next generation, while also enabling gene-passers to do more than just that. For many animals, burrows save and extend lives, while also serving as the places animal families call home.

Keeping this "burrow equals survival" theme in mind, and just in case you are still enthralled with the alligator-crocodile-bird success story of out-surviving non-avian dinosaurs, realize that this is not nearly as impressive as knowing how burrows contributed to the lineage you see reflected in your mirror every morning. Many mammals are fabulous burrowers, and this ability goes back even further into the geologic past than alligators, crocodiles, and birds. Ancestors of these furry vertebrates, called mammaliforms, evolved toward the end of the Triassic Period at about 220 *mya*, which was just after the start of the dinosaurs. The ancestors of mammaliforms, synapsid reptiles, originated even farther back in time, during the Carboniferous Period, more than 300 *mya*.

Once evolved, synapsids, such as Dimetrodon, were terrifically successful, adapting to and dominating land environments throughout the Permian Period (about 300–250 *mya*). Sadly (for them), nearly all went extinct at the end of the Permian, a time sometimes called "The Great Dying" because of how extreme global warming and other factors caused 95% of all species to wave good-bye to their evaporated gene pools. Notice I said "nearly," which implies that a few made it into the next period, the Triassic. From these surviving synapsids, mammals evolved, and their descendants somehow made it past another mass extinction at the end of the Triassic, then were a constant presence throughout the heyday of the non-avian dinosaurs: Surviving, but not necessarily thriving. Then, once the dinosaurs died out in the next mass extinction at the end of the Cretaceous Period, about 66 *mya*, mammals really took off. This success led to our own primate lineage, some of which learned how to control fire, track game animals, identify useful plants, map the heavens, and, finally, flirt with emoticons.

How did mammals and their synapsid ancestors continue to persist and exist after three mass extinctions? One factor they had in common was the ability to make or otherwise occupy burrows. As will be explained later, the synapsids that made it past the end-Permian extinction were burrowers, enabling these toughies to endure the most horrific conditions the Earth could throw at them. Following that, mammals from the Jurassic Period on were (and many still are) burrowers. This makes sense if you imagine yourself the size of a shrew and living in environments where dinosaurs are everywhere. Some want to eat you, while others will carelessly step on you and carry your squashed remains like chewing gum on their feet for days. Oh, you say you live in deep burrows where no dinosaurs can find you or compress you into two dimensions? Yes, that will do nicely. Even better, you now also have the means for escaping global cooling, warming, drought, fires, storms, or other natural nastiness happening in the outside world. Congratulations, shrew-sized mammal: You win the survival sweepstakes, and one tiny branch of your descendants eventually gets to a point where it can discuss how you outlived the dinosaurs.

Burrows do not just start with synapsids and mammals, though, but also go much farther back in time as a tool for survival. For instance, during the Devonian and Carboniferous Periods (420–360 *mya*), lungfishes and amphibians were also digging down and living in burrows. Skeletons of these animals have even been found in their fossil burrows, connecting this behavior with modern-day burrowing lungfishes, as well as with salamanders, frogs, toads, and other amphibians that do the same. Burrowing behaviors enable these water-dependent animals to live in deserts or avoid the worst effects of droughts. Once self-buried, some lungfishes, frogs, and toads can stay underground and become torpid for months or years, popping out once water becomes more plentiful. Granted, lungfishes and amphibians fossilized in their burrows did not survive whatever fate entombed them. Yet enough of their relatives did and then bequeathed burrowing abilities to future generations, which is all that matters in evolution. Moreover, all of these animals descended from water-dwellers that flopped, slithered, crawled, or otherwise landed on foreign shores. How did these aquatic animals manage to overcome the desiccating effects of land environments after emerging from the water? Burrows certainly would have helped.

Much later, vertebrate burrows of all sizes and shapes also provided microhabitats for plenty of other species, which today are best exemplified by gopher tortoises and their homes. These seemingly unimpressive tortoises, which do not get much bigger around than a typical dinner plate, are incredible diggers, hollowing out tunnels that can be more than 10 meters (33 feet) long and 3 meters

(10 feet) deep to keep themselves out of harm's way. Their lengthy tunnels can also have nearly 400 species cohabitating in them, with at least a few of these species having evolved their own specialized niches over many generations of burrows. The underground "rain forests" of biodiversity in gopher tortoise burrows hint at the importance of vertebrate burrows for maintaining life's balance in many ecosystems today.

But enough about vertebrates: What about the real overlords of the earth, such as worms and insects? How about other spineless animals that have adapted to nearly every terrestrial and marine environment? Do modern invertebrates live in burrows, and did their ancestors also live in burrows? Of course they do and did, as attested by anyone who owns a yard, strolled through a park, walked along seashores, or sat on an ant nest. Many of these burrows left a remarkable record of the evolutionary history of animals going back more than 550 *mya*, as they made transitions from surface living to deep burrowing, and as they moved from deep-sea environments to shallower sea bottoms, and from the sea to freshwater ponds and streams, and from the water to land. The bigger picture behind these everyday observations of many holes in the ground, however, is that the long history of these burrowing invertebrates completely altered global environments, from the deepest sea to the highest mountains, and even affected the atmosphere and climate. In short, the entire surface of our planet is built upon one big complex and constantly evolving burrow system, controlling the nature of our existence.

Did humans ever catch on to this fundamental way of life, figuring out that burrowing was an important part of earth history, and that burrowing into the earth was a great way to avoid danger? They sure did. And for that, let's go back about 5,000 years to a place in what is now called Turkey, where people tapped into this deep evolutionary heritage and decided to emulate their burrowing-mammal ancestors in order to survive—an urge that continues through today.

CHAPTER 2

Beyond "Cavemen": A Brief History of Humans Underground

Safely Below Anatolia

Our hotel room was a cave. Fortunately, my wife, Ruth, and I had been fully informed of this when booking it, hence our expectations synched with reality. Upon entering, we were delighted to see how its light-gray rock walls enclosed two connected chambers; one held a modernly furnished bedroom and the other a bathroom with sink, shower, and tub. The walls were not smooth, but textured, defined by finger-wide grooves and ridges. These traces were evidence of human carving and told us we were not in a natural cave, but one intentionally made to look and function like one. In terms of the latter, it seemed to work. Despite the heat of a summertime sun outside, the interior was cool and comfortable, not requiring any artificial air conditioning. We were very happy to stay there.

This rocky start to the day came after enduring a twelve-hour overnight bus ride from Istanbul to central Turkey. Upon arriving midmorning at the bus depot in the small town of Göreme, we were picked up by a hotel shuttle bus and taken uphill to our hotel. With typical Turkish hospitality, the staff greeted us cheerfully, and told us our room would be ready in a few hours, but first we must eat. The breakfast buffet, laden with olives, fresh tomatoes, cheeses, breads, and fruits, felt like a well-earned reward after our bus-confined journey, and we enthusiastically sampled what it had to offer. Afterward, we sat on a patio and took in our surroundings, some of which we had seen on the bus coming into town. It was marvelous.

This part of Turkey is called Cappadocia, a region defined by its geographic position between the Taurus Mountains to the south, coastal highlands north, the Euphrates River east, and various historical provinces west. Towns in the area include Göreme, Nevşehir, and a few others, but Cappadocia is largely a rural area with farms and pastures. It is also located on a plateau (steppe) about 1,000 meters (3,300 feet) above sea level, a highland formed by tectonic uplift accompanied by extensive volcanism. However, this was not what captivated us as we looked out onto the surrounding countryside. Cappadocia is world famous for its unusual geology which, combined with a longtime human presence, resulted in what is often described as a "fairyland." This might seem like an odd way to describe a real place, but this term was understandably inspired by the landscape and how people have modified that terrain over the past several thousand years. As Ruth and I gazed around and down from our vantage point, we saw thousands of spires, towers, and pyramids rising above ground surfaces, apparently composed of the same gray rock surrounding our hotel room. As we regarded these prominences more closely, we could see many were dotted with rectangular windows, doors, and vestibules. In some we spotted in the distance, tiny figures appeared in or moved through these openings, seemingly affirming a magical kingdom inhabited by wee people who slipped in and out of the earth itself.

Yet, if this was a fairyland, it was one where its pixies had been under constant siege. Based on some of what I had read before coming there, at least a few of these rock-hewn homes were made more than 1,500 years ago by Christians trying to stay hidden from the then-reigning Romans. After the demise of the Roman Empire, invading Arabic forces of the Ottoman Empire gave people more incentive to dig into the local rock formations. And dig they did. Not only did the Christians living in this area make homes for themselves and their churches, but they also carved out vast underground cities capable of holding thousands of people.

Before saying much more about these elfin moles and their human history, it is probably best for me to back up just a bit, chronologically speaking. Fair warning, though: Because I'm a geologist, "just a bit" means a few million years. So let us consider some geological time units. For instance, think of the Miocene Epoch, which ranged from 23 to 5 *mya*, and the Pliocene Epoch, which immediately followed the Miocene and lasted until about 2.5 *mya*. (Both epochs are subdivisions of the Neogene Period.) During the latter part of the Miocene and earliest Pliocene, from 9 to 2 *mya*, conditions were hellish in Cappadocia. This hellishness, however, could not be blamed on a lack of Christians back then, but on plate tectonics. Colliding Eurasian and Afro-Arabian tectonic plates triggered extensive volcanism, some expressed as lava flows but most of which blanketed the land as volcanic ash flows and mudflows. The volcanic ash in particular—its mix of minerals corresponding to the igneous rock andesite—is what composed the vast majority of the bedrock in the evocative landscape in and around present-day Göreme and Nevşehir. Once ash solidifies, it forms a rock called ignimbrite, reflecting its fiery origin. Ignimbrites were deposited originally as thick layers of hot ash and other rocks, which welded together to make a sandy rock cemented by minerals and glass.

Nonetheless, the cement holding the rock together was not so strong, which made it softer than most other rocks. Granted, if I picked up a chunk of ignimbrite and threw it at someone, it would likely provoke a loud "ouch!" from my target and a well-deserved larger chunk thrown back at me. But if that same assaulted person then decided not to seek revenge, but instead whittled a little statue from the rock commemorating the event, he or she could easily create such an artwork using simple hand tools. Such differences in rock hardness, influenced by their degree of cementation, result in what geologists creatively call "hard rock" versus "soft rock."

Thus without even knowing the geologic history of the area, people who lived on this plateau of central Turkey must have learned quickly that their bedrock was soft enough for them to fashion it into makeshift caves. This probably was not an accidental discovery, as the landscape itself would have hinted it could be shaped in various ways. For instance, the rounded spires, towers, and pyramids around Göreme looked as if a gigantic master sculptor had worked on them over many years. Yet these forms were actually a result of the soft ignimbrite first getting fractured, then worn away, by water and wind.

At some point after the Pliocene Epoch, weathering and erosion began working on the thick ashfall sediments, which had hardened a little since their deposition, but not enough to resist daily degradation. This breakdown of the rock progressed most quickly wherever it was fractured, as water would have flowed along the paths of least resistance. Over time, water-caused wear evidently made separate but closely spaced "islands" out of what used to be widespread and massive rock bodies exposed at the surface. If someone observed these processes as a time-lapse sequence over a few hundred thousand years, the pillars would have looked as if they were growing taller, like mushrooms popping out of the ground after a rainstorm. Instead, though, they were being "elevated" at the expense of the land around them, which eroded at a faster rate than the isolated rock bodies.

Such collections of rock pillars are called *hoodoos*, which non-geologists (but never geologists) also call "fairy chimneys." Probably the most famous place in the world to see hoodoos is in Bryce Canyon National Park in Utah, but these striking features can form in any place with the right combination of soft bedrock, fracturing, intermittent flowing water, wind, and not enough plant roots to hold in soil. Nevertheless, for a hoodoo to form properly, it needs to be capped by harder rocks, such as those made by lava flows, which are accordingly more resistant to weathering than ashfall deposits. This harder rock prevents top-down erosion, and thus prevents a pillar from eroding down to just a short lump. In Cappadocia, such differences in erosion patterns imparted rounded to pointed caps on some of the hoodoos. This led to undeniably (and impressive) phallic forms in some hoodoos. Bawdy anatomical comparisons aside, these and other geologic features must have awed the first people who settled in this area, which based on archaeological evidence would have been more than 5,000 years ago.

So why dig into your local bedrock, other than to escape occasional invaders? Just a few days of Ruth and me walking to the underground dwellings and churches around and outside of Göreme during summer gave us a clue: It was hot and dry. These conditions ensured that we drank plenty of water in the morning and carried water bottles throughout the day, and we refilled them often. The steppe climate of Cappadocia, however, is not that of a pure desert, but more of seasonally variable one, where summers are hot and arid but winters are cool and wet. In an underground environment, these extremes in temperature and humidity average out, making for agreeable conditions all year round: not too cold or hot, not too humid or dry.

This may sound pleasant in itself, but anyone visualizing how they might live underground for months or years might also think about other biological basic needs. For instance, what about food and water? Today, any concerns about food scarcity are easy to forget in Turkey. As anyone who has spent more than a day there can attest, it is a nation abounding with wonderfully fresh, varied, and delicious food. Was this always the case, specifically for people living in the region of Cappadocia over the past few thousand years? Obviously people could not grow crops in dark places, but neither did they need to be down below all of the time. For water, healthy amounts of water are provided by winter rains and snow in the nearby volcanic mountains, meaning people could have had access to plenty of fresh water in the wet seasons. These water supplies then could be stored during dry times, whether directed downward into cisterns or accessed as groundwater through wells. Abundant water also ensured that people could grow their own grains, vegetables, and fruits during spring and summer, aided by good soils, and then harvest and store this bounty underground during the leaner months. The lush summer soil was richly productive because of mineral-rich sediments supplied by eroded volcanic rocks. In short, Cappadocia was—and still is—a great place to grow crops.

For example, even in the short time we visited there, bountiful apricot trees, many growing just outside the entrances of underground chambers around Göreme, were dropping their juicy fruits on the ground. Just a few bites of these delicious apricots drove home the point that even if you were living in a cave, good food could be mere steps outside your doorway. Apricots, dates, figs, olives, and other fruits could also be dried and stored underground indefinitely, supplementing grains and additional foodstuffs that were also in the right climate-controlled environment for preserving them. The countryside also had enough native vegetation for grazing livestock. Thus fresh milk was available from cows, sheep, and goats that could live just outside as well, which could be used to make cheeses that would last for months. The same livestock giving this milk were also walking larders and could be easily slaughtered, dressed, and brought down below after a quick trip to the surface. Even better, some domestic animals could be kept underground, cutting down on surface forays.

Many of these thoughts came to me during our first day in Göreme as we walked to nearby historic Byzantine churches and relatively small homes that had been hewn out of the local rock. Yet it was not until the second day in this area, with a tour to the former subterranean city of Derinkuyu, that I began to realize the astonishing magnitude of hard work and planning that went into making a functional and livable community for hundreds or thousands of people underground.

Derinkuyu today is a small town of about 10,000 people, and everyone lives where they can easily see the sky every day. Yet at one time a population *twice* that lived underground. The Derinkuyu underground city is the deepest known in the region of Cappadocia, plumbing depths of about 85 meters (280 feet), with five levels between the surface and its deepest parts, all carved out of the rock using nothing more than hand tools and people power. According to archaeologists and historians, this city might have been started well before Christians moved into the area, perhaps as long ago as when the Hittites were there, which was 3,000–5,000 years ago.

Other than living chambers, what else would the inhabitants of a city need to stay happy and healthy while working for extended times underground? Wine and olive oil, of course. For Derinkuyu, this meant creating rooms for pressing grapes and olives, as well as wineries for fermenting grapes. Also needed were food- and water-storage (cistern) areas, kitchens, stables for domestic animals, schoolrooms, and places for worship, with religions depending on who was living there at the time. For air circulation, a 55-meter (180-foot) deep ventilation shaft connected to the surface, and thousands of ventilation ducts emanated from this shaft, ensuring that chambers received air from the outside world. Rather than relying on wells—which could easily be poisoned or otherwise sabotaged at the surface by enemies—water was channeled from groundwater and through subsurface conduits. In a seasonally wet-dry climate, it made sense to tap into this underground water supply, instead of relying on ephemeral surface-water sources. For further comfort, linseed-oil lamps illuminated the darkness, with the oil coming from locally grown plants (species of *Linum* and *Eruca*). A room or two devoted to pressing plant seeds for their oils was helpful for keeping the lights on.

Although many passageways in Derinkuyu were narrow (more on that soon), some of the horizontal tunnels past these stairways opened up into veritable highways. Such widened corridors were used for moving livestock to and from stables. These were sensibly constructed on the first level, as even the most domesticated of animals would have resisted descending too far into the earth. Chambers connected by tunnels also ranged from the size of a Tokyo efficiency apartment to that of a spacious ballroom. These big rooms were held up by pillars, which were simply fashioned out of the stone. Small, embossed rectangular alcoves held oil lamps, which provided light at all levels. All in all, these functional spaces added up to a self-sufficient and relatively sustainable environment for its inhabitants.

Access to multiple levels and chambers in this underground complex was accomplished by shaping inclined stairways from the stone and long horizontal passageways, respectively. Both types of routes vary considerably in width and height, but most of the stairwells are tight, barely accommodating one person at a time. These were also short enough that the average-height people of today have to stoop to prevent unwanted rock–head collisions. However, the claustrophobic character of these stairways was not a flaw, but a feature, as they were designed for defense.

So let's say it's about 300 A.D. and a Roman century of about a hundred men shows up at one of the entrances to an underground city. Once there, the centurion in charge decides it's time to conquer these people, or at least do some good old-fashioned pillaging. Nonetheless, because of the narrow stairways, the Roman soldiers are forced to move in single-file through a constricted space, reducing a mighty, thrusting Roman phalanx to a mere pinprick. Also, the underground layout is deliberately planned as a maze, and could be darkened instantly by residents dousing oil lamps as soon as the soldiers arrived. This creates a situation where the invading soldiers are literally in the dark and clueless about what awaits them around each corner, whereas the defenders know every nook and cranny. As every military tactician or horror-movie aficionado knows, proceeding under such conditions is a really bad idea. All that was needed were a few pointy sticks wielded by made-in-the-shade Cappadocian defenders to put a little damper on this Roman holiday. Given just a first few speared soldiers clogging the passageway, the others would have stopped behind the still-writhing bodies of their disemboweled comrades to pause, reflect, and perhaps reconsider their plundering ways.

Even if these soldiers got past the first few defenders, they then likely encountered thick, massive rounded stones that the inhabitants rolled across the stairways. Such barriers became psychologically more impenetrable with the knowledge that more spear-wielding people were on the other side, ready to protect their families. These stone doors also served to redirect intruders, causing them to double back and turn a different corner, where yet another trap might await them. Given enough losses and failures, our imaginary centurion eventually would have called it a day, gone back to camp, and sought out some surface dwellers to bother instead. Several centuries later (in years, that is), Arabic invaders would rediscover the same disadvantages of attacking these underground cities, and likewise make the wise choice to stay above it all with their vanquishing aspirations.

Everyone probably agrees that getting speared in a dark, narrow passageway is a bad way to die. But it could be worse. How about having boiling oil poured on you just as you enter through a doorway? However unfair it might sound, this tactic was used by the inhabitants of the underground city Özkonak. Although not quite as large or deep as Derinkuyu, Özkonak had ten levels and was similarly designed for sustaining a large population of people underground for a long time. However, its planners made conduits above its tunnel entrances, through which boiling oil was dumped onto home intruders. Özkonak also had another innovation that was apparently unique among the underground cities of Cappadocia, best described as "speaker pipes." Residents used these

5-centimeter (2 inch) wide hollow tubes between levels and rooms to converse throughout the city. Much like today's Internet, this series of tubes probably led to miscommunications, especially if messages were being relayed from level to level. Yet these pipes would have still provided major tactical advantages if the city was assaulted at multiple entrances, as defenders could have passed information very quickly throughout the complex.

Nevertheless, probably the greatest enemy in day-to-day underground living in Cappadocia was boredom. For instance, did these people feel isolated from other communities by living underground? Maybe, but this potential problem was solved by the people of Derinkuyu building an 8–9-kilometer (5-mile) long tunnel that joined with another underground city, Kaymakli. Kaymakli was wider than Derinkuyu but also shallower; still, it had eight levels below ground. Its ventilation shaft was deeper than the one at Derinkuyu, at about 80 meters (260 feet) long, but Kaymakli otherwise resembled Derinkuyu by having many small and large chambers, a food-storage room, and an all-important winery. Having Kaymakli and Derinkuyu connected must have been advantageous to people in both cities from the standpoint of exchanging goods and services, improving their social lives, easing ennui, and of course vanishing from their enemies.

Amazingly, Derinkuyu, Kaymakli, and Özkonak may not be the largest of underground cities in Cappadocia. In 2013, demolition workers in Nevşehir (north of Derinkuyu) discovered entrances to a previously unknown city, which is now considered as a good candidate for the biggest and deepest found thus far in the region. Preliminary studies by Turkish geophysicists suggest that it may be as deep as 113 meters (371 feet), covering an area of about 46 hectares (114 acres). Although its total number of levels is still undetermined, archaeologists have already verified extensive tunnels and stairways linking living chambers, air-circulation shafts, water conduits, kitchens, facilities for metalworking, and wineries. Like other nearby cities, it seems to have been designed for long-haul subsurface living, and other artifacts indicate that Christians used it often as a haven while waiting for the Ottoman Empire to fall. Today, it promises to be one of the greatest tourist attractions in the area, bringing in thousands of more people to go downtown (so to speak) and marvel at its subterranean wonders.

Does this newly discovered city in Nevşehir qualify as a candidate for the most extensive underground living space ever devised by humans, especially for defense? Perhaps for preindustrial times, but it was definitely surpassed by many ambitious defense-minded projects started in the mid–twentieth century. Appropriately enough, these were not encouraged by the threat of invading armies, but instead by the invention of weapons capable of causing our extinction.

Subsurface Surviving in Nuclear Times

It would be easy to look back on the reasoning of Cappadocia's underground city planners, builders, and inhabitants as outdated, a product of their place and time. Yet despite thousands of years passing, the spread of people into nearly every ecological niche on earth, staggering advances in technology, and the ability to change world climate, whenever humans feel threatened, they still look underground for protection. This default behavior manifests itself in private citizens (nowadays called "preppers," as in "preparing") constructing personal shelters, whether intended to survive nuclear warfare or the much less likely threat of a zombie apocalypse. But this behavior also becomes apparent in small desert towns where people dig out their homes to keep comfortable year-round, as well as in modern cities where citizens wish to continue functioning through brutal winters. However, this protective instinct surely reached its zenith when industrialized nations began planning for their governments to survive worst-case scenarios inflicted by human-caused or natural disasters.

For the grandest example of a nation that devoted extravagant resources and human labor to subsurface subsistence of its government in case of war, look no further than the United States in the twentieth century. Although the U.S. was the reigning military superpower after World War II, its government wanted to further ensure its continued existence. This paranoia and keen sense of self-preservation was prompted by the Cold War, which the U.S. waged with the former Soviet Union (and to a lesser extent, the People's Republic of China), with nuclear weapons at the crux of the threat. Whether delivered by bombers, land-based missiles, or missiles shot from submarines lurking just offshore, nuclear weapons were a source of dread, worth worrying about like no other menace.

Countermeasures to possible nuclear war ranged from the ridiculous to the profound. For the former, American citizens who grew up in the U.S. during the 1950s and 1960s may remember "duck and cover" drills conducted in public schools. In these drills, schoolchildren watched training films that showed them how to get low ("duck") and below ("cover") by hiding under their school desks, where they were quickly reminded of just how much gum had been discarded there over the years. This action was supposed to somehow shield them, just in case the Soviet Union or one of its allies lobbed a missile with a nuclear warhead at their schoolhouse. (Just to clarify: "Duck and cover" probably would not have worked very well.)

However, while this ineffectual training was going on, the U.S. government was implementing a more operative strategy that would have allowed at least a few people to survive atomic warfare. The plan was to get beneath the problem of M.A.D. (Mutually Assured Destruction) and, like the people of Cappadocia, dig out massive refuges in the earth. So, starting in the early 1950s, the U.S. began looking for places where key military personnel could be sheltered and maintain communications during and after a nuclear exchange with the Soviet Union.

In 1950, the U.S. military was authorized by the government to start excavating such a facility in Pennsylvania, which became known as the Raven Rock Mountain Complex. Raven Rock, also called "Site R," is located near Pennsylvania's southern border with Maryland and just southwest of the American Civil War battlefield of Gettysburg. Although this might seem a little out of the way for Washington bigwigs, it's actually a short helicopter ride from the D.C. area and only about 10 kilometers (6 miles) from the president's vacation home of Camp David. From 1951 to 1953, the underground part of Raven Rock was hewn out of greenstone, a metamorphosed basalt sometimes erroneously called "greenstone granite." This rock formed originally as lava flows, which were then put under heat and pressure generated by colliding tectonic plates more than 300 *mya*, becoming part of the Blue Ridge province of the Appalachian Mountains.

To make this facility, which is located about 200 meters (650 feet) below the mountaintop, an estimated 14,000 cubic meters (500,000 cubic feet) of greenstone was removed from the mountain interior. This space was hollowed out to fit five office buildings, old-style (i.e., room-sized) computers, communication equipment, air-exchange systems, two water reservoirs and a water-treatment plant, power generators, living spaces, a dining hall (romantically named "Granite Cove"), an infirmary, and amenities such as a convenience store, barber shop, fitness center, and chapel. Office buildings had rock-wall partitions between them; appropriately, all buildings lacked windows, which admittedly saved costs on window washing, blinds, and curtains. A nine-story-tall microwave radio tower juts out from the mountain, but seven of its stories are buried. Raven Rock's capacity was for about 3,000 people, and those not already there would have gained access to the underground facility via four vehicle entrances.

Raven Rock was used for joint communications between the U.S. Air Force, Navy, and Army throughout the Cold War, and still serves as an emergency center for those branches of the military. It was also set up so that the president of the United States, who might be close by at Camp David, could be kept safe there before or after an attack. Along those lines, Raven Rock supposedly served as Vice President Dick Cheney's secret lair soon after the 2001 terrorist attacks in New York and Washington, which he denied, thus neatly verifying he was there. This base is still operational, so guided tours for the public, like the ones conducted of underground cities in central Turkey, are rather unlikely.

Another underground bunker constructed to safely handle national or international disasters is the High Point Special Facility, otherwise known as the Mount Weather Emergency Operations Center. It is located near Berryville, Virginia, about 80 kilometers (50 miles) from Washington. Rather than being connected directly to the U.S. military, this facility was built specifically to ensure the executive and judicial branches of the U.S. government would continue to function after a major disaster. Appropriately, the Federal Emergency Management Agency (FEMA) is in charge of Mount Weather. Excavation of the Mount Weather facility started in 1954 and its interior construction was finished in 1959.

Primary protection of Mount Weather was provided by its overlying geology, which, like Raven Rock, is also part of the Appalachian Mountains. However, the Mount Weather interior was also reinforced against attacks by more than 20,000 bolts, as well as a massive, more than 30-ton door that Cappadocians would have coveted. This interior holds the equivalent of twenty buildings, all connected by tunnels, and include a power generator, media studios, living spaces, dining halls, and a hospital. Perhaps most importantly for sustaining sanitary conditions over more than a few days, it also has large-capacity water reservoirs and its own sewage treatment plant. Although designed to hold several thousand people if needed, private bedrooms are reserved for the president, his or her cabinet, and even justices of the Supreme Court. Mount Weather was kept mostly out of public view until 1974, when a commercial jet crashed into its side because of bad weather. This accident brought in plenty of media folks who quite rightly wondered about the meaning of this odd taxpayer-funded facility beneath the crash site. The ensuing scrutiny resulted in U.S. government officials admitting the existence of Mount Weather, although it still continued to operate under high levels of security and secrecy.

As some citizens of the U.S. might recall from middle-school lessons, the U.S. has three branches of government, yet so far only two of those—the executive and judicial branches—seemed to have been protected by underground facilities. So what about the legislative branch, also known as the U.S. Congress (the House of Representatives and the Senate)? In a move that probably would not receive much support from U.S. voters today, a subterranean place was constructed to save members of Congress. Unsurprisingly, this congressional refuge was not placed beneath a school gymnasium, cornfield, or hazardous-waste dump, but instead was sited directly below a luxury hotel. The bunker, conceived in 1956 and built from 1959 through 1962, was located underneath the Greenbrier Resort in White Sulphur Springs, West Virginia. The host rock for the facility is the Millboro Shale, which is in the Allegheny Front of the Appalachian Mountains; sediments composing these rocks were deposited in an inland sea during the Late Devonian (380–360 million years ago). One code name applied to the facility was "Casper," which presumably referred to a popular cartoon-character ghost, who could easily appear and disappear. Still, considering its intended somnambulistic guests, it just as easily could have been named after the Casper mattress company. Another code name, "Project Greek Island," left little doubt about how legislators imagined their continued existence would play out while the rest of the U.S. was a smoldering radioactive wasteland.

Analogous to its proposed occupants, this facility was shallow, with its first level only about 6 meters (20 feet) below ground, and consisting of more than 10,000 square meters (107,640 square feet) of space. It also had government-provided housing (dormitory), meals (food rations, kitchen, and dining hall), health care (hospital), and broadcasting (television station and other communications equipment), as well as other perks these officials would have denied their constituents just before the latter were unceremoniously nuked. Tragically, no golf courses, boutique wineries, spas, or cigar shops were included in its underground floor plans, reflecting the dire conditions that would have been imposed on its guests. Although the U.S. Congress has 535 voting representatives, "Casper/Greek Island" was also designed to sustain nearly 1,000 people for two months, which would have allowed for a bare-minimum complement of about 500 Washington lobbyists to join their employees. (In all seriousness, though, a former speaker of the House, Thomas "Tip" O'Neill, said that Congress members were told they could not bring their families there, creating a moral dilemma for anyone fleeing an attacked Capitol.)

In the event of a national emergency, 25-ton doors would have blocked vehicle and pedestrian entrances to the underground structure. Further protection was supplied by thick layers of concrete and steel, with an estimated 50,000 tons of concrete poured into the site. Above ground, federal authorities could commandeer the entire Greenbrier Resort if necessary. The West Virginia Wing of the hotel included two auditoriums, each sized to accommodate the House of Representatives and the Senate, and an exhibit hall for holding joint session of Congress. Unfortunately for all voters who like the idea of their elected officials going down a hole and forever vanishing from public sight, a 1992 *Washington Post* article revealed "Project Greek Island" for the needless boondoggle it was, and it was closed in 1995. However, the Greenbrier Resort capitalized on its notoriety, with the staff now conducting guided tours of this Cold War relic.

As an example of how thinking went even further down in the Pentagon, U.S. military planners suggested yet another U.S. government installation in 1962, imaginatively dubbed the Deep Underground Command Center. However, it was never built. This facility, which was supposed to be near the Pentagon and serve the U.S. military, would have more than fulfilled its appellation, as it was supposed to be sited 1,000 meters (3,280 feet) below ground. Had it been constructed, it would have been designed to resist a direct hit from a nuclear bomb, and host 50 to 300 military personnel for more than 30 days while they figured out how to respond to an attack on the U.S. capital or other parts of the nation.

Meanwhile, on the other side of the continent, NORAD (the North American Aerospace Defense Command), located in Colorado Springs, Colorado, but jointly operated by the United States and Canada, was conceived as an early-warning/defense system against enemy missile attacks. However, because NORAD's headquarters was on the surface, an alternate facility (the Alternate Command Center) was housed inside Cheyenne Mountain just outside of Colorado Springs. Protected by about 600 meters (1,969 feet) of igneous rock, this subsurface command center was intended as a blast-proof bunker. This still-impressive (and operational) center possesses: multiple massive steel blast doors designed to endure a 30-megaton nuclear explosion; access and side tunnels; main chambers; a water reservoir; ventilation systems (with air filters); power generators; a medical treatment center; and much more. The entire facility is contained within the equivalent of 15 three-story buildings and protected by more than a thousand giant shock-absorbing springs that keep these structures in place, just in case of natural or bomb-imposed earthquakes. The Cheyenne Mountain NORAD facility is also supposedly the only U.S. military installation guaranteed to withstand the electromagnetic pulse (EMP) emitted by a nuclear weapon, which would disable nearly all electronic communications equipment on the surface.

Massive underground bunkers meant to endure nuclear attacks were not just an American invention, but were also constructed by its main nuclear rival, the Soviet Union. Like the U.S., the Soviet government developed subsurface centers for continuity of its leadership, as well as extensive subways and tunnels underneath Moscow and surrounding areas to connect these centers. A 1991 report by the U.S. Department of Defense noted that one bunker was directly below the Kremlin and another near Moscow State University. In this same report, both bunkers were estimated to be 200–300 meters (657–985 feet) deep and capable of hosting about 10,000 people. The facility next to the university, which was below the Ramenki district of Moscow, was later confirmed and nicknamed the "Underground City." According to some Russian sources, it could have held 15,000 people, although for how long is unknown. With the end of the Cold War, at least one former bunker ("Bunker 42"), which is located about 60 meters (197 feet) beneath the surface, was converted into a tourist attraction. For an admission fee, visitors can see the rooms, tunnels, and equipment that Soviet personnel would have used to survive a nuclear attack, and imagine bombs exploding on the surface, incinerating their comrades. But if this is not enough fun for tourists, they can always hang out in its restaurant and karaoke bar: *perestroika*, indeed.

The People's Republic of China was another nuclear weapon–bearing nation that sought protection through subsurface hideaways. In 1969, following a serious skirmish on the Soviet-Chinese border, Chairman Mao Zedong commanded Beijing residents to prepare for the possibility of conventional or atomic bombs being dropped on their city, an order that could have been stated simply as "Start digging." And dig they did, as approximately 300,000 people excavated more than 20,000 bomb shelters underneath Beijing. These shelters could be accessed through thousands of hidden entrances in homes and other buildings, connected by more than 30 kilometers (18 miles) of tunnels. Similar to Cappadocians from many centuries previous, Beijing citizens mostly used hand tools and human power to make this underground system. Moreover, they also put in thousands of ventilation shafts, located potential well sites, and made food caches for hundreds of thousands of people living in these shelters. To lend a sense of normalcy to this siege mentality, classrooms, restaurants, movie theaters, and other comforts of home were later opened in this city-beneath-a-city.

Following Mao's death in 1976, and as the threat of a Soviet annihilation was replaced by the siren call of capitalism, more and more of these shelters and tunnels were converted into commercial spaces and housing, some of it for long-term renters but also for hostels. Additionally, the Chinese government required all new building projects to incorporate underground structures as emergency shelters, but with the requirement that these also have a commercial purpose. As Beijing commercial real estate boomed in a non-nuclear way up to and after the 2008 Olympics there, many new arrivals began moving into this cheaper underground housing.

Today, as many as a million people live in what has been called *Dixia Cheng* ("The Underground City"), partially for its inexpensive rent, but also to mollify the effects of Beijing's hot and humid summers, which alternate with bitterly cold and dry winters. Subsurface living is further encouraged by dust storms that arrive with every spring. However, it is not this seasonal disturbance that gives residents yet another reason to seek refuge below ground whenever necessary, but rather human-caused air pollution. Now infamous for its severity, Beijing's air pollution is caused by increased industrialization depending on coal as a main energy source, as well as millions of cars replacing bicycles as personal transportation over the past few decades. As is typical in big cities with smog, this pollution becomes far worse with summertime heat, a double whammy that makes an underground environment all the more appealing.

All of the preceding shows the extremes of how governments will create subsurface environments to survive horrific assaults by our own species. Nonetheless, the "underground city" of Beijing also demonstrates another benefit of going below, which is to protect people from environmental threats. This trait, which humanity shares with all other burrowing animals, is one we often choose whenever trying to cope with places where climate alone can bring us great harm.

Winning the Cold (or Hot) War

Before the Cold War started and well after it had ended, people used underground environments to fight environmentally induced cold, heat, or both, depending on where they lived. Although each environment provided an impetus for such measures, the degree to which people adapted their underground homes in accordance with surface conditions varied considerably and ranged from small villages to large, modern cities.

So let's start with a hot spot. One of the most interesting underground towns in the world is also located in one of the most inhospitable places imaginable for humans who do hard labor for a living, Coober Pedy in south Australia. Coober Pedy is a community of a few thousand people in the middle of the central desert of Australia, nearly 900 kilometers (560 miles) north of the nearest sizeable city, Adelaide. Its unusual name comes from a corruption of *kupa-piti*, an indigenous (aboriginal) term that translates as "white man's hole." This naming was inspired by how melanin-challenged people of European descent, once entering the longtime territory of aboriginal people, soon afterward dug into the ground and eventually lived in holes of their making.

What motivated non-native people to move to the middle of a desert, dig below it, and then live there? Opal, and heaps of it. In 1915, when prospectors began looking for earth resources in the Coober Pedy area, they were astonished to find much high-quality opal there. Opal is an iridescent gemstone mostly made of silica (SiO_2), but includes some water in its mineral structure. The opal of Coober Pedy is normally found in sandstone beds of the Bulldog Shale, a geologic formation formed when a shallow sea covered this region during the Early Cretaceous (about 110 *mya*). As the seas receded from this region more than 50 million years later, acidic groundwater dissolved some of the clay minerals and quartz in the Cretaceous rocks, which put silica into solution. This silica then combined with water and precipitated as opal in vast quantities, making the area around Coober Pedy the richest opal deposit in the world. The opal is so abundant that some of it even filled interior spaces of Cretaceous clams, snails, and marine-reptile bones, creating gorgeously sparkly fossils.

However, acquiring this opal for moneymaking purposes meant mining, and in early twentieth-century Australia, most miners dug into the opal-bearing sandstone using hand tools. As they excavated shafts and tunnels, though, they also were reminded daily that they lived in a very hot and arid desert, where daytime temperatures during the summer could reach 40–45°C (104–113°F). Also, as is typical of deserts, temperatures fluctuated considerably between the day and night, giving miners the exhilarating experience of heat exhaustion during the day and hypothermia at night. That is, if they stayed on the surface. What they soon discovered after spending so much time underground was that temperature swings were moderated. Indeed, when not swinging a pickaxe or hammer while breathing in Cretaceous rock dust, miners found the underground quite pleasant. So down they went. Accordingly, people modified abandoned mine tunnels, added rooms by subtracting rock, and otherwise whittled out dream homes where they could rest and recreate in comfort. Even today, despite the subsequent invention of air conditioning, nearly all of the several thousand Coober Pedy residents still choose to live underground. This choice is quite sensible as it saves on potentially exorbitant power bills, while also making a unique hook for attracting tourists who might not otherwise consider visiting an Australian desert for fun. Like the former Cold War bunkers of Moscow and Beijing, Coober Pedy also has underground amenities for visitors, such as hotels, restaurants, and (of course) pubs.

If you then had to name a city with nearly all of the opposite qualities of Coober Pedy, you would do fairly well by naming Montreal, Quebec (Canada). However, even though Montreal is literally on the other side of the world from Coober Pedy, and on the opposite end of the spectrum in both climate (warm, humid summers alternating with very cold winters) and population (more than 3,500,000 people in its metropolitan area), the two cities share a subsurface solution to the environmental challenges faced by their residents.

Winter months in Montreal can be brutal, even for Canadians who take great national pride in thermal deprivation. January temperatures average around –10°C (14°F), and winter snowfalls normally total about 2 meters (6.6 feet), but of course can exceed this average. These conditions also do not take into account wind chill, which can make already frigid temperatures feel more like –25°C (–13°F). Although average highs in summer, at 24–27°C (75–80°F), are far below those of Australia's central desert, high humidity can be discomforting and frequent thunderstorms tend to dampen outdoor activities, too. Considering this situation, Montreal residents would be perfectly justified in huddling at home most of the year, perhaps occasionally venturing outside during the winter for food supplies or going to the airport so they can fly to some place more tropical and exotic, like Toronto.

As a result of its inhospitable climate, Montreal city planners and developers over the past 50 years began having deep thoughts, and these visions were realized in what is now called RÉSO, or *La Ville Souterraine* ("The Underground City"). Montreal was built on glacial clay, sand, pebbles, and gravel from the Pleistocene Epoch (2.5 million to 12,000 years ago), which were deposited either directly by glaciers, by glacial meltwater, or in glacial lakes. Most of this "bedrock" is relatively unconsolidated, which made it easy to excavate for the RÉSO. This complex, covering a 12-square-kilometer (4.6 square mile) area below the central business district (thus qualifying as a real "downtown"), is connected by about 32 kilometers (20 miles) of pedestrian-amenable tunnels linked to the subway system, the Montreal Métro.

Montreal's Underground City began simply as an office building with an underground mall in 1962, but expanded dramatically with further development of the Métro throughout the 1960s. Métro stations were deliberately sited below major office buildings, both to alleviate car traffic and to ease employee access to their workplaces, especially during winter. With expansion of the Métro system through the 1970s to the 1990s came more buildings and shopping malls, accompanied later by apartments, condominiums, hotels, restaurants, art museums, concert halls, cinemas, public plazas, and much more. Universities, always known for having a metaphorical "student underground," also linked to this system, with students, staff, and faculty of the Université du Québec à Montréal (University of Quebec in Montreal) and McGill University using it to get to and from campus buildings. On any given day (but more so during the winters), anywhere from a quarter to a half million people pass through this network. Hence although the RÉSO was an ad-hoc development, and much of it is actually more of an "indoor city" rather than fully below the surface, it still qualifies as the best-developed underground network of any city in the world.

The Subsurface Redemption

Whenever humans have needed to flee an undesirable situation, they have often looked for underground solutions, such as tunneling through soil or rock. Digging one's way to freedom was probably best dramatized in the 1994 film *The Shawshank Redemption*, in which a wrongly convicted man uses a small rock hammer to chip away at the wall of his prison cell for many years, eventually making a tunnel that allows him to escape. A real-life example of a well-dug tunnel being used for a prison break happened on July 11, 2015, when the notorious Mexican drug lord Joaquín Guzmán Loera (nicknamed El Chapo) said adiós to his cell by dropping down into a vertical shaft underneath his shower stall, which connected to a tunnel. The tunnel joined another vertical shaft below a "construction site" 1.4 kilometers (0.9 miles) away from the maximum-security Altiplano Prison, which was immediately demoted to somewhat-moderate-security status. The tunnel, evidently excavated by people under his employ, was marvelously engineered, reinforced by wooden beams and outfitted with adequate ventilation and electric lighting along its length. A customized motorcycle set on rails was likely used to move rocks and dirt out of the tunnel as it was built, but also might have been ridden by Guzmán to speed him along. Guzmán, who was recaptured on January 6, 2016, was quite a fan of subsurface transport of drugs and people. His cartel is credited with the construction of many drug-smuggling tunnels under the border between Tijuana (Mexico) and San Diego (U.S.), demonstrating the ineffectiveness of border walls or other surface barriers when applied against people who think more deeply.

Perhaps the most famous of escape tunnel feats was in 1944, when 76 Allied soldiers burrowed their way out of the World War II Stalag Luft III (German) prisoner of war camp in Poland, a breakout that later inspired the 1963 movie *The Great Escape*. Scientists also recently uncovered physical evidence for another World War II example of people using a tunnel to free themselves from Nazi oppressors. In 1943–44, about 80 Jews from the Stuffhoff concentration camp near Vilnius, Lithuania, were forced to conceal war crimes from approaching Soviet troops by unearthing and burning thousands of bodies from mass graves in the Ponary "killing fields." Because the laborers were kept in a pit, they took advantage of this topographic low, going in to get out. For 76 days, using just their hands and spoons they found with exhumed bodies, they secretly dug a 35-meter (115 foot) long tunnel into the side of the pit. On April 14, 1944 (the last day of Passover), about half of them fled through their hand-hewn conduit. Although most were caught and killed, a dozen made it to the nearby forest and liberation, living to tell of the atrocities they had experienced. In June 2016, an international team of geoscientists and archaeologists, using geophysical tools to scan the ground in and around the pit, detected the subsurface tunnel and finally documented the escapees' traces, evidence of their extreme determination and survival.

To summarize, people ranging from military and government planners in high places with extremely large discretionary budgets thinking about the unthinkable, to early twentieth-century miners, to modern-city planners, to those escaping justice or persecution, were all linked by extant or imagined dismal situations, whether temporary or permanent. Yet time and time again, they came up with a common solution. Their best plan for enduring everything from bad weather to the most heinous inventions in the history of humanity echoed what hundreds of generations of people in Cappadocia had done: Go below.

The Burrowing Imitation Game

With limited time in Cappadocia before moving on to other parts of Turkey, my wife, Ruth, and I had a choice of seeing either of the underground cities in our immediate area—Derinkuyu or Kaymakli—but not both. Somewhat arbitrarily, we picked a tour to Derinkuyu, and we were very glad we did. Not only was it an informative outing (or was it an "inning"?) and an extraordinary place to visit, but it also offered a wonderful moment of synchronicity that occasionally happens to scientists of my ilk.

After spending about two hours underground at Derinkuyu, we had nearly reached the surface, but paused at the first level to learn from our guide about the former stables there. There we intersected with another group near the end of their tour, composed of inquisitive tourists like our group. Yet I certainly did not expect to see anyone in this group I knew, let alone knew well. It was all the more shocking when I turned a corner in a stable chamber and nearly bumped into my longtime friend and colleague, Dr. Renata Guimarães Netto, visiting from Porto Alegre, Brazil. Renata, who also is an ichnologist (someone who studies animal traces), was just as surprised as I was, and once we got over being taken aback, we warmly greeted each other and laughed at our good fortune. I turned to Ruth behind me and said, "Look who's here!", as they had met just the year before at an ichnology conference. Along with such a delightful surprise, though, we soon assessed how this crazy coincidence of Renata and my being in the same place and time (and underground, no less) was almost inevitable.

Like me, Renata is a university scientist and is well known for her research in animal burrows. She is particularly knowledgeable about modern and fossil crustacean burrows, such as those made by burrowing shrimp, crabs, or lobsters, with much of her research done in South America. We were both in Turkey for an international ichnology conference a few days later in the coastal city of Çanakkale, where each of us would be presenting research on fossil burrows. (Ruth and I had front-loaded our trip with the vacation part, which included Istanbul and Cappadocia as our main touristic goals before going to the conference.)

Nonetheless, Çanakkale was another twelve-hour bus ride from Cappadocia, and hence far out of the way for us conference goers. This led to our asking each other, "Why are you here in Cappadocia?" The answer for both of us was a shared interest in the fabulous and world-famous geology of the area, but also how its friable bedrock had offered the opportunity for preindustrial peoples there to modify it extensively, carried out to extremes with underground cities like Derinkuyu. Renata and I then agreed that it was not so strange for an American and a Brazilian ichnologist to encounter each other underground in the middle of Turkey. After all, we both were fascinated by animal burrows and had studied complicated burrow systems. Of course we would gravitate to a place with huge, complex underground burrows, ones that begged for comparison to the crustacean burrows each of us had studied.

This not-really-a-fluke meeting of the ichnological minds raised some good questions. For the people who first started digging into the bedrock of Cappadocia, was this an original idea for them? In other words, did they come up with such plans to make massive and extensive underground communities on their own? Or were they inspired by someone (or something) else? Also, once started, did new generations of people alter their subsurface dwellings as they gained more knowledge and experience, finding out what did or did not work? Or might they have also they learned "best practices" by watching other burrowing animals that lived in the same area?

To better contemplate that last question, let's do a little thought exercise. Imagine a giant ichnologist with the vast knowledge of

Dr. Netto visiting Cappadocia more than a thousand years ago, and with the occupants of Derinkuyu and Kaymakli as Lilliputians (or fairies, if you prefer). From the surface, this grand ichnologist spots a few of the many entrances to these underground cities; with a little more investigating, she then finds the hidden portals. She maps these openings to see if any patterns emerge, such as groupings that might give clues about the how this colony functions. For instance, more openings might be close to fruit orchards or other crops, rivers, or other resources needed for the colony to function properly.

After mapping the surface—while enduring stings from spears and mild burns from boiling oil flung by the city's tiny occupants, who at some point either flee the scene or hide in the deepest chambers—she carefully excavates down to each level. This investigation reveals wide, rectangular or rounded chambers interconnected by thinner horizontal tunnels and inclined stairways, as well as the vertical ventilation shaft with its many lateral branches. She maps each level and eventually compiles these maps to look for vertical changes in the overall architecture of the city.

Once the lowermost level is reached and she is satisfied that it is the deepest, she cuts a vertical slice through one side of the city, then another, and another, all evenly spaced and parallel to one another. These sections help to show how the different levels relate to one another laterally, and whether smaller or larger chambers are concentrated at the top, bottom, or distributed evenly throughout. This sort of painstaking analysis, accompanied by much labeled sketching and measuring, results in a mental image of the three-dimensional form of the city. Because it is a complex structure made over multiple generations, it surely would defy easy interpretations if just examined from the surface alone.

If all of this excavating, mapping, drawing, measuring, and thinking sounds like a lot of work, it is. So let's say you're another giant ichnologist who wants to scoop your rival, while also using an easier method to study the overall form of an underground Cappadocian colony. To accomplish this, you locate all of the entrances on the surface, mix up a huge volume of cement, and pour it down those holes. Yes, this would kill all of the occupants, and you would be stealing the other ichnologist's research ideas, but you reassure yourself that it will be well worth it for a cover article in a high-impact research journal. Anyway, after pouring, you would wait a while for the cement to set, and then dig out the resulting cast. *Voilà!* You would have a beautifully rendered and complicated arrangement of shafts, tunnels, chambers, dotted by entombed Cappadocians and their livestock, as well as purple-stained areas denoting locations of the wineries.

The point of the preceding faux-genocidal scenario is to note that once considerations of scale and humanity are set aside, the underground cities of Cappadocia might resemble those of an ant colony or crustacean burrow system. As we will learn in upcoming chapters, some invertebrates can create such exquisitely complex burrows that they defy glib characterizations of these animals as "simple." However, if you still object to the idea of our species being compared to invertebrate burrowers, then you might think about the underground systems made by closer kin, such as naked mole rats, ground squirrels, gophers, and other small burrowing mammals. This convergence in form related to the function of a mammal burrow system and an underground city might seem remarkable, but it is not. After all, a poorly constructed under-ground shelter, whether made for an individual, family, or com-munity, would quickly result in the deaths of its makers. And death has an annoying habit of preventing genes from getting passed on to the next generation: No evolution for you.

So just to review the traits of a Cappadocian subsurface city, it minimally had the following: more than one access point to resources on the surface; multiple levels with different functions; living chambers; food caches; water storage; defenses; proper ven-tilation for air circulation; proper waste disposal; and places where other species (livestock) could be kept. Well, guess what: Nearly all of these traits are also in other mammal burrow systems. Even better, a burrowing mammal still common in Cappadocia and throughout much of Eurasia that fulfills most of this underground-living checklist is the European badger (*Meles meles*).

European badgers, like their relatives in North America, Africa, and elsewhere are well known as ferocious predators, but perhaps are less famous as burrowers. Like many other members of the weasel family (*Mustelidae*), these badgers excavate and use burrows and excel at it. This is not surprising once you examine their robust claws, arms, and shoulders, which are well adapted for digging out complex and communal burrow systems in soils. These systems have many surface openings used as entrances or exits; some of the openings are used only for leaving the burrow, and never for entering, analogous to always arriving through the front door of a house and leaving from its back door. Tunnels leading to and from openings then connect with larger chambers used as dens, also called *setts*.

Badger setts are used for resting, sleeping, and nesting, and are typically occupied by one badger family. Setts are also placed more than 5 meters (16 feet) away from an entrance, and usually are 1–2 meters (3.3–6.6 feet) deep, which is far enough to make nest chambers easier to defend from anything hankering for a baby badger. (To all aspiring predators of badger babies trying to get past their parents in a burrow: Good luck with that.) However, setts of different families are by no means separated from one another, but linked through more tunnels. Burrow systems also can be reused and expanded over generations, resulting in networks with hundreds of meters of tunnels, and more than a dozen badger families all living together in the same underground city.

Sound familiar? It gets better. These badgers bring in grasses, leaves, mosses, and other vegetation from the outside world to line their chambers, which they discard and replace frequently. They also keep their burrows quite tidy, making sure all waste is deposited outside. Even when a badger dies, its family members will either convert a sett into a burial chamber by closing it off or drag the body outside and bury it there. Once other mammals—such as foxes and rabbits—see such attractive subsurface real estate, these animals might even cohabitate with the badgers, as the burrow systems are big enough that they may not run into one another.

Given all of these parallels in form and function of human and badger colonies, one might wonder if the Cappadocians were actually ichnologists who carefully watched, studied, and mimicked the European badgers living in the steppe of Cappadocia 5,000 years ago. Then, did future generations of humans living in the area continue the traditions of the pioneering human burrowers there or of those of other mammals in the area? I have no idea, and absolutely no evidence to back up such speculations. Nonetheless, it is worth making a parallel comparison to show how two different mammal species living in the same ecosystem arrived at similar solutions when faced with challenges, whether from rivals competing for the same space and resources or from the environment itself. Going a step further, even more remarkable are the parallels in planning used for the underground cities of Cappadocia to the modern, industrial ones used for ensuring the survival of a population.

This brief review of human burrowing as a form of protection and escape alludes to the original phrase of "duck and cover," but implemented as a plan by ducking and covering underground, rather than under a mere desk. However, if one goes back to the original 1951 "Duck and Cover" propaganda film produced by the U.S. Federal Civil Defense Administration, you will see that a human was not used as a symbol for ducking and covering. Instead, a cartoon turtle, nicknamed "Bert the Turtle"—who actually looks more like a tortoise—stars as the animal that, once threatened by an impending explosion wielded by a mischievous monkey, expertly drops to the ground and retracts all of his limbs into his shell. The overt message of the film for 1950s children and adults alike was that Bert the Turtle/Tortoise stayed safe from danger by ducking and covering.

Interestingly, a modern animal provides a much better example of ducking and covering, but uses burrows to protect it and its offspring through all sorts of peril. For subterranean inspiration in this respect, look no further than the southeastern United States and a humble-looking reptile, the gopher tortoise (*Gopherus polyphemus*). Not only is this tortoise one of the most accomplished of all animal burrowers, but it also serves as a landlord for many other animals by creating special environments that drive biodiversity up as tortoises dig down.

CHAPTER 3

Kaleidoscopes of Dug-Out Diversity

Grown-up Non-Mutant Burrowing Tortoises

The hole in the ground was not very wide, but it sure was deep. As I peered down its half-moon-shaped entrance, its walls dissolved into darkness. Yet this hole did not go straight down. Instead, it was built like a ramp, sloping about 30 degrees relative to the low mound of sand where I knelt directly in front of it. Based on what I could see from the first few meters, the half-moon shape continued downward, rounded on top and flat-bottomed, looking vaguely like a freeway tunnel in New York City. This tunnel, however, was held up not by cement or steel, but by compacted sand, and thin plant roots punctuated its rough, corrugated walls. I listened carefully to this inside environment, hoping it would act as a resonating chamber for whoever might be home, but only silence returned. A quick inhalation of breath through my nose took in some not-unpleasant loamy scents, reminding me of gardens past. As I exhaled, my breath condensed into a tiny cloud that dispersed as soon as it met the hole. And that was it. There was otherwise no telling what was down there, what it was doing, how it got there, and why it made such an awesome burrow.

This wonder-invoking burrow was one of many pockmarking a grassy field on St. Catherines Island on the Georgia coast. This was the same island I visit often for its alligator dens, but on this cool January morning I was there with colleagues from Georgia Southern University—Robert K. Vance and Sheldon Skaggs—to study these other burrows. As I stood up and looked around us, I saw that their ecological setting clearly differed from that of the wetlands or maritime forests hosting alligator dens. For one, nearly all of the burrows were adjacent to creamy-white sand piles and separated by low-lying patches of vegetation. The main ground cover in this field was composed of wiregrasses, such as pineland threeawn (*Aristida stricta*), ably fulfilling their nicknames as thin, tall grasses that projected up like bundles of wires. Sharing this field with the wiregrasses were longleaf pines (*Pinus palustris*), which ranged from seedlings to three-story-tall trees. Longleaf pines are likewise aptly named, with 30–40-centimeter (12–16-inch) long, yellow-green needles bursting from their branches, a trait especially apparent in immature trees. If any trees could be described as cute, these would qualify, as trees not much taller than me had rail-thin trunks leading up to a bottlebrush-like crown of lengthy needles that spilled up, out, and down. A bunch of these trees standing together in this field looked like several 1980s rock bands on a reunion tour.

Longleaf pines and wiregrasses are essential parts of an endangered ecosystem in the southeastern U.S., the longleaf-pine forest. This mix of pine forest and grassland savanna used to be the most common environment in this part of North America, stretching more than 3,200 kilometers (1,988 miles) along a broad swath of coastal plain from what is now east Texas to Virginia, dominating most of Georgia and Florida. The pines, wiregrasses, and their associated plants grew quite well in well drained, sandy coastal-plain soils. Such soils marked where barrier islands and other coastal environments had deposited sand during higher sea levels of the past 60 million years.

Longleaf pines can live hundreds of years, which is a good thing for their communities, as old pines are necessary to get young pines started. They outcompete other tree species through a simple strategy: producing lots of seeds in their pinecones and living through forest fires. To do the latter, longleaf pines are fire-resistant, a superb advantage whenever a lightning-sparked conflagration sweeps through their habitat. Such fires clear out pine needles, which makes way for pine seeds to establish themselves in the soils. The temporary elimination of ground-cover plants also eliminates competition, as these taller plants would have prevented seedlings from getting sunlight. Burnt plants also return nutrients to the soil, further helping seedlings to grow. This blend of factors thus gives pine seedlings the means to settle and grow up into adorable "bottlebrushes." Later, as they turn into mature pine trees, they contribute pine needles as the main fuel load for fires, ensuring future generations of seedlings.

What caused fires that were frequent enough to ultimately select for flame-enabling trees and other plants? Let's just say if you have ever yearned to be struck by lightning, then you should move to the southeastern United States. Through a unique set of climatic circumstances encouraged by a location between the Gulf of Mexico and the Atlantic Ocean, lightning strikes are more common in southeastern states than anywhere else in North America. Florida alone has an estimated 1.5 million electrified hits annually, rivaling hurricanes and tornadoes for sheer havoc wreaked. Multiplying millions of lightning-caused fires each year by a few millions of years, the ancestors of modern longleaf pines and other plants either would have gone extinct or adapted to these frequent and rapid ecological disasters. Fire-wielding Native Americans also likely shaped longleaf-pine ecosystems over the past 12,000 years, their effects having magnified an already present threat of fire. This human factor hence further nudged the evolution of the ecological community in a direction favoring fire-surviving plants and animals, a subtle trace of a former Native American presence. Sadly, the arrival of Europeans in the sixteenth century, soon followed by intensive deforestation and agriculture, shrank longleaf-pine habitats to less than 1% of their original realm, leaving only small patches in the southeast. For instance, the tiny parcel I was experiencing on St. Catherines Island was a restoration experiment, placed on a former grazing field and sited on a sparsely populated barrier island to safeguard its success.

Although already full of surprises, another unexpected facet of this ostensibly mundane ecosystem is its biodiversity. In my experience, whenever I do a word-association game with my university students in Georgia and say "biodiversity," they respond with "Amazon rain forest!", "Great Barrier Reef!", "African savanna!", or similar faraway exotic places. Such knee-jerk responses—honed ably by many years of gee-whiz conservation-minded documentaries filled with charismatic megafauna—are not necessarily wrong. Nevertheless, they do expose ignorance of what is much closer to home. For example, in one of the few remaining old-growth longleaf-pine forests in southwest Georgia, more than 1,000 species of plants contribute to its ground cover: This matches the plant biodiversity of some tropical rain forests. Also, because most of these are flowering plants, many of them have pollinators. This translates into thousands more insect species that use these plants for pollen, as well as for other forms of food or living spaces. In turn, many of these insects have predators, such as other insects, spiders, amphibians, birds, and mammals. This food web spreads further into more vertebrates, such as lizards, snakes, and other reptiles, contributing to an astonishingly long list of species all inhabiting a "mere" pine forest, a mosaic composed of a myriad of interlocking and interdependent tiles.

In fact, one reptile had made the marvelous burrow in front of me, and others of its species had excavated nearly all other such burrows punching through the ground of that longleaf-pine eco-system on St. Catherines. It could be credited to the handiwork of one of the most industrious reptiles in the world, the gopher tortoise (*Gopherus polyphemus*). This special reptile was key to encouraging an even greater biodiversity, but one hidden in an ecological dimension that is often neglected by our surface-biased senses.

Gopher Tortoises for the Underground Win

Gopher tortoises are so remarkable that I and other people who study them cannot stop talking about them. As a testimony to my awe for these animals, I once began a presentation about them at a professional meeting with a photograph of a gopher tortoise, and using my best "Randall" voice (narrator of the hilarious 2011 "honey badger" parody video), said, "This. Is the gopher tortoise. Look at him. He's pretty badass." Which he was, as is every other gopher tortoise, because few animals can match them for sheer burrowing prowess while also imparting such a significant effect on their ecosystems. In fact, ecologists quite properly refer to these tortoises as *ecosystem engineers*, placing them in the same elite category as beavers for their ability to shape a habitat and affect so many other species.

Nonetheless, many people who see a gopher tortoise for the first time are not so impressed. Most adult tortoises are the width of a Frisbee, at about 25–35 centimeters (10–14 inches) long and slightly less wide. Like many tortoises, their shells are dull brown-gray on top, but have a daring splash of yellow on their bottom part, also known as a plastron. Interlocking bony plates compose their shells, and overlapping scales and smaller bony parts (scutes) cover their legs and heads. They are quiet animals, at best only emitting a hiss when disturbed, and their eyes and facial expressions reveal only a limited range of emotions. If seen moving about in an open field, they are dreadfully slow, walking only about twice as fast as a texting teenager, but can move surprisingly fast once they sense danger. In short, "badass" might not the first description that comes to mind when you see a gopher tortoise.

Yet when it comes to gopher tortoises, appearances can be deceiving. Take a closer look at a gopher tortoise's front feet and there is the first clue of what makes them such extraordinary animals. Along the outer edges of their thick, meaty front feet are hard ridges, accompanied by flat soles and robust claws that look much like the "teeth" on the business end of a backhoe. These feet connect to powerful arm muscles that provide more than enough strength and stamina to tear through the earth however much is needed. They burrow using alternating strokes of these front feet, ripping into the soil maybe a half dozen times with the left, a half dozen times with the right, and so on. Their hind feet resemble miniature versions of elephant feet, with rounded pads attached to stout, pillar-like legs; they too have claws, albeit not as prominent as the ones on the front feet. These feet mostly stabilize a tortoise as it digs, but also help with kicking back any sediment loosened by the front feet.

As these tortoises burrow deeper and move back and forth in tunnels of their own making, their hard shells act as soil compactors, pushing against formerly loose sand and firming the walls around their burrows. The half-moon profile of the entrance I noticed in every tortoise burrow is more or less an outline of the tortoise's body length and height. The tortoise makes the burrow just big enough to allow it room to move up and down the tunnel, but also so it can turn around while still inside. Sand pushed behind their burrows accumulates as overlapping sheets outside the entrance, forming a sand apron in front of the tunnel.

And down this tunnel goes. Although most burrows are "only" 5–10 meters (~16–33 feet) long, some are 14 meters (46 feet) long. These burrows also can have vertical depths of 6 meters (~20 feet), but normally are 2–3 meters (6.6–10 feet) deep. Just to put their burrows in perspective when compared to the body dimensions of a typical tortoise, this would be like me digging a 35–70 meter (115–230 foot) long tunnel, but with only trowels. On the way down with their digging, tortoises take a gradual right or left turn, adding a twist to their tunnels that might form a half spiral. They finish off their subterranean masterworks with an enlarged chamber, which serves as the main living space for the tortoise. In general, the sandier the soil, the deeper the chamber, and although we would quickly suffocate in such enclosed chambers, tortoises handle them just fine, as they are adapted for low-oxygen conditions. In contrast, more clay in an otherwise sandy soil tends to retain carbon dioxide in a burrow, which ultimately is bad for a tortoise. Accordingly, their burrows are shallower in clay-rich soils, which gives them a better chance to easily reach the surface and put more oxygen in their blood. Given a choice, then, gopher tortoises benefit more from burrowing into sand rather than clay.

It thus makes sense that if you were trying to spot gopher tortoise burrows, whether on the ground or flying overhead, you would look for sand. Because their burrows are so deep, these tortoises must eject large volumes of sandy soil, which then composes the sand aprons. These are 1–2 meters (3.3–6.6 feet) wide, 20–30 centimeters (8–12 inches) tall, and are normally composed of bare sand, as the continued digging and other movements of an active tortoise prevent plants from putting down any roots. This is one of the ways I could tell right away whether a burrow is occupied by a tortoise: Few plants, freshly ejected sand, and crisply defined turtle tracks all say, "I'm home." In contrast, healthy plant growth and no tracks in a former apron signal an abandoned burrow. A lack of vegetation also makes it easy to spot active tortoise burrows from above, as the sand aprons show up as bright white spots in fields of green. This means active burrows can be mapped using aerial drones that fly above them and snap photos, and they even can be spotted on satellite images by using Google Earth. Yes, that's right: Gopher tortoise traces can be seen from space, which, you have to admit, is pretty badass.

Each tortoise burrow only has one entrance, which also serves as its exit. As a result, actively occupied burrows often hold evidence of where they walked and dragged their shells while coming and going. Sometimes I have spotted them briefly just outside their burrows on the sand apron, or poking out of the burrow entrance, warming themselves with a little morning sunshine. If they spot me, though, they nimbly turn around and vanish, leaving behind an outline of their shells and tracks on the sand. Yet tortoises do not emerge just for thermal therapy, but also to eat and mate (and not necessarily in that order). Tortoises are herbivores that take advantage of the wide variety of plants in their habitats, through which they will range far and wide to find their favorites, such as prickly-pear cactus. Hence aerial photos with obvious sand aprons of active burrows also reveal subtle trail networks woven among them, linear paths worn down by frequent forays outside of their havens.

Burrows and trails also give hints of other tortoise activities. For instance, tortoises do not dig just one burrow and stop with that as their dream home. Instead, similar to some baby boomers, they can own so many residences that they lose count of them. Interestingly, male tortoises make as many as fifty burrows, which is two to three times as many as made by females. Tortoises might also move from burrow to burrow, chowing down on plants during their commutes. This burrow productivity means that estimating tortoise populations is never as simple as counting burrows with sand aprons, which would grossly overestimate the number of tortoises in a given area. Usurping and swapping are also common practices, a sort of burrow promiscuity that impels gopher tortoise researchers to tag individual animals to see where they go and how often they change addresses.

It is then suitably easy to imagine how trails between burrows might also be traces of tortoise-boy-meets-tortoise-girl stories. Such tales begin with a male tortoise leaving one of his many burrows in search of love, or at least a female tortoise that will not reject him that given day. Similar to males in other animal mating systems, a male tortoise may have to travel quite far and quite likely is spurned by many possible mates encountered along the way, for what may be perfectly reasonable female tortoise reasons. If a mature and consenting pair of tortoises meets—usually just outside her burrow—the male then commences a courtship display, unleashing a series of alluring head-bobbing moves. If this successfully woos a female, she will come out of her burrow and indicate her interest by bobbing her head, too. That is all he needs to know, so he then bites one of her front legs: In gopher tortoise language, this translates as, "Turn around, baby." The pair then does what nature intended.

However, unlike most birds or many mammals, mated tortoises have little to do with each other once the deed is done, duties are fulfilled, genes are delivered and mixed, the tortoise has come into the station and left, and, well, you get the point. This aloofness also means they do not later relax together in each other's burrows, illuminated by the afterglow, nor do they shack up as a tortoise couple living in blissful harmony. Instead, they go back to their separate burrows. Later, when the time is right for laying eggs, the gravid female tortoise takes advantage of the all-natural incubator sitting just outside her burrow, the sand apron. For her nest, she digs into the thickest part of the apron, lays a clutch of 2–10 eggs, and buries them. Until recently, researchers thought gopher tortoise mothers did not watch out for their eggs, nor even stay in the same burrow next to them. However, secret video footage recently taken of a mother tortoise showed it aggressively defending both its nest and burrow from a rapacious armadillo. Once laid and buried, eggs take about 90–100 days to hatch. Overall temperatures of the sand apron during that time control the sex ratio of the clutch, with warmer temperatures resulting in more girl hatchlings and cooler temperatures producing more boys.

As soon as hatchlings break out of their eggshells, they claw through the overhead sand, pop out of the sand apron, and, just like 1950s U.S. schoolchildren faced with nuclear peril, they duck and cover. They do this by scurrying straight to the nearby adult burrow entrance and going inside, where they can quickly conceal their still-soft shells from snakes, raccoons, hawks, foxes, or other predators who might enjoy some fresh baby tortoise treats. Newborn tortoises are also literally born to burrow, and within mere days of hatching, they may add their own smaller tunnels branching off the main burrow walls or make their own new burrows outside. Most additions to their mother's burrow are concentrated toward the top, making what was originally a relatively simple structure into a more complex one. Once these hatchlings get big enough, they strike out to make homes of their own elsewhere.

When this brief period of vulnerability passes and gopher tortoises grow up into adults with hardened shells, they are difficult to kill. For instance, any animal that wants to eat a gopher tortoise usually ends up hungry, as its intended dinner clams up, retracting all of its tasty bits inside its personal body armor. Indeed, gopher tortoises are great survivors, with some living for more than 80 years. Such longevity implies that if a baby tortoise makes it to an elderly status, it may have dug tens of thousands of burrows during its lifetime. If you take that number and multiply it by, say, a hundred gopher tortoises in the same longleaf pine forest, then this collective effort will have resulted in more than a million burrows. Granted, some of their older burrows may have collapsed, or been filled by sand washed in by thunderstorms, or erased by burrows of newer generations of tortoises, as well as the burrowing actions of other animals, from earthworms to armadillos. Yet the overall cumulative effect of these burrowing tortoises will have been to completely overturn, alter, aerate, and ultimately enrich the soils of that ecosystem. This cumulative activity in turn encourages the high diversity of the local plant community. In short, without these tortoises, the pines and other vegetation in their habitat would not be nearly as plentiful or as varied.

Other than predation of young tortoises or humans messing with them, what threats do they face in their natural habitat? Very few. As mentioned previously, predators are almost powerless against an exposed tortoise. Even if a predator decided to excavate a tortoise, it might expend more calories digging than it would gain from eating a tortoise. Just like Coober Pedy or Montreal, tortoise burrows also moderate potential problems with daily or seasonal variations in temperatures, maintaining even temperatures year round. Likewise, humidity levels stay at comfortable levels. That leaves natural disasters as the only major problems that tortoises might face in a longleaf pine forest.

Quiz time: What natural disasters would be most likely in a longleaf pine forest? If you said "Fires, and lots of them," you get an "A" for paying attention earlier. Yet in the event of a fire, a gopher tortoise would not simply pull in its legs and head and wait until the conflagration passed through its neighborhood. If so, it would then become a surprise item on a Southern barbeque, followed quickly by an argument over which sauce would go best with it. This is again an instance of where burrows aid in tortoise survival, fire after fire. With the first flickering flames, a gopher tortoise will retreat to a nearby appropriately sized burrow, go meters below the surface, and simply wait until the inferno has subsided.

This extremely successful survival strategy of gopher tortoises also points to yet another reason why gopher tortoise burrows are so splendid and why I have paid attention to them both as an ichnologist and paleontologist. These lengthy holes in the ground are not just self-serving, ensuring that tortoises will eat, head-bob, and leg-bite their way into making future generations, but also enable tortoises to be accidental altruists for other animals. These burrows can even direct the evolution of multiple animal lineages, resulting in amazing adaptations to these underground environments. In other words, tortoise burrows do not just save them, but they also aid in the continued existence of a menagerie whose lives would surely end if not for the fortuitous placement and abundance of these subsurface shelters.

The Underground Zoo

Based on a description that only focuses on the wonder-inducing burrows of gopher tortoises, you may have calculated that their apparent animal biodiversity consists just of a gopher tortoise and perhaps (temporarily) its offspring. This would be very much wrong. Gopher tortoise burrows, already worth our admiration for their sheer depth and length, also deserve notice for how they expand the biodiversity of a longleaf pine forest or savanna into a dimension often unnoticed or neglected by mere surface-dwellers: underground.

Because of the voluminous amount of space afforded by their burrows, tortoises can host many roommates, and of a wide variety, consisting of nearly 400 species. Yes, that number is 400, not 40, and it justifies why biologists who survey gopher tortoise burrows are always surprised by what they find in them. This subsurface biodiversity is further magnified by the great number of tortoise burrows, most of which are temporarily unoccupied or abandoned; such burrows then act like empty houses in a neglected neighborhood. Regardless of whether or not the tortoises are home, their burrows still provide many perfectly suitable shelters for other animals, just so long as their entrances remain open. Some of these burrow residents (also called *commensals*) are permanent, and would have a very tough time indeed if gopher tortoises were not around. Such animals are *obligate commensals* in that they live their entire lives in the burrow, never leaving it. In contrast, residents that can check in or out anytime, getting anything they want, are *facultative commensals.*

Of the tortoise-burrow roommates documented thus far, more than 300 are invertebrates (mostly insects) and about 60 are vertebrates. For the vertebrates, more than half are other reptiles (lizards and snakes), whereas amphibians and mammals make up the rest. But perhaps the most famous co-occupants of tortoise burrows are snakes, the most charismatic of which are eastern indigo snakes (*Drymarchon couperi*) and eastern diamondback rattlesnakes (*Crotalus adamanteus*).

The eastern indigo snake is the longest snake native to North America, reaching more than 2.5 meters (8.2 feet), and if you are ever lucky enough to spot one, it is usually on its way to or from a tortoise burrow. This snake is not only lengthy but is also adorned with iridescent blue-black scales. These nonvenomous snakes, which prey on a wide variety of vertebrates (frogs, toads, lizards, birds, rodents, and even other snakes), depend so heavily on gopher tortoise burrows that their health and abundance are directly linked to the number of tortoises in a given area. Tortoise burrows do this by providing indigo snakes with even temperatures all year, keeping them safe from predators, and protecting them against the frequent and intense fires of longleaf pine forests. These snakes also nest in burrows, which conceal both eggs and baby snakes from anything that might eat them.

These same factors apply to rattlesnakes, which often curl up in gopher tortoise burrows to escape seasonal extremes in temperature or natural disasters. Perhaps unexpectedly, indigo snakes and rattlesnakes might live together in a tortoise burrow, but they normally leave each other alone. This is not surprising, though, as indigo snakes include other snakes on their menu and are immune to rattlesnake venom, which creates a sort of détente between these two serpents. As far as we know, both snakes are also innocuous housemates if an adult gopher tortoise is living in a burrow with them, although either probably would not hesitate to eat hatchling tortoises. However, such instances of tenants eating their landlord's children are probably rare.

As mentioned earlier, other vertebrates that live in tortoise burrows are frogs, toads, and mammals; some of these animals live in the burrows year round, whereas others only pop in for an occasional visit. Among the permanent occupants are gopher frogs (*Rana capito*). Gopher frogs are small (2.5–4 centimeters/1–1.5 inches long) but big-headed, and have blotchy brown-gray and warty skin over most of their upper bodies. These frogs depend on tortoise burrows for most of their lives, only leaving burrows occasionally to get food or travel overland to seasonal shallow pools of water to mate and lay eggs. As soon as tadpoles sprout legs and are able to hop on land, they point their weighty heads toward the nearest burrows. This reliance on tortoise burrows is again because of the steady safety they offer these tiny frogs: even temperatures and humidity year round, as well as protection against predators and fires. Humidity control is especially important for frogs, as their skin would dry out too quickly if regularly exposed to arid conditions. Hence tortoise burrows also help adult frogs and other amphibians survive droughts.

Of the roommate animals discussed so far, none of them changes a tortoise burrow; they just live there. Not so for the Florida mouse (*Peromyscus floridanus*), which is cheeky enough make its own additions to tortoise burrows. This burrowing mouse digs out horizontal, mouse-wide looping tunnels off the upper 2 meters (6.6 feet) of the main tortoise burrow. It then adds curving vertical shafts that connect to the surface, giving mice easy entrances and escape routes from their snaky, rodent-eating neighbors. These shafts also give mice the means to still access an abandoned tortoise burrow after its entrance had collapsed, like having secret entries to a vacant warehouse. Mice also may save themselves some digging by following premade tunnels of hatchling tortoises, which they then modify for their own needs, such as for nesting chambers that they use to raise their young. So when you look at a gopher tortoise burrow, think of it having mouse nurseries and day care centers in it, too. Also, just like many other animals living in tortoise burrows, mice use these places as refuges from heat, low or high humidity, predators, and fire.

Unlike the preceding semipermanent residents, most vertebrates use tortoise burrows like a roadside motel, dropping in just for a day or two to rest. For animals that need to overwinter, they may just stay seasonally. However, these burrows become highest in demand whenever a fire starts in a wiregrass savanna or longleaf pine forest. Suddenly, tortoise burrows are the places to be, and anyone who can fit down these holes will seek sanctuary in one. Fires can thus cause unusual mixtures of animals in a burrow, such as snakes, lizards, rabbits, foxes, armadillos, skunks, opossums, and chipmunks. Such disaster-induced groupings must get socially awkward whenever they result in predator-prey pairs getting together. Nonetheless, the odds are slightly better for prey animals that shack up with their predators, rather than getting cooked in the blaze outside a burrow. Also, some tortoise burrows are large enough that rabbits and foxes may not run into each other.

Other burrow inhabitants include nine-banded armadillos (*Dasypus novemcinctus*), which take over abandoned gopher tortoise burrows and modify them according to their needs. Like tortoises, armadillos carry personal body armor as overlapping plates of bony skin, but differ by having rounder profiles. As a result, I can always tell an armadillo-altered burrow from an unaltered one made by a gopher tortoise by the elliptical cross-section of the former and the half-moon shape of the latter.

As mentioned earlier, most invertebrates living in a gopher tortoise burrow are insects, although a good number of worms, spiders, ticks, and other non-insect animals might call them home, too. What is unexpected about these insects, though, is the number of them that evolved exclusively to fit the microenvironment of a gopher tortoise burrow. For instance, burrows host a number of dung beetle species that are well suited for using tortoise dung, such as the little gopher tortoise scarab beetle (*Alloblackburneus troglodytes*), the gopher tortoise copris beetle (*Copris gopheris*), the punctuate beetle (*Onthophagus polyphemi polyphemi*), and smooth gopher tortoise beetle (*Onthophagus polyphemi sparsisetosus*). This little beetle-run sanitation service neatly solves the problem of a tortoise having to go all of the way up the burrow to use an outdoor latrine, while also keeping its burrow relatively tidy. Such tidiness is important not for aesthetic reasons, but for health, as less dung also means fewer flies in the burrow, as well as fewer intestinal parasites lurking in the dung. Of course, dung beetles are always happiest when they get to feed crap to their kids, too. (Confession time: I love dung beetles.) Dung is used as provisions in brooding chambers, where mother beetles lay their eggs; these eggs hatch into hungry grubs yearning for some excremental goodness. These species of dung beetles are so specialized that they are only found in tortoise burrows, suggesting that they may have evolved in harmony with tortoises as mutualistic partners, united by poop and burrows.

While these beetles serve as a tortoise's cleanup crew in their burrows, other beetles act as exterminators. These insects are the gopher tortoise hister beetle (*Chelyoxenus xerobatis*) and equal-clawed gopher tortoise hister beetle (*Geomysaprinus floridae*), as well as the gopher tortoise rove beetle (*Philonthus gopheri*) and western gopher tortoise rove beetle (*Philonthus testudo*). All of these beetles are predators, seeking out and eating eggs, larvae, pupae, and adults of insects or other arthropods living in a tortoise burrow. These beetles thus act like pest control, decreasing the number of flies, lice, ticks, and other arthropods that might spread disease or otherwise adversely affect the health of a tortoise. Again, these beetles live exclusively in tortoise burrows and apparently evolved in these subsurface environments with their vertebrate hosts.

Not all flies in a tortoise burrow are annoying vermin that must be destroyed, though. Some flies also help with dung removal, whereas others rid a burrow of unwanted insect guests. For waste control, the gopher tortoise burrow fly (*Eutrichota gopheri*) also partakes in tortoise dung, using their droppings as a food source for its larvae. Other flies are insect predators, lending air support to their ground-dwelling compatriot beetles. One of these is the gopher tortoise robber fly (*Machimus polyphemi*). To get a clue what robber flies do, they are also called assassin flies. To justify such a frightful nickname, these flies forcefully seek out and attack their prey—often other flies—by grasping them, piercing them with a needle-like proboscis, and injecting a cocktail of neurotoxin and enzymes that both paralyzes and predigests their wretchedly hapless victims. In short, gopher tortoise robber flies are handy insect allies for decreasing the number of "bad insects" in a tortoise burrow. Interestingly, gopher frogs likely include dung flies on their menu, pointing toward yet another strand in this complex underground food web.

Still not impressed with how gopher tortoise burrows have nudged branches on the evolutionary tree of insects to point in certain directions? Well then, how about moths that only live in tortoise burrows, some of which only eat dead tortoise skin? Moths, often regarded as the dark-adapted versions of their more colorful butterfly cousins, actually have adapted to a wider range of ecological niches than butterflies, among which are those in gopher tortoise burrows. For instance, the tiny gopher tortoise acrolophus moth (*Acrolophus pholeter*) is well suited to these burrows, living in perpetual darkness and joining other insects in its consumption of tortoise wastes. The gopher tortoise noctuid moth (*Idia gopheri*) performs a similar function as a "litter eater," eating organic detritus that otherwise would accumulate in this enclosed environment. These moths are then aided by what is arguably the most specialized of all insects in a tortoise burrow, the tortoiseshell moth (*Ceratophaga vicinella*).

The tortoiseshell moth deserves our special attention because it is the only insect in North America known to eat keratin, and only keratin. Keratin is a structural protein essential to the outermost layer of our skin, as well as skins of other mammals, birds, reptiles, and amphibians. More specifically, it provides necessary support for fingernails and toenails in humans as flexible but still strong non-mineralized ("non-bony") tissues. In gopher tortoises and their turtle relatives, keratin forms an extra-tough layer on top of their bony plates, giving them additional protection, like wearing a leather jacket over a bulletproof vest. As anyone who has had a roommate annoyingly clip fingernails and toenails and leave clippings on the floor can attest, keratin can last a long time. Hence an insect in your home that would dispose of all fingernail and toenail clippings would be quite handy. For a tortoise, though, the tortoiseshell moth performs its most valuable ecological function when that tortoise dies.

Once a tortoise expires and goes to that Great Wiregrass Pasture in the Sky, whether it is in a burrow or not, the keratin on its shell becomes moth larvae chow. In the typical life cycle of a tortoiseshell moth, it lays its eggs on the tortoise shell, which may be literally belly-up. The larvae that hatch from those eggs then dig a silk-lined burrow 3–10 centimeters (about 1–4 inches) down into the ground below the shell. These burrows not only give the larvae immediate access to their food source, but also protect them from predatory insects while they are still in their larval state and during pupation. Adult moths that emerge from the burrows accordingly seek out more tortoise keratin, which is essential for the perpetuation of their species. These scavengers then help to break down the shell and cycle a former tortoise's molecules back into the ecosystem of its former burrow.

The burrows added by tortoiseshell moths onto the larger burrow of a gopher tortoise also point to how burrowing insects add even more ichnological complexity to the overall form of any given gopher tortoise burrow. Insects in the much larger burrow of a tortoise may also be digging into its walls, drilling narrower tunnels and shafts, like short rootlets branching off the taproot of a tree. For example, camel crickets (*Ceuthophilus latibuli*) do not necessarily have to live in tortoise burrows, but they often do, and when there they excavate tunnels into the walls. Many beetles, including dung beetles, also burrow as part of their life cycles. Now imagine these smaller burrows being added onto those of juvenile gopher tortoises, as well as those of Florida mice, and this seemingly simple "big hole" becomes much more complex, holding thousands of nooks and crannies.

Such burrow complexity was verified partially by the research I was doing with the scientists from Georgia Southern University on St. Catherines Island. For this work, we recorded surface features of the burrows, such as widths and heights of the entrances. Along with this, though, we did something different by applying advanced technology, namely a ground-penetrating underground radar (GPR) unit. This device—also used by scientists who discovered the World War II escape tunnel in Vilnius, Lithuania—did the neat trick of creating digital representations of the tortoise burrows and the burrows of their roommates. This not only avoided disturbing the burrows and their occupants, but also saved us researchers from having to use shovels to expose (and hence destroy) the burrows.

A GPR unit looks much like a lawnmower, with an upright handlebar on one end, which is connected to a boxy part with four wheels. The difference between it and a lawnmower, though, is the small computer console on the handlebar used to program, display, and store data collected by the boxy part. This portion of the GPR device sends microwaves into the ground, which, fortunately, are too weak to cook gopher tortoises or any other animals that might be in the burrow. The microwaves, acting like underground sonar, reflect off and refract in substances with different densities, such as air (in a burrow) and soil (around the burrow). The data collected by the GPR unit are then downloaded into a computer, which converts the microwave data into a series of colorful two-dimensional maps, nicknamed "slices," like horizontal slices of a layer cake. Even better, a computer can then take this series of slice-maps and render them into a three-dimensional rotating image. This was what my Georgia Southern University colleagues were trying to do, and herpetologist Veronica Greco ably steered us to all of the good places to find tortoise burrows.

It worked beautifully. We used the GPR unit on a large, deep, and magnificently expressed burrow made by an adult tortoise, which I described earlier. We then later tried it out on a field where a bunch of younger, juvenile tortoises had dug their smaller, shallower, and more closely spaced burrows. Based on presence or absence of fresh sand aprons and tracks, we could tell that some of these burrows were active, whereas others were unoccupied. First, we walked the GPR unit over a carefully measured area to get maximum coverage of the large adult burrow below the surface. Later, Sheldon Skaggs downloaded the data, put it into a computer program, and made a rotating 3-D image of the burrow. Just as we expected, the single tortoise burrow went down at a steep angle, but twisted to the right, or clockwise if viewed from above. This main burrow also had a smaller, looping horizontal tunnel branching off the main shaft near the top of the burrow. We surmised that this was either the burrow of a Florida mouse, a juvenile tortoise, or a combination of the two. Upon seeing such results, we triumphantly declared "Science!"

However, when we did the same procedure with the juvenile tortoise field, its results surprised us, causing us instead to say, "Science?" The juveniles, because they did not travel far from their mother's home burrow, had dug their burrow entrances fairly close to one another, but definitely separate: Think of university faculty not wanting to have their offices directly adjacent to one another. This separateness did not continue below the surface, though, as the inclined burrow shafts intersected one another. Consequently, the 3-D image of these burrows showed a jungle-gym network of open burrows, whether occupied and unoccupied, crisscrossing and otherwise making a complicated structure. Had I seen such a burrow complex preserved in the fossil record, I would have immediately interpreted it as one made by mammals, not reptiles, let alone tortoises, and juvenile tortoises at that. Lesson learned. All in all, though, we were fairly happy with our results and confident that these would be helpful both to conservation biologists—who would be thrilled to have a new way to study tortoise burrows without bothering their owners—and to paleontologists, who might find similar structures in the fossil record.

However, as is typical in ichnology, other scientists studying gopher tortoise burrows soon bettered our "cutting-edge" and "state-of-the-art" research. We started our field work on St. Catherines Island in January 2011, and I presented our preliminary outcomes to my geologically inclined peers at the annual Geological Society of America meeting. I proudly projected the 3-D rotating models of the tortoise burrows and discussed the composite nature of the burrows-within-burrows, and was encouraged by audience "oohs" and "ahhs." More nonverbal approval was indicated by nodding heads, which I was mostly sure were not from napping, but affirming. Along with this professional exposure of our research, I published a blog post about it on my Web site, albeit without the GPR profiles, as we were saving that for a peer-reviewed paper.

Then, life interfered with our finishing that article in a timely way. Meanwhile, two other scientists, unaware of the work we had done, coauthored a superb article on—you guessed it—the use of GPR for subsurface imaging of gopher tortoise burrows. The article, published by Al Kinlaw and Mark Grasmueck in 2012 in the journal *Geomorphology*, was a well written and richly illustrated study of gopher tortoise burrows using this same newfangled technological tool we used. Unlike our study, though, it was performed on tortoise burrows far south of our field site, in north-central Florida.

Although some scientists at this point might have become exasperated, perhaps even shook fists at the sky and shouted "Curses, foiled again!", we were actually elated and somewhat relieved. For one, because a herpetologist and a geologist had done this study together, they demonstrated the sort of cooperative work we felt needed to be done to properly study these burrows. (Most of us working on the St. Catherines Island project were trained as geologists, only wishing we were as cool as herpetologists.) For another, these scientists took a lot of pressure off us. Now we did not have to explain why this new technique worked to skeptical peer reviewers, but instead could just cite this new article and say, "Hey, they did it, it worked for them, and it worked for us, too." Lastly and perhaps most importantly, we were now free to focus our study more from ichnological and geological perspectives, rather than a conservation biology angle, which was the focus of the 2012 study. In short, we could ask, "What would these burrows look like if preserved as trace fossils?" and suggest that these beautiful 3-D rotating models of tortoise burrows would give geologists and paleontologists the right search images for recognizing burrows made by these tortoises' ancestors, thus giving us better insights on how these animals evolved, survived, and even thrived throughout the geological past. Accordingly, we started rewriting our article, which we hope to have published by the time you read this sentence.

What did Kinlaw and Grasmueck find? Their results were impressive, with their main conclusions stated in the title of their article: "Evidence for and Geomorphologic Consequences of a Reptilian

Ecosystem Engineer: The Burrowing Cascade Initiated by the Gopher Tortoise." This was a significant article for all interested in gopher tortoises, but also had broader implications related to the potential kaleidoscope of diversity in each tortoise burrow. First of all, the article was important because it had lots of pictures depicting the previously secret underground architecture of gopher tortoise burrows. The old adage of pictures expressing meaning more efficiently than words was certainly true here, as the images Kinlaw and Grasmueck produced from the technological pairing of GPR units and computers were stunning.

Nevertheless, as cool as it was to actually see the forms of these burrows, the authors made a more holistic point that emphatically connected these remarkable reptiles' burrows as change agents, both ecological and evolutionary. They stated that gopher tortoises were ecosystem engineers enabling other, smaller animals to add their burrows. This is the "burrowing cascade" part of their title, which I will discuss in more detail soon.

First, let's talk about geomorphology, which is both in the title of the article and comprises the name of the journal where it was published. Geomorphology literally means "study of earth forms [landforms]." It is typically regarded as a subdivision of geology, but also has been co-owned with geographers. Geomorphologists have traditionally classified and interpreted the origin of every bump, dimple, hill, mountain, valley, basin, or pothole on the earth's surface. Much of this science was done from the perspective of physical processes forming these land features, such as the actions of wind and water. This, however, is wrong. Once biologists and geomorphologists realized that many landforms were actually formed by organisms, and especially animals, their formerly separate disciplines began to merge. Along these lines, a book that helped to synthesize biological and geological approaches was one with the unwieldy title of *Zoogeomorphology: Animals as Geomorphic Agents*, written by David Butler and published in 1995. As a sign of how interest in animal-modified landscapes developed since then, Kinlaw and Grasmueck's article was one of 16 articles in a 2012 special issue of *Geomorphology*, co-edited by David Butler and Carol Sawyer and organized around the themes of zoogeomorphology ("study of animal-earth landforms") and ecosystem engineering by organisms in general. Hence now it is common knowledge among both ecologists and geologists that animals not only change landscapes but also can affect the entire direction of an ecosystem.

What Kinlaw and Grasmueck found was that the tortoise burrows exemplified this principle of ecosystem engineering, but in the hidden landscape below ground, where geomorphologists and ecologists alike might not normally look. The GPR- and computer-generated images of tortoise tunnels showed that they not only twisted to the right or left as they descended, but also sometimes spiraled or split into separate tunnels. Just as we found with our Georgia burrows, these Florida burrows also cut across older generations of long-abandoned burrows. This led to the question of whether or not the splitting burrows were made by the same tortoise or more than one at different times. As if this was not already complicated enough, these burrows had many narrower tunnels popping out of their sides. These additions were Florida mice burrows, and they were marvelous. Instead of just going out straight out and curving, they made complete loops that started and ended with the main tortoise burrow, looking like little bows tied onto this animal-architectural present. Although the researchers did not detect many finer-scale insect burrows with this technology (someday, we will), these were also implied.

The images thus documented how gopher tortoises, through their burrows, initiate incredibly complex underground environments, which then enable the burrows of many other animals, from Florida mice through insects. For imagining this ecological effect as a "cascade," think of the gopher tortoise burrow as a small, meandering stream emptying off a cliff as a trickle. Then as more animals add their tributaries to this base flow (more burrows, and lots of them), the trickle eventually becomes a mighty waterfall, one worthy of artsy photographs, honeymoons, and barrel riding.

But for the waterfall to exist, it first needs a stream; otherwise the other tributaries will have no place to join and contribute their flow and instead will evaporate well before reaching the cliff.

Just to continue this analogy just a bit further, imagine thousands of waterfalls in this fantasy environment and how each would have been started by the same species of animal. After all, one tortoise can make many burrows during a normal lifetime, so many tortoises sharing an area would have a multiplicative effect on that ecosystem. Just to do some math, let's assume that at any given time an individual tortoise might have 20–50 burrows in a longleaf pine–wiregrass community. If a hundred tortoises are in that community, 2,000–5,000 burrows could be there. To these burrows, add older ones made by previous generations of tortoises in all stages of life (juveniles to adults) that still have open entrances or otherwise afford some space below ground for other species. Attach another variable to the equation: hatchling tortoise burrows dug into the sides of female tortoises' burrows. Florida mice burrows, looping off the main tortoise tunnel, provide yet another variable. Finally, the most complicated part of the equation are the insect burrows, some of which branch off the main tunnel, the juvenile burrows, the mouse burrows, and those of other commensals.

Given such a mathematical exploration, we soon realize that what was first regarded as banal, half-cylinder tunnels that just turned to the left or right are actually better described as fractal patterns, requiring a fractal analysis or similarly complicated mathematics to assess their true volumes. The final step in our imaginary model is to multiply all of the burrow volumes by the number of burrows (previous and present) in that area. Suffice it to say that the volume of underground habitat would be enormous, far exceeding the surface area of the longleaf pine forest, and with the potential to host a great number of species. In fact, the surface habitat owes its existence to what is below, as the effects of the burrowers and this underground ecosystem have molded the plant and animal communities above.

The take-home message from numerous studies of gopher tortoises is that if you erase tortoises and their burrows from a long-leaf pine forest, it is still that ecosystem, but one far less diverse and adaptable to change. Indeed, some researchers have not only proposed that gopher tortoises are ecosystem engineers, but their absence in their preferred ecosystem can cause a different kind of cascade: a collapse. An ecological collapse—sometimes called a *trophic cascade*—is what happens when a certain species, such as a plant or animal, is removed from an ecosystem. This removal causes many other species depending on that plant or animal's ecosystem functions to start dying before they can reproduce, thus cutting the guide wires in a food web and setting off a wave of extinctions. Just like pulling out the crucial block while playing a game of Jenga, an ecosystem can quickly crumple and become a ghost of its former self.

Considering what we now know about gopher tortoises and their impacts on so many other species around them, it is no wonder why they are protected under U.S. federal law. Legal issues aside, if people purposefully kill tortoises, tear up their habitats, or otherwise mess with these important reptiles, they inflict a huge amount of collateral damage on a coevolved heritage that may have taken millions of years to produce, all of it facilitated by tortoise burrows. This ecological awareness also connects to the question of how such codependent evolutionary relationships began in the burrows of tortoises or other animals. To answer those questions, it is time to go to the past.

Deep Time with Deep-Burrowing Tortoises and Turtles

Modern gopher tortoises did not invent burrowing themselves, but instead had it bestowed on them by the genes and natural selection of their ancestors. When did these ancestors evolve such incredible burrowing abilities, and when did ecosystems begin to evolve in conjunction with such engineers? Unfortunately, because we do not have enough evidence from the fossil record, such questions cannot be answered in full yet. Based on fossil tortoises, though, we know the immediate ancestors of gopher tortoises lived in the southeastern part of North America during the past two million years and were abundant until humans started altering their habitats in just the past few hundred years.

Close relatives of gopher tortoises also lived in the southwestern part of North America, and today that is where all four of the other extant species of *Gopherus* live: the Mojave desert tortoises (*G. agassizii*), Texas tortoise (*G. berlandieri*), Bolson tortoise (*G. flavomarginatus*), and Sonoran desert tortoise (*G. morafkai*). All of these species are burrowers, and like their southeastern kin, their burrows host many species of animals, increasing the biodiversity of their desert habitats. Presumably their burrowing abilities mean their most recent common ancestor—which likely lived more than a few million years ago—also burrowed and had similar ecological effects in its environments. In fact, fossils identified as the genus *Gopherus* go back much further in time to the earliest part of the Oligocene Epoch, about 33 *mya*.

The oldest known fossil tortoises are from nearly twice that time span, found in Paleocene deposits in Asia, dating back to 60 *mya*. Interestingly, fossils of their immediate turtle ancestors in Asia are from nearly 70 *mya*, which was during the Late Cretaceous and just before the mass extinction that wiped out the non-avian dinosaurs. Since the Paleocene, descendants of these ostensibly slow-moving land animals somehow dispersed from Asia to five continents and many isolated islands. Some became gigantic, such as the Miocene to Pleistocene tortoise *Megalochelys atlas*, which probably weighed more than 500 kilograms (1,100 pounds). Tortoises are also among the longest-lived of all animals, with a few estimated to have lived more than 180 years. In short, these have been very successful animals, worthy of our admiration for feats other than burrowing.

Still, despite paleontologists finding and describing many fossil tortoises, we do not know for sure how many of them burrowed or when they began burrowing. Even more disappointing, no one has yet identified fossil tortoise burrows. Fortunately, knowing that modern tortoises make so many burrows per tortoise, it is perfectly reasonable to assume trace fossils of these burrows should be much more common than their body fossils. To find these burrows, though, paleontologists would have to learn and apply search images supplied by studies of modern tortoise burrows, such as GPR-generated reproductions of them. Furthermore, these fossil burrows will probably be filled with sedimentary rock differing from the original soft sediments the tortoise dug into: for instance, a mudstone (formerly loose mud) filling sandstone (formerly loose sand). Such fossil burrows will definitely *not* be open holes with fossil tortoises at their ends.

While keeping such preservational differences in mind, a fossil tortoise burrow also should have a half-moon cross-section matching the body length and overall profile of its tortoise maker. This shape might continue into the rock at a 20–40-degree angle relative to the former ground surface, and if it can be followed far enough, it should twist to either the right or left, and perhaps spiral downward. Fully preserved burrows then should end in a large, oval mass of rock, which would be the main living chamber of the tortoise. The buried remains of the tortoise tracemaker might even be inside this rock. However, considering what we know about modern tortoises and how each one can make hundreds or thousands of burrows, the odds of a paleontologist finding a skeleton in one are low.

Sounds simple, right? Not so fast. In my ichnologically driven mind, I want that purported tortoise burrow to have plenty of commensal burrows attached to it. So that half-cylinder of rock should also have lots of little branches poking out of it, perhaps some looping, perhaps some bunched together, but ultimately conferring a complexity to its overall form. This entity should be a composite trace fossil, a Frankenstein monster of a burrow made of many parts, but stitched together by a menagerie following the lead of a single creator.

However, even if paleontologists find and properly identify such trace fossils of tortoises and their commensals, they will not be satisfied with just that. After all, the evolutionary history of tortoises is not very long when compared to the entire evolutionary history of turtles, terrapins, and their relatives, which all belong to the clade Chelonia. Chelonia stretches back nearly 250 million years, whereas tortoises are a relatively recent innovation, consisting of land-dwelling descendants of what used to be freshwater turtles. The early evolution of turtles was likely during the Early Triassic (about 240–245 *mya*), and although their exact evolutionary origins are still disputed, they are now considered to have descended from Diapsida (diapsids), the same clade containing lizards, snakes, crocodilians, and dinosaurs. Also, as we learned earlier, dinosaurs include birds, so these are diapsids, too. Based on a combination of fossils and genetics, evolutionary scientists estimate that ancestral clades of turtles, crocodilians, and dinosaurs split from one another about 260 *mya*, just before the greatest of all mass extinctions at the end of the Permian Period. *Eunotosaurus*, a 260 million-year-old reptile from Middle Permian rocks of South Africa, backs up this supposition. Its skull and the rest of its skeleton show the transitional stages of development in a turtle's now-standard body armor, including bony plates covering parts of its belly.

The discovery of the oldest undisputed fossil turtle, just announced in 2015, also helped to fill a few gaps in what used to be a mysterious origin story for turtles and their descendants. This turtle, named *Pappochelys* ("grandfather turtle"), is from Middle Triassic rocks of Germany, from about 240 *mya*. The next oldest turtles are *Odontochelys semitestacea* ("toothed turtle with a half shell," from China) and *Proganochelys* (from Germany), both of which are from the Late Triassic (about 220 *mya*). Both fossils are terrific examples of so-called "transitional fossils," in that they show many of the traits evolutionary scientists would expect to find in a primitive turtle. Like *Eunotosaurus*, these fossils had only partially formed bony plates covering their lower surfaces, not on their tops, but were much further along in developing into what we now call a "turtle." Bottom plates were originally interpreted as protection against predators from below as these proto-turtles swam.

Perhaps most intriguingly, though, is the idea that all turtles (tortoises, too) owe their legacy to burrowing. In a 2016 study, paleontologist Tyler Lyson reexamined *Eunotosaurus* and again noted its wide ribs, from which its bony bottom plates grew. But he also discerned that its head was blunt and flat on top, like a shovel. Also, its front feet had large claws, robust fingers and were bigger than its rear feet, and its thick shoulder and forearm bones had attachment sites for hefty muscles. The wide ribs toward the front of its body reinforced the arms and shoulders, providing further support for them. Yet another piece of evidence was the bony rings around this proto-turtle's eyes, which were small enough to suggest that it was adapted to low-light conditions. Granted, this odd arrangement of shoulders and ribs had some costs, such as inhibiting breathing and slowing its movement on land. What was the evolutionary advantage, then? On the basis of these traits, Lyson and his seven coauthors concluded that *Eunotosaurus* was a burrower. This ability then later served well for turtles diving into aquatic environments, as digging muscles can more easily evolve into swimming muscles. Unfortunately, a *Eunotosaurus* burrow has not yet been recognized, let alone one with a *Eunotosaurus* in it, but this intriguing hypothesis that all modern and fossil turtles owe their start to burrowing is now out there, waiting to be tested further.

Given that beginning, turtles took off and evolved beside the dinosaurs throughout the Mesozoic Era, with the greatest diversification of turtles during the Cretaceous Period at about 100–120 *mya*. This is when warmer conditions allowed turtles to move into high-latitude environments, such as the formerly polar environments of Antarctica and Australia. Yes, that's right: Turtles lived in polar environments, and although their bones are still rare in Antarctica, they are fairly common in Cretaceous rocks of Victoria, Australia, when that continent and Antarctica were still alongside each other. This seemingly strange move for what were (and still are) cold-blooded reptiles could be partly credited to the warmer temperatures of the Cretaceous. Yet these polar turtles still had to deal with freezing, dark winters. This calls into question how they survived these trying conditions every year, and for millions of years. What special qualities did they have that allowed them to adapt to and diversify in such hostile environments?

The easiest answer I have—which you probably already figured— is that these Cretaceous turtles burrowed. Many modern turtles live at high latitudes, such as wood turtles (*Glyptemys insculpta*), which are as far north as Nova Scotia (Canada). Other turtle species occur in New England, which also qualifies as a place where basking in warm sunshine is severely limited during winter. Hence these high-latitude turtles are adapted to dig into lake and stream bottoms and bury themselves as a way to cope with cold winters. This strategy also works well during too-hot summers, or times of drought, again showing how burrows are multi-tools of survival. Once buried, turtles go into a trance-like state of hibernation called *estivation*, in which they minimize their breathing and all other metabolic functions. With these modern examples in mind, in 2009 I was not surprised to find turtle-sized and shaped burrows in 105 *mya* Cretaceous sedimentary rocks of Victoria, Australia. The trace fossils, preserved in sediments deposited by formerly polar rivers and sharing the same environments as their bones, accordingly led me to wonder if overwintering turtles made them. If so, turtles were already using burrowing and estivation as survival techniques more than a hundred million years ago.

Here is the main point to think about with regard to turtles, and by extension tortoises. The immediate diapsid ancestors of turtles made it past the most awful extinction event in the history of life, which was about 250 *mya*. The descendants of those diapsids, some of which became true turtles, then survived at least two more mass extinctions in the Mesozoic Era. The first of these was at the end of the Triassic Period, about 200 *mya*. This extinction event killed off many reptiles, including some that had been evolutionarily very successful for tens of millions of years. Yet turtles persisted. The second mass extinction at the end of the Cretaceous at 66 *mya* is famous for ending the reign of the non-avian dinosaurs, but turtles again endured. This extinction also took out many large marine reptiles, with only a few carrying on after this great snuffing. Which big-bodied marine reptiles made it past that extinction and are still the most massive of marine reptiles? These would be sea turtles, which include leatherback turtles (*Dermochelys coriacea*), weighing as much as 900 kilograms (nearly 2,000 pounds).

Just how did the ancestors of both land- and sea-dwelling turtles manage to make it through three heinous events in the history of life and still be around today? If you thought I was going to say "burrowing," you would be right, but it gets more complicated than that. After all, though a few turtles use burrows for estivation and others for homes, most turtles do not live in burrows at all. For example, nobody is suggesting that Cretaceous sea turtles dug out submarine burrows to wait out an upcoming apocalypse. So rather than making homes, turtles have used underground environments in another way to save themselves and future generations: by female turtles burying their eggs.

Despite more than 200 million years of evolutionary history, no turtles give birth to live young, and females of nearly all species must burrow deep enough to at least to hide a clutch of eggs. They also must do this on land, as watery environments have a nasty habit of drowning their eggs. For this, mother turtles use their rear legs to dig into mud, sand, forest litter, or other substrates to conceal and incubate their eggs, such as gopher tortoises that use sand aprons outside their burrows as incubators. Alternatively, turtles that do not bury their eggs and instead lay them on the surface take a bigger risk of these getting eaten, stomped, or otherwise maltreated. Female sea turtles in particular are excavating masters, with each mother turtle lumbering up on land dozens of times per summer, leaving tractor-like treads in their wakes. Once in a suitable place with the right sand, these turtles carve out deep, vase-like egg chambers with their rear flippers, deposit more than a hundred eggs in them, cover these, and go back out to sea: Crawl, lay, repeat. All seven species of modern sea turtles do this, including leatherbacks. In terms of their evolutionary history, trace fossils of sea turtles, nests and trackways link this nest-digging behavior back to the middle of the Cretaceous, nearly a hundred million years ago. Furthermore, the 2015 discovery of *Desmatochelys padillai* in Early Cretaceous (120 *mya*) rocks of Colombia pushed the origin of sea turtles even further back in time, implying that these sea turtles also must have dug nests on land then.

Although we may never know all of the factors that allowed for the persistence of turtles and their kin, I have little doubt that this female-only burrowing greatly increased the survival of sea turtles and land turtles alike, helping them to succeed land-bound dinosaurs and giant marine reptiles. In short, burrows and burrowing not only helped some adult turtles avoid terrible conditions outside, but also protected their unborn future generations.

Evolving Dinosaur Roommates

While considering the major groups of vertebrates found living in gopher tortoise burrows—amphibians, reptiles, and mammals—an attentive reader may have noticed the absence of one of the most diverse of all vertebrate groups: birds. This should not be all that surprising, though, as we know most birds today make their livings on or high above the ground. Birds also have not been around as long as turtles in their evolutionary history, having only arisen about 160 *mya*. As a result, they may not have had as long to evolve as roommates with turtles and tortoises. Lastly, birds are decidedly unexpected candidates for digging their own burrows next to those made by amphibians, reptiles, and mammals, especially when one recalls how their arms were originally meant for flying and not for digging. To the very reasonable thinking of many people, then, birds belong in the sky, not in the ground.

Nonetheless, as most kids (and a few of their parents) will happily tell you, birds are dinosaurs, having descended from feathered theropod ancestors during the Jurassic Period. Yet if you asked these same kids or parents to name a burrowing dinosaur from the

Mesozoic past, you will likely get a blank look. That is, unless they just watched a specific episode of the PBS series *Dinosaur Train*, in which case they may blurt out the name *"Oryctodromeus!"* while beaming with the sure knowledge that they are right.

What is *Oryctodromeus*, and how does it relate to burrowing dinosaurs? If dinosaurs did burrow, what is the evidence for this? How common was this behavior during the Mesozoic Era, when dinosaurs seemingly ruled every environment above the ground? Also, could burrows have helped some dinosaurs escape from the same types of environmental pressures faced by other burrowing animals, such as predation, droughts, fires, and more? If dinosaurs did burrow, did any other animals live in those burrows with them, as with gopher tortoises? And what about their living descendants, birds? Did any of these dinosaurs evolve the means to burrow while still retaining flight (or not), and if so, how do they burrow? Lastly, why ask so many rhetorical questions? Because all of them link to the realization that birds were the only dinosaurs that made it through a mass extinction 66 *mya*, credited to a meteorite smashing into the earth. Following this horrific event, burrowing might have been one lucky strategy that permitted some birds to survive and then diversify and occupy nearly every environment imaginable today, including underground.

CHAPTER 4

Hadean Dinosaurs and Birds Underfoot

Tasmanian Angels

The penguins surrounded us under the cover of darkness. Fortunately, they did not seem to mind that we were in the middle of their bedroom community, so they simply moved toward and around where we sat. The nighttime sky was far too overcast to see them clearly, and we had purposefully kept our flashlights off and stayed perfectly still so as not to disturb them. All we sensed were small, shadowy figures shuffling by us and we heard their bodies brushing against low-lying coastal shrubs. Nonetheless, these whispery sounds were overpowered by raucous, rasping, and wheezing calls they made to one another. Their eerie chorus enveloped us, bearing witness to a time long before our primate ancestors decided that coming down from the trees was a good idea. Like a wave, the birds passed us by, and each voice—one by one—muffled, softened, and then ceased altogether, as if they had slipped into another dimension. It was an ethereal experience that left us wanting more.

My wife, Ruth, and I were on the eastern coast of Tasmania, Australia, and we were experiencing a nightly ritual performed by the fairy penguins (*Eudyptula minor*) of that area: They were coming home. Fairy penguins, also known as little blue penguins, spend their entire day in the sea, swimming, eating, and cavorting with one another. As is typical of penguins, they reserve their "flying" for the ocean, where they perform breathtaking maneuvers while hunting for fish, squid, or other seafood. They do this as flocks from dawn to dusk, and as the sun begins to set, some stay at sea, where they might hang out for a few days. Others, however, swim back to the land and waddle ashore, where they settle in for a night or two of rest, or stay longer to perform necessary species-perpetuating duties, such as mating, nesting, and raising young. Colonies of fairy penguins carry out these familial functions in many places over a very broad area of the Southern Hemisphere, frequenting the shallow waters and beaches of southern Australia, New Zealand, and Chile.

If too many documentary films and cartoons about the 1.4-meter (4.5-foot) tall emperor penguins (*Aptenodytes forsteri*) of Antarctica have colored your perception of penguins, fairy penguins will surprise you, and in a good way. Once they stand up on a beach, their overwhelming cuteness may very well compel you to babble an infantile string of nonsensical monosyllables while dissolving into a puddle of goo. First of all, these are the smallest of penguins, with adults reaching about 30 centimeters (12 inches), one third the height of the *Star Wars* droid R2D2. Second, their backs are composed of nearly iridescent-blue feathers, their bellies are white, and their cheeks have just a hint of blush, a plumage contrasting with that of the severe black-and-white tuxedo outfits worn by their southernmost relatives.

Further adding to their charm, fairy penguins hold their thin black wings out from their sides as they walk with webbed feet, looking as if they are performing a balancing act; which in effect they are, because they would easily topple over at the slightest push. (Please don't do this, though. Remember: overwhelming cuteness.) All of these traits are endearing enough in any given penguin, but when multiplied by hundreds, all of them baby-stepping out of the surf together and looking like one big happy family, it is enough to elicit squeaks and squeals from even the most hardened anti-penguin cynics. No wonder, then, that a longtime fairy penguin colony on the seashore of Phillip Island in Victoria, Australia, has become a huge tourist attraction. Throughout each year, hundreds of thousands of fellow biped admirers gather nightly to watch the hundreds of these birds ambling up the beach in a "penguin parade."

As a result of their diminutive profiles and slow, short-legged gaits, one might say that fairy penguins do not inspire existential dread in their enemies. Accordingly, they are vulnerable to animals that look straight past their endearing qualities and imagine them as dinner. Unlike alligators, these penguins do not possess huge mouths filled with teeth, nor do they wear body armor like gopher tortoises. Given this lack of defenses, they would surely become quick and tasty meals for anything that had an urge to eat a little penguin. This is where fairy penguins again differ from our default image of emperor penguins, which can travel overland for hundreds of kilometers in the open and during daytime. Emperor penguins can get away with such blatant, in-your-face public displays because they have no predators on land. In fact, other than humans brave enough to do field research in Antarctica, they are the largest land animals on that continent. They have no evolutionarily imposed fears of other animals, nor any reason to be cautious.

This is yet another instance where burrows come to the rescue. The entire reason why fairy penguins come onto land with the setting sun is that they are returning to burrows. Fairy penguin burrows do all of the good things that burrows fulfill for other animals: They protect adults and children alike, provide safe nesting environments, keep them warm (or cool) at night, and otherwise give them stable, cozy places to relax after several days of open-ocean swimming and fish-eating.

Burrows of a fairy penguin colony are numerous but spaced apart from one another, showing some respect as neighbors. Burrow entrances are wider than tall and big enough to allow two penguins to pass each other on the way up or down these 50–100-centimeter (20–40 inch) long inclined tunnels. Tunnels end with an expansive chamber and floors adorned by grasses or seaweed: like shag carpeting, but more stylish, while insulating. Penguins dig burrows by scratching with their feet and eventually use the rest of their bodies to push aside soil, although they are known to take over abandoned burrows of previous penguins or other animals. Male penguins do most of the primary burrowing or renovating of old burrows and then use the material evidence of their nursery-making abilities—along with courtship calls and dance moves—to woo females. Meanwhile, they fight off male rivals who dare to approach their hard-earned property, leading to fierce beak-to-beak and flipper-slapping combat. If a lady penguin is suitably impressed by a gentleman penguin and his intended home, she consents to his amorous entreaties, penguin carnal desires are fulfilled, and soon littler little penguins are on their way.

Other than in the ground, penguin burrows also might be deliberately sited below rocks or logs along a shoreline. In some places, humans have reinforced penguin burrows by adding wooden vestibules to their entrances. Male-female pairs normally occupy a burrow throughout a breeding season, but burrow exchanges can happen afterward as seasonal time-shares. Penguins do not mate for life and often swap spouses once the kids are raised and out of the burrow. Unlike all other penguins, fairy penguin mothers can lay two or three clutches of eggs in a breeding season, although each clutch consists of just one or two eggs. Once the eggs are laid in the burrow chamber, parents take turns sitting on the eggs to incubate them for five weeks before hatching. Chicks stay in the burrow chamber for the next three to five weeks, where both parents bring them food and do the avian equivalent of changing their diapers by removing their feces from the burrow chamber. When the chicks are big enough to venture outside of the burrow, they stick close to their parents for another couple of weeks before officially "fledging." Of course, for penguins this does not mean they walk or fly away from home like most other birds. Instead, they are now free to swim in the nearby ocean and find food or mates on their own, perhaps competing with their parents for both.

Ruth and I had previously read about fairy penguins in our *Lonely Planet* guidebook on Australia, mentioned in a chapter simply titled "Tasmania," so these small birds were on our mind before arriving in the coastal town of Bicheno, Tasmania. Although this was a vacation trip and not intended as some ichnologically inspired adventure, we definitely had natural wonders on our agenda, which on the island of Tasmania abounded. Already we had been thrilled by dramatic vistas encountered during the drive from the north to the east of the island, and although glad to have reserved a week there, we were wishing it were two.

Once we arrived at our motel and checked in, the proprietor told us about guided tours that would take us to a local favorite penguin haunt. Nevertheless, we balked at this idea, partly because we were on an academic budget, but also because we preferred to do something a little wilder and preferably not in the company of other humans. He graciously gave us a tip on how to experience the penguins in the area without having to pay for a guided tour. Using a map, he pointed to where we could drive, park, walk to a place along the shore where their burrows were located and just wait for them to come to us.

We arrived at the spot just after the sun went down, parked the car off the road, and walked down a coastal trail to the area. Having never been there before, we were unsure if this were the right spot. However, the sun had already slid below the horizon and our light was fading, so we settled down. Because we had previously seen penguin-wide holes in the sandy ground of other coastal spots, we sat on rocks well above the surface. This consideration ably spared the penguins the indignity of encountering a human butt–occupied burrow at the end of their long day at sea.

Because we were in Tasmania in April, the antipodal fall had already begun there, and the air chilled rapidly as the daytime glow diminished. We waited, and for a while not much happened until we heard them approach, walk by, and go underground. In retrospect, the dampening of their calls after they moved past us told us when they had entered their burrows, presumably as male-female pairs. These pairs also called out to maintain contact with each other, using sound instead of sight as an invisible tether to stay connected.

As special as this moment with these delightful ground-dwelling penguins might have been as a nature-loving experience, though, it also provoked many scientific questions I have pondered since. For instance, did the immediate ancestors of these penguins also burrow? What about other penguins: Is burrowing restricted to just fairy penguins, or do different species share this behavior? What other modern bird species burrow, and why? When did some birds evolve this seemingly atypical behavior for an animal lineage bestowed with wings? And what about theropod dinosaurs, the immediate ancestors of birds? We know that *Tyrannosaurus rex* and its tiny-armed kin—as well as many other dinosaurs—would have been poorly equipped for digging out burrows. Yet at least a few dinosaurs somehow managed carve out an underground niche during the demanding days of the Mesozoic Era, perhaps setting a precedent for their avian descendants.

The Dinosaurian Underground

As my shovel sliced into the ground, I wondered about the dinosaur burrow below. A photograph attached to an e-mail sent to

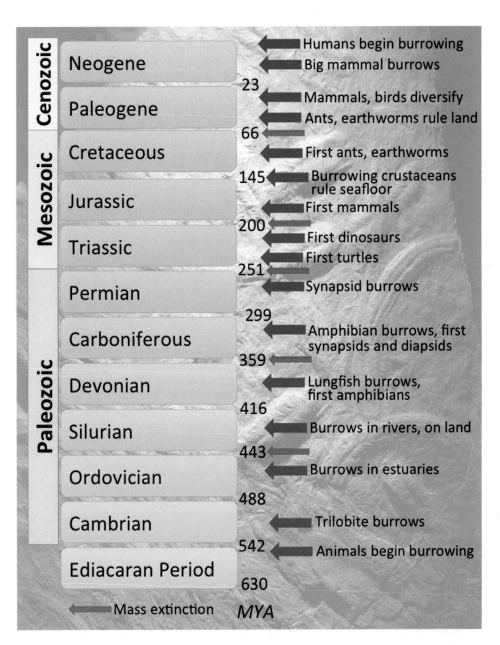

FIGURE 1. A brief summary of animal burrowing through time, from the Ediacaran Period through today. Geologic eras on left, periods on right, *MYA* = millions of years ago, and red arrows indicate times of mass extinctions in the geologic past.

Figure 2. Burrow of American alligator (*Alligator mississippiensis*) in pine forest of St. Catherines Island (Georgia, USA), indicating former presence of a freshwater wetland. But does the lack of a wetland necessarily mean the lack of an alligator in this burrow? Notebook (scale) = 13 centimeters (5.1 inches) long. (Photo by Anthony J. Martin.)

Figure 3. Alligator den entrance with hatchling alligator (LEFT) and overprotective mother (RIGHT), St. Catherines Island (Georgia). (Photo by Anthony J. Martin.)

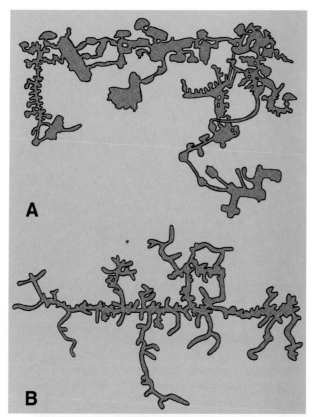

FIGURE 4. A comparison of mammal burrow systems. (A) Basic floor plan of underground city Derinkuyu, Cappadocia (Turkey), made by multiple generations of humans (*Homo sapiens*). (B) Burrow system of naked mole rats (*Heterocephalus glaber*) up-scaled to match size of Derinkuyu, but with tunnel widths greatly exaggerated. (Figure by Anthony J. Martin, with (A) modified from map posted outside main entrance of Derinkuyu, and (B) modified from figure in: Bennett, N.C., and Faulkes C.G. (2000), *African Mole Rats: Ecology and Eusociality*, Cambridge, U.K., Cambridge University Press.)

FIGURE 5. One of the many passageways in the underground city of Derinkuyu (Turkey), with stairs leading from one level to another. (Photo by Anthony J. Martin.)

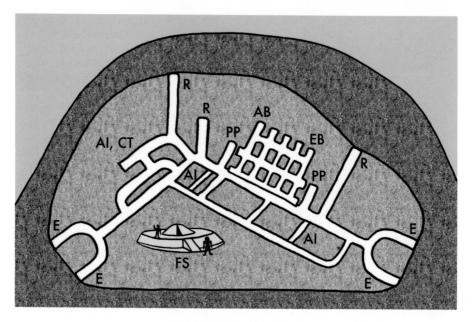

FIGURE 6. Possible schematic plan of "Site R" (Raven Rock Mountain Complex), southern Pennsylvania. Image based on various online sources, which in turn were derived from a Freedom of Information Act request. Key: AI = air intake; AB = A Building; CT = cooling tower; E = entrance; EB = E Building; PP = power plant; R = reservoir; and FS = flying saucer, which may or may not be hidden in the facility. (Figure by Anthony J. Martin.)

FIGURE 7. Burrow entrance and sand apron of gopher tortoise (*Gopherus polyphemus*), St. Catherines Island (Georgia); scale (notebook) is about 12 centimeters (4.7 inches) wide. (Photo by Anthony J. Martin.)

FIGURE 8. View down a gopher tortoise tunnel, showing overall elliptical cross-section and corrugated walls; tunnel is about 30 centimeters (11.8 inches) wide and runs straight for about 2 meters (6.6 feet) before turning to the left. St. Catherines Island (Georgia). (Photo by Anthony J. Martin.)

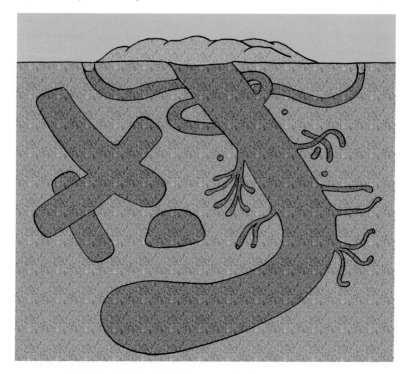

FIGURE 9. Hypothesized complexity of a typical gopher tortoise burrow, with many smaller burrows added to main tunnel by commensal animals, such as mice and insects. Also note intersecting and partially collapsed tortoise burrows to the left. (Figure by Anthony J. Martin.)

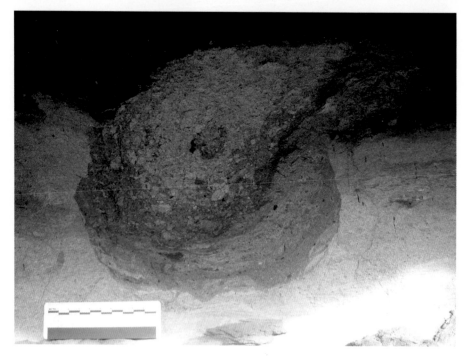

FIGURE 10. Fossil burrow attributed to a turtle in Early Cretaceous (105 *mya*) sandstone in Victoria (Australia), viewed in cross section; scale in centimeters. (Photo by Anthony J. Martin.)

FIGURE 11. Little penguins (*Eudyptula minor*) emerging from a burrow on Bruny Island, Tasmania. (Photo by J.J. Harrison.)

FIGURE 12. Artist conception of an adult burrowing dinosaur *Oryctodromeus cubicularis* and its two offspring in a den during the mid-Cretaceous (about 95 *mya*). (Artwork by Mark Hallett.)

FIGURE 13. Burrowing owl (*Athene cunicularia*) in California, which does not burrow so much as occupy burrows made by other animals. (Photo by Alan Vernon.)

FIGURE 14. Cross-sectional view of bird burrows at the same scale, represented by silhouettes of each bird. From left to right: little penguin (*Eudyptula minor*), Atlantic puffin (*Fratercula arctica*), brown kiwi (*Apteryx australis*), and European bee-eater (*Merops apiaster*), with bee in its beak. Scale = 1 meter (3.3 feet). (Figure by Anthony J. Martin, but two of the birds (little penguin and brown kiwi) are based on figures from PhyloPic.)

FIGURE 15. Burrow of either a rough-winged swallow (*Stelgidopteryx serripennis*) or belted kingfisher (*Megaceryle alcyon*), depending on who is home: Cumberland Island (Georgia). (Photo by Anthony J. Martin.)

me several months before conveyed hints of its form, properly piquing my curiosity and pulling me out of the urbanity of Atlanta, Georgia, to the western U.S. The burrow itself had been buried for about 95 million years, a long time before shovels were invented. Yet now, thanks to a combination of tectonic uplift and weathering, it was close enough to the surface for us humans to find and excavate it with our tools. Sage scents wafted by on the wind and, in between scoops, I looked around at the nearby pine forests and rolling, high-plains grassland nearly everywhere else, then up at an expansive blue sky hosting white, fluffy clouds. You might say I was in a country where the sky was big: Some people just call it "Montana."

My longtime friend and colleague, paleontologist David (Dave) Varricchio of Montana State University, had brought me to this part of southwestern Montana because of the potential dinosaur burrow beneath our feet. We were there in September 2005, about seven months before I would be sitting in Tasmania listening to penguins in the dark. Like many discoveries in paleontology, this one happened because of luck meeting preparation. Earlier in 2005, Dave led a field crew through that area of Montana, where they were prospecting for dinosaur bones in the Blackleaf Formation. The Blackleaf is a geologic unit of sandstones and shales, composed of sediments deposited in and beside rivers that coursed through that area about 95 *mya*. When these paleontologists wandered through this part of Montana, this formation had only yielded a few dinosaur bones, none of which had been identified. As a result, its poorly known fossil content meant it was ripe for investigating and might even hold the remains of dinosaur species new to science. The age of the rocks was also significant, having formed in the middle of the Cretaceous Period, which lasted from 144 to 65 *mya*. This is when dinosaurs were undergoing important transitions in their ecological and evolutionary histories. For instance, at the time when flooding rivers dropped the muds and sands of the Blackleaf Formation, birds had already split from their feathered theropod ancestors, flying and otherwise occupying new ecological niches for about 50 million years. Herds of relatively new clades of herbivorous dinosaurs, such as duck-billed ornithopods and big-horned ceratopsians, had spread throughout the western half of North America, albeit halted in their expansion by a mid-continental inland sea. Another 30 million years remained before a mass extinction, a mishap of global proportions that rid the earth of all ornithopods, ceratopsians, theropods other than birds, and broad swaths of other life forms.

Hence the field crew was in mostly unknown territory, paleonto-logically speaking, and anything they found could be scientifically significant. They felt somewhat vindicated when one member of the group, Yoshi Katsura, spotted a few dinosaur bones poking out of a white sandstone in the ground. A little digging around the rock revealed more of a skeleton, with its bones perhaps all belonging to the same dinosaur. What dinosaur? That was a bit of a mystery, as it could only be identified as a small ornithopod. Still, because it was the most complete skeleton discovered thus far in the Blackleaf Formation and possibly represented a new species, Dave decided it was worth collecting. This involved chipping away at the soft mud-stone around and underneath the sandstone holding the skeleton, and then covering it in plaster to protect the precious bones. Once the specimen was properly encased, the field crew put the specimen on a stretcher, walked it out of the field area to a vehicle, and took it for a ride back to Bozeman, Montana. After arriving in Bozeman, the bones were expertly extracted from their rocky matrix in a lab at the Museum of the Rockies.

Before all of that happened, though, Dave noticed something odd. As the crew dug into the mudstone around the sandstone containing the bones, he saw how the sandstone was more like a rounded, oval mass, rather than forming part of a continuous layer. Even stranger, this oval portion connected to a tubular structure filled with the same type of sandstone. This sandstone cylinder maintained a regular width throughout, dipped from a higher point, and twisted to the right, then to the left before joining the oval part with the dinosaur bones.

This weird sandstone body later turned out to be the first dino-saur burrow known from the fossil record. The originally hollow burrow had been naturally cast by sand-bearing water—probably from a nearby river—that filled it 95 *mya*, preserving its shape in such a way that we could study it in detail. Several months after its discovery, when Dave brought me to the site and we spent two days chopping away the mudstone around the burrow, its full form matched much of what we see in many modern vertebrate burrows. It descended vertically for about a meter (3.3 feet). When I looked at it from above, it had a straight length, turned abruptly to the right, continued another length, and turned to the left, where it had once connected with the sandstone mass with the bones. This mass represented the main burrow chamber, where the dinosaur lived and may have died.

As Dave and I dug around this discovery, I also predicted that a burrow this large should have had commensal animals living in the burrow, just like a gopher tortoise burrow. If so, we might expect to find smaller burrows joined to the main tunnel. Sure enough, later that day we found a cluster of pencil-sized burrows at one corner of the twisting form, and a banana-wide single burrow at a lower corner. For me, these trace fossils helped to complete a paleoecological picture of the large burrow as a place used both by the dinosaur and other animals well suited for digging and living together under-ground 95 million years before we did our own digging.

The burrow was not the only significant part of this discovery, as the dinosaur inside was indeed a new species of ornithopod. Dave, Yoshi, and I named it *Oryctodromeus cubicularis* ("running digger of the den"), and yes, its name reveals how we thought this was the same dinosaur that dug its own burrow. It was about the size of a Labrador retriever (and perhaps slightly smarter than one), but also had anatomical traits backing up our interpretation of it as a burrower. For one, its snout bones were fused, giving it a nice little spade-shaped face. Secondly, the bones in its shoulder girdle had attachment sites for large muscles, similar to those in armadillos or other modern burrowing animals. Lastly, when compared to related ornithopods, it had an extra vertebra in its hip. This addition would have helped stabilize it while digging with its front feet or shoveling dirt out of the way with its snout.

However, just to be good scientists, we remained skeptical of this apparent good fortune by asking ourselves, "How could we be wrong?" For example, one possible error was assuming this burrow actually held the remains of its maker; after all, plenty of animals take over other animals' burrows, just to save time and energy. For instance, I had seen this happen on St. Catherines Island with armadillos moving into old gopher tortoise burrows, altering these premade holes to better fit their bodies. Also, gopher tortoises and other animals might unwittingly host many temporary roommates, especially if a fire or other natural disaster drove those animals to seek shelter. Hence the "squatter hypothesis" for this dinosaur could not be discounted without rigorously testing it. Another potential source of error was if the dinosaur's proportions showed that it was too big for the burrow, which I sometimes call the "Cinderella model." If the dinosaur turned out to be too wide or long for the burrow to accommodate it, then it likely never went inside while alive, but instead had its bones washed into and deposited into the cavity, perhaps long after it died.

Fortunately, we had a tool to test our idea, the scientific equivalent of a spear and magic helmet, which some people call "math." During the same year that *Oryctodromeus*'s bones were discovered (2005), zoologist Craig White published an article demonstrating a mathematical relationship (correlation) between body mass (weight) and area covered by an animal burrow's cross section. This mathematical relationship makes good biological sense for many reasons. For one, most animals will not make an oversized burrow, in which they would expend too much energy for too little gain. A double-wide burrow also would have the disadvantage of losing its ability to control temperature and humidity, while enabling larger predators easier access to whatever might be living there. Conversely, an animal would be less inclined to dig a too-small burrow, which would make every day the one where you have to squeeze into those old jeans or little black dress that fit just fine when you were in high school.

Thus natural selection would have resulted in animals constructing burrows that were size-appropriate: not too big and not too small, and tight enough to maintain the burrow climate while also excluding unwanted dinner guests. In his article, White showed how this correlation between body mass and burrow area held up well for a variety of burrowing animals, from invertebrates—insects, spiders, and various crustaceans—to vertebrates, such as fish, frogs, lizards, and mammals. The only animals that deviated from this norm, expressed as a correlation line on a graph, were vermiform ("wormy") animals, which fell below the line, and birds, which were above the line. Why? I'll just say for now that there are very good reasons for these apparent aberrations, which I will happily discuss later.

So we did a little math. Dave and I had taken a series of width measurements along the length of the sandstone "tube" that represented the natural cast of the presumed *Oryctodromeus* burrow. We then used these to estimate the areas of an average burrow cross-section area. The number we got was about 1,000 square centimeters (~1 square foot), which was too tight of a squeeze for Dave or me (and especially Dave), but fine for most five-year-old dinosaur enthusiasts. Using a correlation line from White's article as a guide, we figured out what mass (weight) corresponded to this burrow area. However, to test its accuracy, we also put our calculated area into a formula to figure the probable body mass of the burrow maker. Its inferred mass was about 25 kilograms (55 pounds), or slightly less than, well, an average-size Labrador retriever. This overlapped with a weight range of 22–32 kilograms (48–70 pounds) Dave estimated for this dinosaur, which he based on bone lengths and widths that in turn correlated with probable dinosaur body size. The burrow cross-section matched the dinosaur. The tunnel segments on the twisting burrow also agreed with Dave's estimated torso length for this dinosaur. This meant it could make each turn—right, then left—without having to flex its body into some improbable yoga-like contortions. In short, this burrow fit this animal.

This already darned good news became damned good when the preparator at the Museum of the Rockies, while scratching away at the sandstone in her lab, began finding smaller limb bones mixed in with the larger ones. Dave already had no doubt that the larger bones were from an adult dinosaur. But once he saw smaller bones belonging to the same species, he realized these were youngsters. Based on limb-bone proportions, he estimated that they were about 55–60% the size of the adult. The co-occurrence of juvenile and adult bones in the same burrow chamber implied something larger than just a small dinosaur using a burrow: This burrow was also used for raising its children. In the world of burrowing vertebrates, this is very common behavior in mammals, called denning. (Remember the Asiatic badger and its setts?) Birds, such as fairy penguins and a number of other species, perform a similar behavior by laying and tending to their eggs (laying and brooding, respectively) in their burrows, then using these safe places to feed and care for their hatched offspring before they fledge.

To summarize, we had an adult specimen of a new species of ornithopod dinosaur, along with two of its partially grown offspring, and all three together in a burrow that was very likely a den. Although we did not know for sure if the burrow also served as a nest, nor whether or not these dinosaurs died together in the burrow, all of the evidence told us that the adult (perhaps with the help of another adult) had dug the burrow, and they had lived there. This further pointed to a previously unknown life habit for dinosaurs. In 2005, paleontologists knew that most dinosaurs lived on the ground, and a few lived in trees well above the ground, but had no good clues that any had gone below ground.

Nonetheless, despite all of these good data and modern analogues, we had a tough time publishing our article reporting this new species of burrowing, denning dinosaur and its burrow. One prominent journal rejected it immediately, and then another did the same, leaving us wondering if the paleontological community was not quite ready for this new idea about dinosaurs. Finally, a third journal gave us a chance and assigned other paleontologists to review the article. This ultimately resulted in its acceptance and publication, and in early 2007 it made a little splash in the popular media. *Oryctodromeus* had become part of the dinosaur family, and now people knew that at least one species of dinosaur had adapted to underground living. Since then, paleontologists have also found more *Oryctodromeus* remains associated with appropriately sized burrows in the Blackleaf Formation, bolstering this originally radical idea.

From 2006 to 2007, dinosaur burrows were on my mind in other ways, thoughts inspired by a striking structure I saw in Cretaceous rocks on the other side of the world. In the first half of 2006, I had a rare sabbatical from my university teaching position that allowed me to travel to and stay for a few months in Victoria, Australia. While there, I worked with the famed paleontologist wife-husband team of Patricia (Pat) Vickers-Rich and Thomas (Tom) Rich. After I found a few previously unknown Cretaceous trace fossils in rocks exposed along the Victoria coast—including tracks made by large carnivorous dinosaurs—Pat, Tom, and other people there gained some confidence in my ichnological skills. They decided to send me to other spots along the coast, looking for more trace fossils in the Cretaceous rocks, but especially for tracks. In 2006, dinosaur tracks were exceedingly rare in that part of Australia, and had only been discovered in two places before then, meaning it was well past time to find more.

One of those field excursions took us to a godforsaken place called Knowledge Creek, which filled my companions and me with the knowledge that we never wanted to go back there again. Slippery slopes, overgrown paths, freshwater leeches in the eponymous creek that danced excitedly at our warm-blooded presence, and a steep climb out of the site all made us regret our decision while assessing the meaning of life. Anyway, our crew, composed of a mix of Museum Victoria and Monash University personnel along with my wife, Ruth, was well on its way to Donner Party status when we finally made it to the coastal outcrop. Its spectacular cliff faces and ledges of buff sandstone, conglomerates, and shale were already enthralling to a geologist like me. Still, I was even more wowed once I turned a corner and spotted something that looked very much like the dinosaur burrow I had seen just eight months before in Montana.

The structure, viewed in cross-section, was a formerly twisting tunnel, with a cylindrical part that descended vertically about a meter (3.3 feet) before connecting to a lower and wider oval part. The tube part was filled with brownish sandstone that contrasted in color and grain size with the gray, coarser-grained rock surrounding it. Its lower portion had a sandstone-conglomerate fill that became finer upward. The tube turned to the right, and then to the left, maintaining a nearly constant width throughout until it reached its end, where it expanded into an oval part. While there, I did some quick measurements of the structure and furtively took photos of it, telling only Ruth what I thought of its possible identity. (At that time, the dinosaur burrow in Montana was still a deep, dark secret, because my colleagues and I had not yet published our findings.)

Against my better judgment, I went back to Knowledge Creek twice over the next few years, thoroughly documented the structure and what I thought might be a few other partially preserved ones next to it and wrote a research article about them. In this article, published in 2009, I claimed not only that these were dinosaur burrows, but also that they were the most ancient in the geologic record at 105 *mya*, 10 million years older than the dinosaur burrow in Montana. The most likely dinosaurs to have made them in this area of Australia were small ornithopods, whose remains were the commonest of dinosaurs in the rocks; these dinosaurs were also similar in size and form to *Oryctodromeus*. However, I did not find any bones in the supposed burrows, let alone those of an adult and two juveniles. This absence admittedly weakened my explanation for the peculiar structures.

Still, I pointed out how, even if these were not dinosaur burrows, dinosaurs *should* have burrowed in this part of Australia at that time. Why? Because 135–100 *mya*, Australia was still attached to Antarctica, and the latitude for the southern part of Australia was at about 75–80 degrees south, close to what was then the South Pole. In other words, the dinosaurs and other animals living in these former river valleys were in polar environments. Also, most of the dinosaurs were too small to have migrated, which inspired Pat Vickers-Rich and Tom Rich in the 1980s to ask a basic question about them: How did they make it through the winter? One suggestion they made in a 1988 article—in an almost offhand comment—is that at least some of these dinosaurs overwintered in burrows.

Although no one has since found similar structures in the vast Cretaceous outcrops of Australia, I later interpreted other fossil burrows there that were likely related to animals using them to make it through cold, dim winters more than a hundred million years ago. Among these were burrows of turtles (mentioned earlier), lungfishes, and possibly mammals. Bones for all of these animals have been found in Cretaceous rocks of Victoria, so it is reasonable to suggest that some might have been burrowers. Although we do not know for sure the lifestyles of these animals, we are fairly certain that these turtles, lungfishes, and mammals did not migrate annually, either. Again, this compels us to ask how these animals survived polar winters.

Since the Cretaceous Period, Australia has drifted northward and is now much closer to the equator, so animals there do not have to deal with such conditions. Yet the harsh desert interior composing much of Australia today means that modern turtles, lungfishes, mammals, and other animals use burrowing to endure seasonal or multi-year droughts there. In other words, burrowing continues to be a way for animals to survive the extremes of environments in Australia, as the continent shifted from icy cold Cretaceous polar winters to the stiflingly hot present-day desert summers.

However, one important point to keep in mind about small, burrowing, herbivorous ornithopod dinosaurs is that they were *not* the ancestors of modern burrowing birds: Those would be theropod dinosaurs. The list of other dinosaurs that may have burrowed to make nests or dens is short, and does not include theropods. This exclusive underground-dinosaur club consists of: *Oryctodromeus*; possibly *Orodromeus*, an ornithopod relative from the Late Cretaceous (about 75 *mya*) of Montana; and *Psittacosaurus*, which is an Early Cretaceous (125–120 *mya*) ceratopsian (horned) dinosaur from central Asia. However, the evidence for *Orodromeus* as a burrower is mostly circumstantial. These ornithopods, which were slightly smaller than *Oryctodromeus*, are typically found in compact masses of rock. A concentration of their bones in formerly small, tight spaces implies that they were preserved in their former burrow chambers, which were later filled with sediment after they died. *Psittacosaurus* is interpreted as a potential burrower based on just one (albeit amazing) discovery: an assemblage of 24 juveniles of the same age all packed together. Such an unusual, well-preserved, and dense population of young dinosaurs is attributed to the Cretaceous equivalent of a mine-tunnel disaster. In this scenario, the dinosaurs were all occupying a burrow chamber—possibly a den, or at least a bowl-like depression presumably dug out by the adults—and were buried alive when it collapsed. However, other researchers suggested these juveniles were buried by a volcanic mud flow.

Nevertheless, given that our fine feathered friends of today are descended from Mesozoic theropod dinosaurs, we must look to those dinosaurs for clues on the evolution of burrowing birds. Some of these theropods were small, and most of were carnivorous; some lineages produced the largest land predators in earth history. Given size as a limiting factor, we can be reasonably sure that massive theropods such as *Giganotosaurus*, *Spinosaurus*, and *Tyrannosaurus*—all Cretaceous dinosaurs—did not burrow. Alternatively, smaller theropods might have had good evolutionary reasons to develop burrowing as a strategy for survival, whether to evade those aforementioned huge predators, or to escape difficult environmental conditions, such as polar winters, fires, summer heat, or other life-threatening situations.

So do we have any evidence of burrowing theropods from the Mesozoic Era? Not yet, although if paleontologists someday discovered the bones of a small theropod in a burrow of its own making, perhaps accompanied by bones of its young, none of this would surprise me. I also would be elated for the paleontologists involved in such a discovery, and decidedly more so if I were one of them. Still, because we do not yet have such evidence, for now we can only conclude that the evolution of burrowing in certain ornithopod dinosaurs and birds evolved independently from one another. This would have happened through a convergence of burrowing behaviors naturally selected in their respective lineages, and understandably so for all of the life-preserving qualities listed thus far.

When did burrowing begin in birds, if not when still in the non-avian-theropod state of their evolutionary history? Probably the best way to address such a broad question is to look at more modern examples of burrowing birds and examine why these birds burrow. Following that, we can then peer back into their evolutionary pasts for clues about environmental conditions imposed on birds that might have caused these animals—so well known and revered for their mastery of flight—to seek refuge in the ground.

Surviving as Birds Below

Most people do not think of birds as burrowers and are often surprised to learn just how many bird species use burrows as an integral part of daily life. For instance, we began this chapter with a specific example of fairy penguins living in burrows. Yet they are by no means an exception, as more than half of the twenty or so penguin species live in burrows. Other burrowing birds include, but are not limited to, puffins, shearwaters, kingfishers, swallows, bee-eaters, and at least one species of parrot.

Given more than 10,000 species of birds, their awe-inspiring diversity, and their adaptations to nearly every terrestrial environment, the evolution of burrowing in a few makes sense as a survival strategy. Most importantly, burrowing in birds is not necessarily done to save adults from premature deaths, but rather to save their eggs and chicks. When it comes to survival, all birds reproduce by laying eggs; thus no bird lineage could have passed on its genes without first laying eggs and ensuring that they hatched.

Assuming that the theropod-dinosaur ancestors of all birds likely made their nests on the ground, and that egg predators became more common during the Mesozoic Era, the immediate ancestors of birds would have had three choices: (1) Continue laying eggs on the ground, but either disguise them, stand (or sit) guard over them, or both; (2) build nests in inaccessible (and usually high) places, such as in trees or cliffs, or on isolated islands protected by crocodilian godfathers; or (3) tuck the eggs in deep burrows, where egg predators could not see or smell them. Although paleontologists have not yet identified fossil bird burrows, let alone Mesozoic ones, the presence of burrowing birds today tells us that birds developed burrowing sometime in the past 160 million years, and used it successfully as a way to protect the next generation.

Of all modern birds that live underground, probably the most famous in the world is the burrowing owl (*Athene cunicularia*). Interestingly, this unduly endearing little owl, unlike alligators, gopher tortoises, or many other animals, actually does little of its own burrowing. Instead, it tends to take over and modify pre-existing burrows, and usually those made by appropriately sized mammals, such as prairie dogs (various species of *Cynomys*) or ground squirrels (species of *Spermophilus*). Burrowing owls are already appropriately sized for getting into and down most of the burrows they choose, but they use their feet to scratch out a better home if needed. This minimal burrowing ability, however, is not to denigrate burrowing owls, which should be treated with unquestioned and absolute adoration. After all, in every other respect these birds are remarkable for how well they have adapted to using underground spaces for starting and raising their families. They also use their usurped burrows as staging grounds for bringing home dinner.

Despite their subterranean lifestyle, burrowing owls can fly and migrate, ranging from the midcontinental and southeastern parts of North America to the Caribbean and throughout the eastern part of South America. As a result, they have minor regional differences in their appearances and behaviors, and accordingly ornithologists have split them into many subspecies based on where they live most of the time. What all burrowing owls have in common, though, are large yellow eyes and long legs. They are also among the few owls that are active during the day, yet retain night vision good enough to hunt in the dark. They seek, acquire, and eat a variety of animals, such as insects, frogs, lizards, snakes, other birds, and small mammals, including mice and moles. Similar to *Oryctodromeus*, which used just its two hind legs for moving about, burrowing owls can run well. When hunting, burrowing owls use a combination of flying and running to catch their prey.

Among the insects on burrowing owls' menus are dung beetles, giving these owls the incentive to develop tools, albeit of a scatological variety. Owls exploit dung beetles' weakness for fresh, sweet mammal feces by grabbing chunks of it with their beaks, carrying it to their burrow entrances, and plopping it down as bait. This practice not only attracts dung beetles to the burrow, but also helps to teach their offspring how to hunt them. Before they can fly, baby burrowing owls stay in and close to their home burrow, which means mom and dad need to bring them meals, most of which are already dead and decidedly immobile. The arrival of small, active prey to the burrow supplies an opportunity for the chicks to start sharpening their hunting skills, and with something nonthreatening. (In contrast, parent owls bringing home live baby rattlesnakes would constitute some "tough love.")

Burrowing owls that live in the southeastern U.S. overlap well in their range with gopher tortoises, which also prefer well-drained sandy soils. Again, this makes sense for both nesting and raising youngsters, as such soils help to ensure a burrow chamber does not retain water and is aerated enough that little lungs can breathe. Adult and baby owls can also survive horrific forest fires by staying underground, just like tortoises. When nesting below, burrowing owl egg clutches can have as few as two eggs, but normally range from five to ten. Once the eggs are laid, the female incubates them in the burrow for about a month while the male brings her food. Upon their hatching, he has to fetch more groceries for the female and their offspring until both parents start teaching them how to get their own meals. Fledging takes about 40–45 days, after which the baby owls become semi-adult owls and go off to seek out prey animals and scrape out a living by taking over preexisting burrows.

Among other charismatic burrowing birds are Atlantic puffins (*Fratercula arctica*), well known for their robust and colorfully striped beaks, stout bodies, and bright orange webbed feet, all of which lend them a clownish appearance. These birds are native to the North Atlantic, living along the coasts of Maine, Newfoundland, Greenland, Iceland, Norway, and Great Britain. Like fairy penguins, they eat fresh seafood (mostly fish), which they catch while swimming in the ocean. Unlike all penguins, though, they can fly, although for only short distances. Once they are done fishing and looking for some puffin lovin', and later have to deal with the parental consequences of avian amour, they rely on their burrows. Puffin burrows are impressive for their depth and sheer abundance, perforating cliff sides near their favorite fishing spots. Burrows are normally a little less than a meter (3.3 feet) long, but some can be as much as a meter below the surface and 3 meters (10 feet) long. Like all other bird burrows, their tunnels end in an expanded chamber that is capable of accommodating both parents and their chick. Yes, that singular use of "chick" is intentional, as puffin mothers lay only one egg at a time, which means they devote all of their attention to making sure it hatches and grows up into a fledgling puffin. Also, puffins are mostly quiet birds, but once in their burrows, they get less inhibited and make distinctive noises, sounding like a profane union between an electric razor and miniature cow.

Interestingly, puffin parents are typically monogamous, but only remain faithful to each other because of their burrows. Once they make a burrow or choose an abandoned one, they stick with it in successive breeding seasons. Such behavior is called *site fidelity*, in which animals—such as crocodilians, sea turtles, and birds—come back to the same place to nest each year, and perhaps over generations. Regardless of whether they are making a new burrow from scratch (so to speak) or improving an old one, both parents are involved in making it a home for themselves and their only child.

Puffins make or alter burrows with their beaks first, then their feet. Using their thick beaks like spades, they chop into the soil to loosen it, and then push it back and behind them with their feet. The nesting chamber is lined with a blend of grasses and feathers, making it a warm, cozy place for their upcoming puffin chick. Once a puffin palace is complete and the egg is laid, the parents share burrow duties, taking turns with incubation, finding food for their chick, cleaning the nest, and acting as a sentry. Puffins are well known for their elaborate walking behaviors, performed just outside of their burrows. These burrows again fit the theme of not only providing a predictable place to call home, but also protection against predators, such as greedy gulls, as well as offering considerably nicer indoor environments during nasty North Atlantic weather.

Nonetheless, puffins do not always live in homogeneous colonies, but often have to share subsurface space with another burrowing bird species, the Manx shearwater (*Puffinus puffinus*). Despite its species name, Manx shearwaters are not puffins, but they also use burrows for their nesting. Despite this shared preference for underground nesting, Manx shearwaters could never be mistaken for puffins, having much narrower, tubular, and grayish beaks, basic black plumage on top and white on bottom, and greater wingspans. Shearwaters are marvelously efficient seafarers, employing a rigorous blend of flying and diving to catch fish, and are capable of migrating more than 10,000 kilometers (more than 6,200 miles) from the North Atlantic to Brazil and back. However, these adaptations come at a price, as Manx shearwaters are slow and clumsy on land, making them easy pickings for pitiless seagulls and other predators. So burrows once again come to the rescue, providing personal safety for both adults and their offspring.

Shearwaters sometimes take over preexisting rabbit burrows, but are more than able to dig their own, with some more than a meter (3.3 feet) long. Like puffins, shearwater parents often reuse burrows for nesting. Similar to maintaining a summer home, they come back to and tidy the same burrow each year. As mentioned previously, adults seek shelter in burrows, but they mostly do this at night after a full day of flying and swimming. If making new burrows, they might crash into a nearby puffin maternity ward, causing the puffins to start over with a new burrow. As a result, burrows of these two very different birds may intersect each other when nesting, comingling individual burrows. Hence instead of nesting grounds, puffins and shearwaters have nesting undergrounds.

Puffins alone can form enormous colonies of more than a million pairs of birds, which can accordingly generate more than a million active and abandoned burrows. So when Manx shearwaters add their numbers to these puffin colonies, these two birds collectively qualify as ecosystem engineers in shore-side soils, completely changing the character of their habitats. If such burrows were to fossilize, paleontologists would be greatly challenged in discerning which burrows were made by which species.

Other burrow-dwelling birds include those that live near burrowing owls in the southeastern United States: rough-winged swallows (*Stelgidopteryx serripennis*), bank swallows (*Riparia riparia*), and belted kingfishers (*Megaceryle alcyon*). However, in contrast to burrowing owls, these three bird species dig their own burrows and do so on vertical surfaces, such as bluffs or natural levees along rivers. They begin their burrows by pecking into banks soft enough to break apart, yet firm enough to hold the shape of a burrow without it collapsing. Once started, they then use a combination of beaks and feet to extend the burrow farther into the bank for about a meter or two (3.3–6.7 feet), keeping it more or less horizontal and consistently wide until the very end.

At that end, they hollow out a chamber allowing enough room for two adult birds plus their upcoming brood.

I have encountered such burrows in soft sandstones of coastal and riverine outcrops of the southeastern U.S., and they are quite distinctive. These bird burrows have a mostly circular or oval profile, but oftentimes are grooved on the bottom, the latter scored by their feet dragging along the burrow bottom as they fly in and out of the burrow entrance. The horizontal orientation of the burrow allows the swallows or kingfishers to quickly disappear into and reappear from their burrows. Burrows are placed high on bluffs for a reason, keeping both the adults and chicks out of harm's way from predators and well above water bodies adjacent to these bluffs. Where these birds are similar to burrowing owls is that they also can be economical and save the labor of digging a new burrow by using an old one, or stealing a burrow actively used by another species.

One of the most important examples of modern burrowing birds, though, are those known collectively as bee-eaters, aptly named because of their impressive bee- and wasp-eating abilities. Bee-eaters consist of 27 species, nearly all under the genus *Merops*, and they are geographically widespread, with species in Europe, the Middle East, Asia, and Australia. Considering their distribution, they also live in a wide range of environments. However, what they all have in common—other than deftly snatching stinging insects out of the air—is burrowing, which they use for nests and raising their young. Indeed, ecologists who have studied bee-eaters also regard these colorful, elegant birds as ecosystem engineers, owing to their prolific burrowing.

Despite their relatively small sizes (think wrens or sparrows, rather than crows), bee-eaters are incredibly industrious diggers. Like bank swallows and kingfishers, nearly all bee-eaters dig their burrows on vertical banks, with only a few going into horizontal surfaces. Burrow making is a tag-team effort, with male-female pairs working together to build their little underground day care center. When making a burrow, they first employ their long, curved beaks to break up whatever soil they deem suitable for their needs (more on this geological astuteness soon). They then scratch at the sediment with their feet to shape their horizontal tunnels, which can be quite deep. For example, European bee-eaters (*Merops apiaster*) make 1.5–2-meter (5–6.6-foot) long burrows, and other species make similarly lengthy tunnels compared to their body lengths. Nest chambers of European bee-eaters at the end of each tunnel are spacious; if filled with water, they would hold about 3.5 liters (about 1 gallon): This roominess is as it should be, for these chambers must be big enough to hold both parents and as many as six chicks.

Bee-eaters sometimes reuse their burrows, but normally make new ones every year. For reasons only bee-eaters know, they also sometimes stop digging and leave partially excavated burrows near their final one. Although bee-eaters are very often colonial nesters, this large number of "starter holes" creates a false impression that many more birds are present than the actual population. Regardless of this home-building fickleness, bee-eater burrows reaffirm the advantages they offer to their makers. For example, in arid environments, burrows maintain high humidity levels, nearly constant temperatures year round, and of course protection against predators. This defense is especially effective for bee-eaters placing their burrows on steep banks, as bee-eater eaters then have to climb up the slope, somehow pick an occupied hole out all of the ones perforating the bank, and make their way a meter into that hole without being attacked by protective parents.

Despite bee-eaters' minuteness, the massive amounts of digging by their colonies on riverbanks, bluffs, and other surfaces significantly alters the ecological character of these places. Even more remarkable is how bee-eaters are not only ecosystem engineers but also geological engineers, as they pick nesting sites based on soil firmness and clay content. Both of these factors are important for successful nesting, showing why these birds are so discriminating about their geological materials. For instance, soil firmness is essential for a burrow to keep its shape, particularly as a bee-eater pair actively cuts through it. If the sediment is too soft, then the burrow roof falls in. This means lots of wasted energy for the parents and—in the worst-case scenario—eggs or chicks trapped under a pile of collapsed sediment, which is less than desirable for passing on genes and evolving. On the other hand, too-firm sediment is also detrimental. Although a hard, rocky substrate would certainly prevent roof collapses, bee-eater parents trying to work it would end up with bent beaks and broken claws. Hence these birds look for an ideal of "not too soft, not too firm" when selecting nesting sites. The amount of clay minerals in the sediment is also important, as too much clay would seal tiny spaces between sand and silt grains, decreasing ventilation and the ability of a bee-eater burrow to drain excess water. As a result, bee-eaters prefer soils that are sandier rather than clayey.

All of the aforementioned birds have notable digging and sub-surface abilities. But perhaps the most surprising of burrowing birds to most people is a parrot. Parrots comprise nearly 400 species that live mostly in subtropical and tropical environments, a diversity that reflects many adaptations and ecological niches. Moreover, despite parrots' common portrayal as wisecracking shoulder-bling for pirates, they are also regarded as among the most intelligent of birds, bested only by crows and ravens for their cognitive abilities. So given these two facts about parrots— diversity and smarts—it makes sense that at least one of them, the Patagonian conure (*Cyanoliseus patagonus*), evolved a life that was best done underground.

Known by Argentines as *loro barranquero* ("burrowing parrot"), these birds—with green feathers on top and splashes of orange and yellow underneath—consist of four subspecies, with most in Argentina and smaller numbers in Chile and Uruguay. Interestingly, Patagonian conures have the largest breeding colonies of any parrots, which, like swallows, kingfishers, and some other burrowing birds, are manifested as nesting holes in cliffs. How big are these colonies? Some seaside cliffs may host as many as 50,000 parrot burrows, enough to change coastal environments in ways comparable to puffin and shearwater colonies living far north of them. Like other burrowing birds, these parrots mainly use their burrows for nesting and raising chicks. Although these parrots live from the foothills of the Andes to steppes to forests to grasslands to seashores, they all require firm substrates for digging burrows, such as poorly cemented sandstones or limestones.

Pairs of parrot parents dig horizontal tunnels that are 0.8–3.5 meters (2.6–11.5 feet) long, and like good little mining geologists, their burrows often follow bedding planes of the host sedimentary rock. Burrow entrances are oval, with a horizontal width of 15–50 centimeters (6–20 inches) and a vertical height of 10–25 centimeters (4–10 inches). Tunnels made by different parents can crisscross one another, but apparently they are amiable neighbors, with no evidence of parents taking over nest chambers or tossing out eggs of another pair. These pairs also stay the same from year to year, as burrowing parrots are monogamous. First-time couples tend to construct new burrows, whereas couples that have been around the cliff a few times are more likely to reuse, widen, or otherwise alter older burrows. Each burrow has a large nesting chamber at its end, where the mother parrot lays 2–5 eggs. After eggs are laid, parental labor is divided so that the mother parrot incubates eggs for about 25 days while the father does all of the food shopping. Chicks take about two months to fledge, during which both parents feed them by working together. However, even after their offspring are more than capable of flying away from home, their parents seem to have a tough time letting go and continue to feed them for another four months.

Sadly, generations of these parrots are being shortened as people shoot, poison, or otherwise kill them. This animosity is provoked by parrots' foraging forays on farms. Looking at it from a burrowing parrot's perspective, it makes sense to look for food on agricultural lands, as these not only provide sustenance, but also replaced natural habitats where they ate native foods. Also, as is typical of animals labeled as "pests," their reputations precede them and their crop-destroying abilities may be exaggerated. Yet another problem facing these parrots is poaching, where people seek out nests, capture and cage nestlings, and sell them into the illegal pet trade. Such abuse is unfortunately all too common for parrots and their cockatoo relatives worldwide, as 30–40% of their species are threatened by extinction. It is even more sobering to know that the one species of parrot that developed the means to burrow—a process that may have taken millions of years to evolve—is nonetheless still not safe from habitat alteration and humans.

For those of us who never thought much about burrowing birds and always assumed that birds either nested in trees or on the ground surface, the preceding review of them should be enlightening. Perhaps most remarkable is that burrowing emerged as a reproductive strategy in animals that use their arms to fly. In this respect, burrowing penguins are somewhat understandable, as they restrict their "flight" to watery environments. Yet for those burrowing birds that can soar above the Earth any time they like, going farther down into it seems contradictory to their nature. Just how did this behavior become genetically programmable and naturally selected over the course of geologic time, and when?

Burrowing Birds through Space and Time

When considering the evolution of modern burrowing birds— penguins, owls, puffins, shearwaters, kingfishers, swallows, bee-eaters, and parrots—it might be tempting to propose they all share a common ancestor, a great Grandmother Burrowing Bird of the Mesozoic. If so, such a genealogy would help to solve the mystery of when burrowing began. Alas, most of these birds and their ancestors are not closely related to one another, and the best we can say for now is that they all belong to a broad clade called Ornithurae. This clade, however, contains two more inclusive ones, Neognathae and Palaeognathae, which may provide a few clues in our quest to figure out when the birds first burrowed.

Neognaths comprise the vast majority of birds that would be recorded by any birder attempting a "Big Year," in which she or he may try to identify as many species of birds as possible in one year. Most neognaths are songbirds, also known as *passerines*. Passerines are appreciated daily by bird lovers for their visual beauty and spirited antics, and also for their delightful melodies, which can be heard anywhere from urban jungles to actual jungles. In contrast, most palaeognath lineages went extinct at the same time as the non-bird dinosaurs, about 66 *mya*. The only modern birds belonging to this clade are the flightless ratites—ostriches, emus, cassowaries, rheas, and kiwis—as well as tinamous; the latter consist of about fifty species of small grounddwelling birds in Central and South America.

Do any modern palaeognaths burrow? Despite the popular depiction of ostriches burying their heads in sand, they do not, nor do any of their large-bodied relatives, such as emus and rheas. For a ratite burrower, we instead need to look at kiwis, which are also among the most evolutionarily improbable of birds.

Native to the two islands of New Zealand, kiwis are represented by five species under the genus *Apteryx*, all of which are covered by brown, downy feathers, and are more or less chicken-sized, ranging from 35 to 45 centimeters (14 to 18 inches) tall. Although kiwis are flightless, bearing only vestigial nubs of wings, they descended from winged ancestors that flew to New Zealand around the start of the Miocene Epoch at about 20 *mya*. Unlike all other birds, kiwis have long, narrow beaks with nostrils on their tips, indicating a ground-dwelling lifestyle. As omnivores, kiwis apply their keen sense of smell to find seeds, insects, worms, and small vertebrates in forest soils, and their beaks can detect vibrations from burrowing prey. Such sensory abilities compensate for their tiny eyes, which are not as needed because they hang out in burrows during the day and forage at night. Perhaps the most startling kiwi adaptation, though, is in the females of each species, which can hold and lay an egg one-sixth of their body weights: For perspective, imagine a human female bearing and then giving birth to a five-year-old kid. (On second thought, don't.) As a result of this outsize parental responsibility, kiwis normally lay only one egg, which they incubate in burrows. Once the egg has hatched, the parents raise their chick in the same burrow until it is ready to go out and seek food on its own.

Since arriving on the isolated landmasses of New Zealand, kiwis apparently traded flight for burrows, which they use as underground roosts, nests, or dens. Given their tiny wings and sensitive beaks, kiwis must use their feet to scratch out hollows in New Zealand forest floors and other environments. For the brown kiwi (*A. australis*), their burrows are about 12–14 centimeters (5–6 inches) wide and 1–1.5 meters (3.3–5 feet) long, with vertical depths of 20–50 centimeters (8–20 inches). For the little spotted kiwi (*A. oweni*), burrows can be as long as 2.5 meters (8.2 feet), a big feat for such little feet. Burrows end in an enlarged chamber that can accommodate a pair of kiwis, although these tunnels point slightly upward along their lengths, twist to the right or left, or make U-turns. One presumed purpose of such complications, seen in burrows of gopher tortoises and many other animals, is to confuse predators. Yet before humans arrived on New Zealand with their dogs, cats, rats, stoats, and other invasive mammals, kiwis had no natural enemies other than other ground-dwelling birds that occasionally snacked on their enormous (and thus tempting) eggs. Oddly, kiwi burrows used solely as nests and dens tend to be shallower than adult-only roosting burrows. In such instances, burrows may function primarily to keep kiwis warm and otherwise out of the elements, rather than protect them against other animals. Also, paleontologists do not know if flighted ancestors of kiwis were burrowers or not, meaning current burrowing behaviors might have only developed during the last 20 million years or so.

Meanwhile, across the Pacific Ocean, tinamous of Central and South America are similar to kiwis in their habits and habitats by living mostly on the ground and in forests, though some also are in grasslands. However, one of the biggest differences between kiwis and tinamous is that tinamous can fly (albeit not very well); more often than not, they run when confronted by threats. Tinamous are also more active during the day and prefer to rest at night. Perhaps a more important ichnological distinction between kiwis and tinamous, though, is that the latter do not burrow for their nesting. Male tinamous instead make nests on ground surfaces.

One of the more peculiar behaviors of tinamous compared to most birds is that males can mate with more than one female, and all females who have enjoyed a male's company are welcome to use his nests. However, what's good for the gander tinamou is good for the goose tinamou, as females can also mate with a variety of males. In humans, such polyamorous arrangements would likely result in some rather awkward family reunions, but they seem to work very well for tinamous. Moreover, the males of these avian swingers have parental obligations, as the males incubate the eggs and take care of the chicks produced by their various mates. Still, no burrows are involved in all of this reproductive behavior, leaving us wanting for an evolutionary echo of it in tinamous and other palaeognaths. Although tinamous are definitely going low socially, kiwis are the only examples of palaeognaths literally doing so.

Given that living palaeognaths provide few insights on when burrowing began in their lineage, perhaps we can look at the evolutionary history of neognaths to estimate when burrowing may have evolved in these birds. This would be especially helpful if we found that burrowing neognaths originated close to the extinction of most palaeognaths at the end of the Cretaceous. Of the burrowing neognaths then, the most intriguing in terms of evolutionary timing are penguins. The oldest known fossil penguin is from the earliest part of the Paleogene, about 62 *mya*, only a few million years after the mass extinction that took out the dinosaurs. Just before then, though, penguin ancestors originated in the Late Cretaceous around 70 *mya*. Although we currently have no information on whether or not the immediate ancestors of today's waddlers burrowed, it is nonetheless interesting that they had some quality that allowed them to pass on genes after one of the worst natural disasters in the history of life.

Psittaciformes, the clade that encompasses modern-day parrots, also probably originated in the Late Cretaceous. Likewise, clearly identifiable fossil owls have been discovered in rocks from about 60 *mya*, implying that their ancestors existed in the Late Cretaceous, too. However, kingfishers and bee-eaters, which both belong to the clade Coraciiformes, are geologically younger, with their fossils dating to the middle of the Eocene Epoch (about 40 *mya*). Puffins similarly belong to a group (Alcidae), whose oldest fossils are in Eocene rocks. Shearwaters are in Procellariidae, a broad group of related seabirds that may or may not have begun in the latest part of the Cretaceous; but shearwaters as a genus (*Puffinus*) have only been extant for the past 5 million years or so. In conclusion, some clades with burrowing birds can be connected to a time just before or just after the Late Cretaceous, but not all. Such admittedly circumstantial evidence thus suggests that burrowing evolved independently in many bird lineages both before and after the end-Cretaceous event.

Now, this is where ichnology could help to clarify when birds first went underground. Earlier I said that fossil bird burrows have not yet been recognized from the fossil record. But how would a paleontologist identify and justify that she or he found one? One clue to identifying an ancient bird burrow is what it would *not* look like, which is a burrow made by a reptile, mammal, or other animal. Remember that neat little graph of animal-burrow cross-sectional areas matching body masses of their burrowers? This graph works great for most animals, except for those with wormy bodies (such as worms) and birds. Bird burrows, especially those made on banks, have considerably larger areas when compared to the body sizes of these birds. What do birds have that make them special among all burrow-making vertebrates that they need wider burrows? A one-word answer should suffice: wings. For instance, bank and rough-winged swallows fly in and out of the entrances of their burrows, meaning the burrows must be wide enough to allow for at least half-folded wings. These burrows also have grooves dug into the floors of the burrows, caused by the many times their bird occupants dragged clawed feet along these surfaces while coming and going. As mentioned earlier, penguin burrows also might be widened enough to allow each of a penguin pair to pass each other.

Given what we know about modern bird burrows and burrowers, I am confident that paleontologists will someday correctly identify fossil bird burrows. Perhaps we will even find burrows with bird skeletal remains in them, paralleling what we have found with the Cretaceous dinosaur *Oryctodromeus* in its burrow. Even better would be a burrow with fossil eggshells and hatchlings in it, confirming how such burrows were used for nesting and raising young. And while I'm having such ichnologically inspired dreams, how about Late Cretaceous bird burrows with such fossil evidence, and then some early Paleogene bird burrows, suggesting that this was how birds survived a mass extinction 66 *mya*? Underground nests certainly would have been one way for the ancestors of all modern birds to ensure the survival of future lineages, which ultimately would have led to the bird-shaped landscapes of today.

Birds and Mammals, United in Time (and Survival) by Burrows

The dependence of burrowing owls on burrowing mammals is an interesting one to note from the perspective of geologic time. As mentioned before, birds have been around since the Late Jurassic, for about 160 million years. In contrast, mammal ancestors (mammaliaforms) go back further, showing up in the fossil record just after the most primitive dinosaurs in the Late Triassic Period, at 220 *mya*. The common ancestor of mammaliaforms goes even deeper into the past with synapsid reptiles, which originated during the Carboniferous Period, around 320 *mya*. Synapsids then proliferated, diversified, and occupied most ecological niches of land environments by the end of the Permian Period, from 300 to 250 *mya*.

At least some of these Permian mammal ancestors lived in burrows of their own making, which we know because paleontologists have found synapsids in their burrows. Somehow a few synapsids persisted after the Great Dying of the end-Permian circa 250 *mya*, a mass extinction that rid the earth of nearly 95% of its species. Again, some of these synapsids have been discovered in their burrows, preserved in sediments laid down during the earliest part of the Triassic after the extinction of so many other animals. Later in the Triassic, these synapsids were succeeded by different synapsids that became our more immediate ancestors: mammaliaforms. These synapsids then bested yet another mass extinction at the end of the Triassic at about 200 *mya*, and of course mammals—which originated in the Jurassic (~160 mya)—made it past the next mass extinction at 66 *mya* that took out the non-avian dinosaurs.

Evidently these small, furry critters burrowed throughout the rest of the Mesozoic under the feet of the dinosaurs, a lifestyle we infer based on their small sizes and trace fossils from the Jurassic and Cretaceous resembling modern-mammal burrow systems. Indeed, subterranean living may have been a default way of life for mammals before the extinction of the dinosaurs. Afterward, mammals spread onto the earth's surface and eventually occupied every environment, including the world's oceans and skies.

Consequently, I hereby propose a burrow version of the "Which came first, the chicken or the egg?" riddle, which is, "Which came first, the chicken or the [mammal] burrow?" The simple answer is that mammal burrows preceded dinosaur burrows by more than a hundred million years, and birds by about 60–70 million years. This means that regardless of when the first bird evolved the means to live in burrows, mammals had already been underground long before then. This also implies that the first small theropod-dinosaur ancestors of birds may have taken over old mammal burrows when they needed to hide or make nests. Indeed, modern tinamous, representing the ancient lineage of palaeognath birds, do exactly this sort of opportunistic behavior today by ducking (in an avian way, no less) and covering in burrows: not in their own burrows, but those of mammals. Such observations lead to wondering whether the first burrowing dinosaurs or birds might have been less like kiwis and more like burrowing owls, taking over and modifying already existing mammal burrows rather than making their own.

So when exactly did mammals start burrowing, and when did their more reptilian synapsid ancestors begin living underground? Could burrows have helped synapsids survive the worst mass extinction in life history about 250 *mya*, and allowed their descendants—mammals—to make it past yet another mass extinction 50 million years later, then the most recent one 66 *mya*? In short, do we and all other mammals alive today owe our evolutionary heritage and existence to burrows? The definite answer is, probably.

CHAPTER 5

Bomb Shelters of the Phanerozoic

Time Tunnels

The *Lystrosaurus* woke up to a nightmare. As she climbed the ramp of her twisting burrow, she sensed something different in the air. It was warm, with charcoal smells accented by burnt flesh. When she poked her head out of the sunlit burrow entrance and looked around, all of the familiar landmarks she used to navigate her territory—the *Glossopteris* tree to the left, the well-worn trail cutting through foliage in front, the scent post she and others had marked with their feces to the right—were all erased. Much of the landscape was blackened, dappled only by splotches of variegated lichens on rocks and green ferns sprouting from reddish clay. The latter was hardened, expressed as mud-cracked polygons merging with a horizon formerly covered by a vast lake. A meandering river that once emptied into the lake was also gone, replaced by dunes of dry, shifting sand. In the distance, a volcano smoldered as it always had, but its flanks had thickened, draped with layers of dark, ropy basalt. An odd quietude presided over this vista, a soundscape devoid of the usual raucous calls exchanged between others of her kind, as well as the voices of reptiles, amphibians, and insects.

In her immediate area, several other *Lystrosaurus* should have already been out before her. In previous springs, members of her clan would have been foraging on new shoots, ambling down to the lake for a long drink and a parasite-ridding mud bath, or getting down to business with mating rituals. Later, many more would have joined the early wakers, each lumbering out of holes in the ground over the course of several weeks and eventually amassing a hundred or more in their colony. Yet this time, only the stark white bones of a few remained, as did their tracks, some with dried fractures radiating from their claws. She was alone.

Like her mother and siblings in the previous two years, she had overwintered in her burrow, slowing her breath and otherwise reducing her metabolism to a minimum during those months of cooler temperatures. She also did not eat or drink during that time. This abstinence resulted in some weight loss, but a springtime bounty of seed ferns and nearby water bodies normally alleviated both of those problems. Instead, a denuded and desiccated terrain greeted her. Even worse, her family members were nowhere to be seen, nor smelled, nor heard, and they apparently had not shared the burrow chamber with her, either. This was also not quite right, as her kind was gregarious, with families often crowding into a burrow chamber throughout the year. If outside, they traveled overland or waded into the nearby lake or river, seeking fodder together. Although adults were more than a meter long, these group behaviors conferred safety against large toothy predators, while also making it easier for adults to guide naïve juveniles back to their home burrow at the onset of danger, from bad weather to rapacious killers.

Given such an overwhelming amount of negative stimuli in this outside environment, she almost backed up and down to the comfort and security of her burrow. This aversive decision would have been perfectly understandable in any case because it was, after all, a marvelous burrow. Its entrance was slightly taller than wide, but wide enough to admit its adult occupants, who also helped make it. Its ramp went down at a gentle angle, but turned right, then left, making a descending semi-spiral. It ended well below the surface with an expansive chamber large enough to hold a family of two parents and two offspring. Although air in the burrow held less oxygen farther down, *Lystrosaurus* was adapted for it; if oxygen levels were ever to plummet outside, she could handle this, too. Despite these burrow benefits, she felt the tug of other biological urges, which eventually pulled her out of what could only become a pit of despair. Having taken no sustenance all winter, she was both hungry and thirsty. She would have to abandon her burrow and go on a journey.

Because all visual cues of nearby environments had been altered so radically during her slumber, she relied on her nose for direction, depending on the wind to bring her messages from afar. Her flattened face did not readily reveal this innate skill, but a series of interconnected sinuses inside her skull received minute chemical cues, which were then processed by a relatively large olfactory lobe into simple but meaningful messages: Food. Water. Fire. Danger. Others. However, the arid air currently held few expressive molecules, which made this quest a little more difficult. Nonetheless, stout of body and short of legs, her tail straight behind her, she began walking.

Although she did not have a memory from which she could retrieve specific details of days past, she had a general feeling of where to go and what to do that was partly based on previous experiences and instinct. Given this mix of inference and the innate, she ate the surviving ferns near her now abandoned burrow, then began walking north, the same direction her clan had moved whenever resources had become scarce before. This choice was confirmed throughout the day as she received more scents funneled down the former river valley and nearby ravines, all of which promised nourishment. As she strolled, the changing terrain did not signal that she was traveling downhill, the differences too slight to detect. This topographic change, however, was important for the subsurface flow of water, which eventually would gather where she and other animals could use it to continue their lives and make more of their own kind.

Vision was not her strength, but what she had helped, and nearly everything she saw was new to her. The former lakeshore and river valley led to the broad floodplain of another river, also dry and barren, its levees and former oxbows festooned by crescentic dunes. Filthy shades of gray, black, beige, and red dominated the ecological palette, with no green in view. Among the sights she encountered after abandoning her former home were scattered and dismembered corpses of other *Lystrosaurus*, as well as remains of other animals she had never seen, living or dead: *Ictidosuchoides*, *Moschorhinus*, and *Tetracynodon*. Like *Lystrosaurus*, these land-dwelling animals were therapsids, representing one branch of a diverse clade of reptiles called synapsids. Unlike her, though, these three ate other animals for a living. Of them, *Moschorhinus* was the apex predator and often successfully preyed on *Lystrosaurus* and the other two therapsids. Yet none of these bodies had moved under their own power for more than a year, their disarticulated and gnawed bones telling tales of both scavenging and more than occasional cannibalism. Once you eat your own more than others, the end of your species is nigh.

Surrounded by such grimness, her chance find of several leafy plants growing between the bones of a former *Lystrosaurus* imbued a feeling approaching pleasure. These plants, standing nearly as tall as she was and bunched together by their clonal growth, required just one more test before she would indulge in their bounty. She inhaled sharply, and the chemicals delivered to her small brain registered "not toxic." She ate by cropping the leaves and stems with her beak, then ground her jaws forward and backward to process their parts before swallowing. These vegetative bits included seeds that had evolved to pass unharmed through her gut and come out the other end the next day with fertilizer on top. Little did she and other still-living *Lystrosaurus* know, but they were helping to reseed the devastated landscape.

After such a fortifying meal, she relaxed a bit, but this respite was interrupted by a surprise: another *Lystrosaurus*. This one had watched the newcomer from the top of a nearby ridge, gathering enough information to make a decision whether or not to approach. Like the other traveler, she had similarly arisen from a winter slumber and left behind her burrow only a few days before, wandering about while looking for the stuff of life. But this one was older and larger; hence her bulk commanded instant respect from the visitor. At first they appraised each other from more than two body lengths away, until the larger one emitted a low-frequency greeting call. The smaller *Lystrosaurus* returned a similar call, confirming that they belonged to the same species. The larger one then slowly approached the young traveler, who backed up slightly but otherwise held her ground as they exchanged sniffs. They rubbed heads to further exchange scents, a final ritual that evaporated any tension that might have triggered "fight or flight" in either. Because *Lystrosaurus* was inherently a social animal, this meeting of two lone individuals strengthened them both.

From there, they walked together, but with the elder female leading the way. Her experience took them more to the northwest, a direction the younger one would not have picked. This change in strategy soon delivered, as they discovered more plants, consisting of a small, isolated stand of *Pleuromeia*. This plant, taller than each of them, had a single stem topped by a cone and simple leaves, and its base was wide, connected to well-developed roots. They ate everything, including the roots, which the elder dug up with her powerful front legs. This was an action the younger *Lystrosaurus* had seen her parents do only a few times when they had foraged for her. Beetles and other insects that had burrowed into the soil near the roots scattered as the therapsids destroyed their former homes, and more than a few were accidentally consumed as the pair ate. Once satisfied, they rested for a while, waited until the sun set, then moved on. They walked through the relative coolness of the night, with the elder frequently stopping to check on and wait for her companion.

Soon after dawn, the clear day became hotter than the previous two and the air became more still. Given such conditions, they would soon have to find both shadows and water. The treeless land promised none of the former, so they would somehow have to make their own shade. The older *Lystrosaurus* solved this problem, too. By mid-morning, she began burrowing into a hillside, making an entrance that faced away from the maximum arc of the sun. Although nowhere nearly as skilled or as strong as the other, the younger *Lystrosaurus* joined her, and soon they were taking turns shoveling aside the hardened topsoil to reach the more friable sediments below. Within only a few hours, they had made a good "starter burrow." The tunnel was much shorter than normal, and only turned once to the right before connecting with an expanded chamber. But this was not going to be a permanent home: more like a lean-to shelter than a mansion. Still, the coolness and high humidity of the burrow contrasted nicely with the harsh conditions outside. They crawled down the tunnel, snuggled together in the chamber, and stayed there for the rest of the day.

In the middle of the night, the elder *Lystrosaurus* roused her new traveling companion, and up they went to the surface for more questing, a full moon lighting their way. Water would be essential for their survival from here on in. The meager plants they had eaten thus far supplied them with a small amount of water, but both were dehydrated after a winter of burrow habituating and would need much more. Soon into their walk, the milder nighttime air carried organics that caused them both to stop, inhale again, and change direction accordingly. The elder linked these scents to a water body, and her apprentice likewise recognized them, a sensation bestowed by her former lakeside home. Thus properly beguiled and beckoned, they went down a gulch, which had funneled both the wind and odors their way.

Fortunately for them, but unfortunately for the *Moschorhinus* at the bottom of the ravine, these air movements carried their aromas up and away from the pond where he waited. Millions of years of natural selection in his lineage had contributed to a predatory instinct to settle down by ponds, lakes, or rivers, where eventually his meals would come to him. This strategy was especially apt during lean times, and this was one of them. Despite his massive bulk—easily twice the size of the older *Lystrosaurus*—and his robust skull, bearing sharp incisors and long, stabbing canines, he had not killed anything for weeks. He had stayed alive only by scavenging the few therapsid bodies left in the area with any muscle or skin on them. He was emaciated and needed fresh meat soon.

Lacking an earlier alert, the arrival of the two *Lystrosaurus* on his right startled him. Giving a short bark, he jumped back from his lying position, but recovered quickly as he assessed their profiles as "food." At the same time, the pair of *Lystrosaurus* recognized his profile as "death," unceremoniously declined his invitation to dinner, and began running along the edge of the pond. The *Moschorhinus* accordingly began pursuing them. As soon as he bounded close enough to capture the younger *Lystrosaurus*, the elder turned abruptly to the left. She then jumped into the pond, the splash distracting the *Moschorhinus* enough to pause and look to his left as she therapsid-paddled her way toward its deeper center.

This was all the younger *Lystrosaurus* needed to escape his grasping claws. She likewise plunged into the water, following the moving shape ahead of her. He was not so keen on entering the pond, having almost gotten mired in its muddy bottom several times before. He paused along its edge and watched the two figures recede, each defined by V-shaped ripples highlit by the moon. Thwarted this time, he lay down to wait for something else to come along.

The other side of the pond was far enough away that the *Moschorhinus* could not see them there, so once the two *Lystrosaurus* were on land, they shook themselves off and inspected their immediate area. Discerning no more danger, they turned around to the shoreline and drank. Once satiated, they walked a little more than ten body lengths away from the pond, found a dune, and together scratched out a shallow hollow in its side. Here they rested into the early morning; when the sun arose, they roused themselves and dug a little deeper. Again, this shelter was not intended to serve as a permanent home, but would effectively hide them from other land predators while decreasing exposure to the oppressive heat outside. They had also picked a good site for provisions, as a variety of wetland plants were nearby, their stems and leaves jutting from the water. Following their morning renovations, they indulged in this bounty, the most food they had encountered thus far on their sojourn. As they ate, flying insects hummed around the plants. Although most lived at least part of their egg-larva-pupa-adult life stages in burrows, the plants provided both food and habitat for them, too.

The *Lystrosaurus* emergency bunker proved worthy in other ways later that afternoon when a fierce thunderstorm rolled in, delivering high winds, lightning strikes, and rain dumped onto the baked landscape. This was unusual, as sandstorms had been more frequent than rainstorms in the past few decades. With so little vegetation to slow the water flowing down the arroyos, and with so few roots holding down the soil, the runoff was both voluminous and muddy. This spelled trouble for the elder and younger *Lystrosaurus*, as the pond level began to rise quickly from the sudden input of water and sediment and approached their burrow. The brief lag between lightning and flood gave them just enough time to pop out of their temporary refuge and head for higher ground. A slurry of mud and sand filled their newly dug shelter, as well as burrows of varying sizes and depths made by previous generations of vertebrates large and small, casting them for a distant future.

The onset of calmer conditions triggered instinct, and it was time again to do what they did best, which was to dig, and dig deeply. Given the ample supply of food and water, and apparent absence of predators, it seemed appropriate to settle down at this locale for as long as these resources lasted. Their more permanent burrow, made well above the pond level, took several days to construct. It plumbed down more than five body lengths of the larger *Lystrosaurus*, the tunnel describing a complete whorl on its way down to a wider end chamber. Given this stable home environment, a routine emerged, with daily life based on cycles of the sun and moon: nighttime drinking and foraging, some morning and twilight basking, but otherwise resting together in the burrow chamber. The practice changed as spring became summer and the days became longer, hotter, and drier. This seasonal shift necessitated more time in the burrow than before, the cooler and moister climate inside providing instant relief from the extremes outside.

It was only a matter of time before the repeated movements of these two pioneers in such a small area concentrated their scents and transmitted their presence to more *Lystrosaurus*. These increased numbers caused positive feedback that brought in eleven during the first lunar cycle, then more than double in the next. They sometimes arrived as male-female pairs, or sometimes as the remainder of family units. Almost none arrived alone, and if they did, they often died quickly, too weakened by the terrible travails endured on their way there.

As their numbers increased and their burrowing with it, they began to change this place more to their advantage. For instance, the newcomers added their burrows to generations of abandoned homes made by defunct therapsid residents near the water's edge. These shafts and tunnels, which had perforated the crusty ground and often intersected one another, effectively worked like pipes, linking nearby groundwater sources with the pond. This connectivity helped to replenish the pond, requiring less dependence on increasingly rare surface runoff. The collective action of more *Lystrosaurus* wallowing and feeding in the shallow water around the pond also expanded its boundary, as trampled and foraged areas filled with water. The plant community, well adapted to disturbance and bearing spores and seeds spread by *Lystrosaurus* feces—many of which were conveniently deposited in the pond—spread and thrived through the summer and into the fall. Insects hosted by these plants accordingly became more numerous, their sounds mingling with the low-pitched calls of the *Lystrosaurus* colony.

Soon the biodiversity of the pond increased from within, as water seeped into the sediments surrounding estivation burrows of lungfishes, amphibians, and the crocodile-like *Proterosuchus*, signaling to all "time to wake up." Each lungfish had sealed itself in a mucus cocoon after wriggling into previously moist mud; the cocoon and burrow doubly protected them from previously dry conditions outside. Large salamander-like amphibians had tried a similar strategy, but sometimes crowded into submerged burrows, which they closed to prevent evaporation. Dire necessity even resulted in strange bedfellows, where amphibians occasionally shared a burrow with a therapsid. *Proterosuchus*, however, was not into sharing, and these reptiles became torpid in burrows by themselves. Only a few had survived the shrinking of their water hole, and once awakened they were not only voracious, but also thrust to the top of the food web. Within only a few weeks after emerging from their dens, they had eaten most of the lungfishes and amphibians in the pond. One *Proterosuchus* was even lucky enough to nab a newborn *Lystrosaurus* that made the mistake of leaving its home burrow and strayed too close to the predator lurking just below the water surface.

Despite these occasional losses, the pond and its surrounding area lacked large predators. The *Moschorhinus* that stubbornly stayed by the pond waiting for food to arrive eventually died, and no others of his species replaced him. This situation, combined with a bountiful water supply, increased plant resources, and no herbivorous competitors, translated to an Eden-like paradise for the *Lystrosaurus* colony. Soon it numbered in the hundreds, with the original cofounder of the colony—the younger *Lystrosaurus*—contributing six offspring to this crowd. The male who sired all of her children had arrived with his mother during the first few weeks of her stay there. However, he had injured his front right leg in a fall during the long walk there, so she did most of the digging for their nesting burrow. This was fine, as she only needed his gametes anyway, along with a little help in maintaining their burrow and raising each pair of youngsters, who were spaced over the course of three years.

The elder *Lystrosaurus*, who had been such a superb mentor for the younger one, eventually disappeared from the colony. Soon after the hatching of the first egg clutch from her protége, she walked out into the darkness of a moonless night, never to return. Whether she detected more tempting resources, another colony, or was simply following the winds to new experiences is unknown. Regardless, she ultimately was responsible for the cofounding and survival of this colony, which became part of a larger survival story for *Lystrosaurus* and related therapsids, then and now.

The evolutionary key to the survival of *Lystrosaurus* and its close relatives is how they and their ancestors had dug into the earth and used burrows to survive, while also changing their local ecosystems in ways that benefited them. Over millions of years, they and their ancestors had unknowingly prepared themselves for an overwhelming global disaster that spread like a plague, wiping out nearly all other species. Given the right combination of genes, behavior, and environmental pressures, their burrowing abilities were favorably selected, affording them the means to live through dramatic seasonal and climatic changes. This selection, however, also bestowed an ability to survive fires, droughts, thunderstorms, and other natural disasters during the pervasive awfulness. Because of their burrowing habits, they could even tolerate less oxygen in the earth's atmosphere. They were, in essence, prehistoric "preppers."

Meanwhile, their burrowing also altered ecosystems, created new avenues for surface and subsurface water to flow, and disturbed soils such that certain plants were selected for these conditions, which then laid a foundation for local food webs. The surviving *Lystrosaurus* became ecosystem engineers, changing terrestrial conditions for themselves while also improving the lot of other species. The few other animals that could adapt to this new *Lystrosaurus*-dominated world did, and went along for the ride. It also helped if these animals could secure themselves in burrows and wait for heinous conditions to subside just long enough to pass on genes to the next generation.

Throughout all of the southern supercontinent of Gondwana and the rest of the world, over just a few tens of thousands of years, a new time started. Other environmental calamities and mass extinctions happened after then, but none was as horrific as the one endured by *Lystrosaurus* and its kin in those years. When entering their burrows during the last gasp of the Permian, they had no inkling that they would exit in a new era. More than 250 million years later, distantly related descendants of a few burrowing synapsids that lived among the many *Lystrosaurus* would give this age of renewal and recovery a name. They called it the Mesozoic Era.

A Quick Natural History of Lystrosaurus and Its Burrowing Kin

Lystrosaurus was such a fine example of a survivor that its picture deserves to be in a dictionary next to the word "survivor," ably replacing any association with, say, a contrived reality-TV series. Represented by about nine species (maybe more, maybe fewer), *Lystrosaurus* mostly lived in the southern continents of Africa, Antarctica, and India, and during the latest Permian Period through the earliest Triassic, with the Triassic as the first period of the Mesozoic Era. In the Permian, at about 280–250 *mya*, the now-separate continents of Africa, Antarctica, and India were united with South America and Australia as one giant interconnected landmass, Gondwana.

North of Gondwana was Laurasia. It was composed of what would later become North America, Europe, Asia (excluding India), and a few more lands. Interestingly, although *Lystrosaurus* was mostly Gondwanan, at least a few made it to Asia, with specimens found in China, Mongolia, and Russia. This seemingly aberrant location for *Lystrosaurus* probably happened because Laurasia and Gondwana were also linked during the Permian. This bonding formed a massive supercontinent, Pangea, surrounded by a great ocean, Panthalassa. The Permian interconnection of these now far-flung landmasses thus better enabled *Lystrosaurus* and other terrestrial animals to disperse widely, undaunted by oceans or other major geographic barriers. Indeed, fossils of *Lystrosaurus*, *Cynognathus* (another therapsid), a seed fern (*Glossopteris*), and a freshwater reptile (*Mesosaurus*) in parts of former Gondwana helped support the idea of "continental drift" in the early twentieth century. In this scenario, these organisms were all close to one another in a united Pangea, but their fossils would be found thousands of kilometers apart after the continents drifted apart from one another. In the second half of the twentieth century, this fossil evidence for "continental drift" contributed to modern plate tectonic theory, with *Lystrosaurus* as a key piece in the puzzle showing how the continents were formerly joined.

So just who was *Lystrosaurus*? To answer that question, a family tree can give some perspective. All species of *Lystrosaurus* were synapsids, and one of the main evolutionary traits shared by synapsids is a pair of temporal foramens, with one on each side of its skull. If you are perplexed by the term "temporal foramen," just ease that momentary tension by placing your fingertips on either side of your head just behind each eye, and massage your temples. The slight depression you feel on each side is a temporal foramen, which is a hole between the skull bones there. This means you are a synapsid, as are all mammals. Platypuses, echidnas, kangaroos, mice, squirrels, honey badgers, elephants, lemurs, pandas, whales, dogs, your neighbors, and even Internet-starring domestic cats are all bonded evolutionarily through right-left temporal foramina (plural of foramen), among other traits. In contrast, diapsids, which belong to the clade Diapsida, have two temporal foramina on each side of their skulls. Diapsids are represented by archosaurs, among which are crocodilians, lizards, snakes, turtles, birds, and non-avian dinosaurs. Keeping in mind how clades are conceptually like Russian dolls, with each descendant clade nested inside a "mother" clade, Synapsida also has Therapsida within it. Hence it would be erroneous to call *Lystrosaurus* and its therapsid cohorts "reptiles," as they more closely share a common ancestry with mammals. Nonetheless, *Lystrosaurus* and other therapsids are sometimes misleadingly termed "mammal-like reptiles," which we should just stop doing right now.

Synapsids originated from other egg-laying land vertebrates in the Late Carboniferous a little more than 300 *mya*, but probably split off their evolutionary tree just before diapsids. The oldest known fossil synapsids are *Archaeothyris*, *Clepsydrops*, and *Echinerpeton*, and all three lived in land environments during the Carboniferous in what is now Nova Scotia (Canada). Not too long after that (geologically speaking), fossil therapsids show up in the middle of the Permian Period, at about 270 *mya*.

Therapsids succeeded other synapsids called pelycosaurs, a loosely defined group (not a clade) of animals that dominated Early Permian landscapes. One of these pelycosaurs, the sail-backed *Dimetrodon*, is the bane of dinosaur paleontologists everywhere, as adults (but never children) erroneously call them "dinosaurs." Adding insult to injury, these well-meaning but ignorant people then place *Dimetrodon* toys in the dinosaur bin of natural history museum stores. Ironically, *Dimetrodon* is more closely related to the people misnaming them than to dinosaurs. Regardless of this identity crisis, *Dimetrodon* was a formidable predator. It reached lengths of about 5 meters (16 feet), had small and large pointed teeth (hence its name, *Dimetrodon*: "two measures of teeth"), and a broad, tall sailback supported by long vertebral spines. Yet it also was a synapsid and had other traits that link it much more closely to mammals than diapsids.

If the gentle reader will bear with me for a moment, a little therapsid taxonomy is necessary before getting back to something that really matters, like burrows. Within Therapsida are two other clades, Dicynodontia and Cynodontia, which split from a common ancestor. Dicynodontia (dicynodonts) includes *Lystrosaurus* and its close relatives, whereas Cynodontia (cynodonts) consists of mammals and their close relatives. Cynodontia is very important. Why? Because of all synapsid clades that arose during the latter part of the Paleozoic Era, it was the only one that survived into the latter part of the Triassic Period. Today, mammals are the only living descendants of cynodonts, and all dicynodonts likely died by the end of the Triassic. Another key point about dicynodonts to keep in mind is that most probably made and lived in burrows. This is not only indicated by the "smoking gun" association of dicynodont skeletons in fossil burrows, but also by anatomical traits suggesting that digging was part of an evolutionary heritage for these synapsids.

I purposefully did not describe the physical appearance of *Lystrosaurus* in the preceding fictional narrative, as I suspect at least a few people swayed by superficial standards of beauty may have started cheering for the *Moschorhinus*. Let's just say that *Lystrosaurus* outwardly looked like the product of an unholy threesome between an iguana, a pig, and a naked mole rat. Granted, other paleontologists and I find it exceedingly winsome, but we also do not expect most people to share our astute tastes. Anyway, if you only caught a quick glimpse of a *Lystrosaurus* before it dashed down its burrow, you would describe it as having a flat face, short limbs, wide torso, and a small, stiff tail. Its name, assigned to a fossil skull described in 1870, may have alluded to its spade-like face, as *Lystrosaurus* means "shovel lizard." Its other facial features were those that only a mother *Lystrosaurus* would adore, with an abbreviated snout (think *ET*), a horny beak, and wide-set eyes. Some skulls have prominent tusks (canines) in the upper part of their jaws, but these were the only teeth known in these synapsids. These fearsome accouterments may have been used to intimidate rivals or impress potential mates.

Body sizes varied with species, as some *Lystrosaurus* were as small as a half meter (2.5 feet) to more than 2 meters (6.6 feet) long, but their average length was about a meter (3.3 feet). Smaller sizes may have benefited *Lystrosaurus* species during times of ecological stress, as bigger bodies require more food; this would have been more so if *Lystrosaurus* were endothermic ("warm-blooded"). At the same time, *Lystrosaurus* had another advantage over most other vertebrates in its ecosystems: It grew up fast. Studies by Jennifer Botha-Brink and other paleontologists have revealed that *Lystrosaurus* bones were greatly vascularized, but in their earliest stages lacked annual growth rings. This shows that juveniles developed rapidly and reached sexual maturity early, perhaps within just a few years. This in turn allowed *Lystrosaurus* to spread more quickly than other animals that might have occupied the same ecosystems.

Nonetheless, the most noteworthy anatomical attributes of *Lystrosaurus* are its forelimbs. From its shoulder to its front toes, stout bones reinforced what must have been massive musculature and connective tissues. These appendages were not just made for what wildlife biologists call the three "F's"—fleeing, fighting, and mating—but also for digging. We also know that *Lystrosaurus* burrowed because their burrows have been positively identified: Which is to say, these burrows had *Lystrosaurus* skeletons in them, just as the small Cretaceous dinosaur *Oryctodromeus* was found in its den. This is where its name ("shovel face") was unintentionally appropriate, as this was an animal well adapted for shoveling dirt. Their burrows were similar to those of modern gopher tortoises, forming tunnels that twisted to the right or left on their way down, ending with an enlarged living chamber.

These attributes connect with my two main reasons for giving *Lystrosaurus* so much love and admiration here: (1) It was a burrower; and (2) it and other dicynodont and cynodont burrowers survived the worst mass extinction in the history of life, the end of the Permian. Coincidence? Maybe, maybe not. Of course, burrowing ability by itself would not have favorably selected *Lystrosaurus* during a mass-extinction event. Still, *Lystrosaurus* was the only burrowing vertebrate that eked into the Triassic. Accompanying it on this subterranean survival story were *Thrinaxodon* (a cynodont), and *Procolophon*, a reptile that looked like a modern lizard, but with a skull like a turtle. Species of both animals are in Permian-Triassic rocks of Gondwana, with *Procolophon* in South America, Antarctica, and South Africa, and *Thrinaxodon* in Antarctica and South Africa. *Diictodon* is also worth mentioning, as it was a burrowing dicynodont that lived in South Africa, having made spiraling burrows more than 1.5 meters (5 feet) deep. Sadly, though, it only lived during the Late Permian. Still, some of its dicynodont relatives did make it into the Triassic, suggesting that they were burrowers, too.

Thrinaxodon is a fine example of a burrowing cynodont, having both the means to burrow and lots of burrows to show for it. In fact, *Thrinaxodon* was the first discovered example of a burrowing synapsid from the earliest part of the Triassic. *Thrinaxodon* remains have been found in their burrows, which then also helped paleontologists to better interpret their burrows without bones. One *Thrinaxodon* burrow also contained an unlikely roommate, a salamander-like amphibian called *Broomistega*. The amphibian skeleton shows it was injured, so it may have been seeking refuge in the burrow with the *Thrinaxodon*. The *Thrinaxodon* then either tolerated its presence or had just died before the *Broomistega* moved in and also died there. *Procolophon*, given its short, stocky limbs with stubby fingers, is also interpreted as a burrower. Unfortunately, it has not yet been found in a burrow, nor have geologists interpreted probable *Procolophon* burrows.

With *Lystrosaurus* and so many other good examples of burrowers at the end of the Permian and earliest Triassic, it is well worth exploring how they and other underground dwellers may have used such shelters to make it past the most disastrous extinguishing of animals on land. *Lystrosaurus* was an extreme example of this, having seemingly taken over nearly empty ecosystems during the earliest Triassic and inadvertently throwing a big post-apocalypse party for itself. According to some paleontologists, *Lystrosaurus* may have constituted 95% of all land vertebrates for a short time, a homogeneity approaching that of an exclusive country club. How did this happen, and how could burrowing have provided these and other animals some advantage? Probably the best way to answer this question is to examine the cause of the end-Permian mass extinction, often labeled by paleontologists as "the time that life almost died."

The End of the Permian Was Really Awful

Given current interest in the "Sixth Extinction" happening today, in which many species are on the brink of dying out, along with climate change, pollution, ocean acidification, and other environmentally themed doom-and-gloom topics, we can always console ourselves by saying it could be worse. Short of our waging all-out nuclear war, the end of the Permian demonstrated how the Earth could inflict far more damage than anything we mere humans have done (so far).

Let's begin with plate tectonics. Other than an occasional wayward meteorite, plate tectonics is responsible for nearly everything that happens on the surface of the earth, and this was certainly true during the Late Permian. In what is now Siberia, volcanism dominated the last million years or so of the Permian, caused by either a series of volcanically active hot spots (like Hawaii or Yellowstone, but much bigger), and/or rifting of the continental crust there. Regardless of its cause, this volcanism unleashed huge amounts of magma in what geologists term as *flood basalts*. These sheets of lava flowed across Siberia and deposited an estimated volume of several million cubic kilometers of basalt, now called the Siberian Traps. These deposits were thick, but also widespread: Imagine the lower 48 states of the United States covered in black rock. Artists who attempt to illustrate the Siberian Traps basalt flows often depict scenes with volcanoes belching broad rivers of orange-red lava, obliterating all ecosystems in their path. Such representations may remind people of the setting for the climactic light-saber battle between Obi-Wan Kenobi and Anakin Skywalker (soon to become Darth Vader) in *Star Wars: Revenge of the Sith*. (Or more likely inspire fantasies of immersing Jar-Jar

Binks in hot magma.) Film criticism aside, such visualizations are meant to convey how nothing—not even the hardiest of bacteria—could live on a surface that regularly received pulses of 1,000°C (more than 1,800°F) liquids.

Yet it was not this local heat nor magma that triggered a mass extinction in all of Pangea and Panthalassa, but the pronounced and long-lived heat caused by the massive release of greenhouse gases and the global warming that followed. The sources of the gases were both on land and in the sea, with volcanism providing one and the seafloor supplying the other, a deadly one-two combo that started a trophic cascade like no other since.

The first stage of this ecological meltdown was the eruption of the Siberian flood basalts. These eruptions were accompanied by immense amounts of carbon dioxide, which, as all but a few billionaires know, is a greenhouse gas. Greenhouse gases are like blankets, retaining more heat energy (infrared radiation) than is reflected off the earth's surface. Thus as this volcanism pumped more carbon dioxide into the Permian atmosphere over the course of a million years, and ecosystems could not absorb or bury it quickly enough, the Earth got warmer. This warming was already bad enough, but the flood basalts made matters worse by burning Carboniferous coal deposits lying near the surface. Yes, that's right: Permian flood basalts in Siberia were about 250 million years ahead of us upright, semicivilized primates in setting fire to these carbon-rich deposits. This burning added even more carbon dioxide to the atmosphere, which likely reached levels far exceeding what humans are contributing through our activities today.

Yet another problem caused by the coal burning was one humanity also experienced after doing too much of the same, which was acid rain. Coal contains varying amounts of sulfur-bearing minerals, which when burned combine with oxygen to form sulfur dioxide. This compound in turn hooks up with atmospheric water to form sulfuric acid. With these processes in mind, geoscientists have proposed that freshwater lakes, streams, and soils became more acidic toward the end of the Permian, making life more difficult for plants or animals that could not handle such caustic conditions. The degradation of plant communities would have been particularly troubling, as these communities not only formed the base of nearly all food webs on land, but also produced oxygen from photosynthesis. Take away most plants, and animals would have been left with less food and lower-quality air to breathe.

These broad-scale changes in the earth's atmosphere were accompanied by local changes in Siberian ecosystems, which understandably were not well suited for hot lava flowing over them. Flood basalts would have smothered and covered former forests, lakes, streams, deserts, and anything else in their path, turning these formerly diverse environments into blackened and lifeless vistas. Relentless lava flows, spread over such a broad area and piling onto older layers of basalt, would have effectively squeezed out plants, animals, fungi, or anything else that preferred to live in places cooler than a thousand degrees.

The long-term emission of massive amounts of carbon dioxide from volcanism on land was enough in itself to trigger global warming. Then the seafloor made it worse. As ocean temperatures climbed, frozen deposits of methane in deep-sea areas of Panthalassa began to thaw. Like carbon dioxide, methane is a greenhouse gas, but even more effective in holding heat: like using a thick wool blanket to keep warm instead of a thin nylon one. So as soon as this formerly solid methane thawed and released big bubbles from the seafloor, it immediately went up and into the Earth's atmosphere, adding a new and warmer blanket. As a result, heating of the Earth's surface accelerated. This positive feedback loop then fed on itself, in which higher air temperatures meant increasingly warmer seas, which thawed out more methane. Conditions went from a mere greenhouse to a hothouse, rendering hell on Earth.

Given such an already terrible situation, how could it get worse? Well, how about decreasing oxygen in the world's oceans? Oceanic circulation stagnated toward the end of the Permian, leading to the expansion of oxygen-poor zones, which moved from deeper water into shallow-water environments. Lower amounts of dissolved oxygen would have shortened the lives of any animals that used gills to filter oxygen from the water, killing many invertebrates, fish, and more. This was bad.

Again, how could it get even worse? Okay, how about the oceans absorbing enough of the carbon dioxide to form carbonic acid, which along with acid rain made the oceans more acidic? Carbonic acid, a typical ingredient in carbonated soft drinks, is also formed in natural systems by mixing carbon dioxide with water. Very simply, if more carbon dioxide is in the water, carbonic acid is easier to make. Ocean acidification would have hindered marine algae and animals from making shells or skeletons, or weakened their hard parts by partially dissolving them. Acidic waters were especially lethal for marine algae, which like land plants produce lots of oxygen from their photosynthesis. With both land plants and marine algae taking a hit, oxygen levels in the atmosphere would have plummeted, too. This was also bad.

That should pretty much do it, right? Nope. How about depleting the ozone layer? The weird mix of gases in the Late Permian atmosphere likely caused such a depletion, which would have allowed more short-wave ultraviolet (UV-B) radiation to reach the Earth's surface. Ozone, which is composed of three oxygen atoms (O_3) instead of the normal duo we inhale daily (O_2), is abundant enough in the upper part of the Earth's atmosphere (stratosphere) to form a thin layer. This layer blocks or otherwise inhibits UV-B radiation from reaching the Earth's surface, where it could do some serious damage to life. For instance, thanks to the discovery in the 1970s that human-made chlorofluorocarbons (CFCs) eroded the ozone layer, we began to realize its planetary importance and took steps to stop this. Increased UV-B would have caused more frequent occurrences of skin cancer in humans, but it also would have resulted in more genetic defects, disrupting the normal physiology and reproduction of many organisms. Modern experiments with UV-B radiation demonstrate that it tends to stunt plant growth and weaken their defenses against diseases, and many marine algae are hurt by it, too. Ozone depletion and bombardment by UV-B radiation toward the end of the Permian would have constituted yet another blow to the foundation of continental and marine ecosystems, severing links in worldwide food webs. This was very bad.

Could anything top all of these problems? Sure. How about all of the continents getting together as Pangea at the end of the Permian, which made an already awful environmental situation utterly dire? This continental coziness slowed the movement of moisture to the interior of Pangea, which formed extensive semi-arid land environments. But with climate change, these environments turned into deserts that would have made the Sahara look like a lush tropical rain forest. Having all of the continents together also decreased the number and areas of coastlines, which meant far fewer shallow-marine habitats for animals, which translated into less marine biodiversity. Considering that all of the previously described nastiness that descended upon the world during the last 50,000 years or so of the Permian, most life in those inland and shallow-marine environments did not stand a chance. As soon as a few ecologically important species in continental and marine environments died out, the rest would have quickly gone with them, a domino effect of ecological collapses that traveled around the globe again and again.

All of this planetary awfulness may sound depressing, especially because so many of the environmental problems suffered by life at the end of the Permian sound so familiar: global warming, climate change, acid rain, ocean acidification, ozone depletion, mass extinctions, and more. Considering that these pervasively dreadful conditions lasted for a minimum of 50,000 years, it is a wonder that anything made it through at all. Just how did any animals on Pangea or in Panthalassa live long enough to reproduce a next generation of their species? Then how did their offspring—often vulnerable and otherwise unready to cope with harsh circumstances—also grow up and pass on their genes, too? Especially on land, what could have shielded *Lystrosaurus* and other animals from extreme temperature fluctuations, acid rain, low oxygen, UV-B radiation, collapsed ecosystems, and more?

Given such horrible circumstances, one might say it was enough to make an animal crawl into a hole . . . and survive.

The End of the Triassic Was No Picnic, Either

The recovery of life on Earth following the end of the Permian was slow, but marine and continental ecosystems alike were thriving by the end of the succeeding Triassic Period. This revival happened over about 50 million years, resulting in remarkably diverse communities on land and in the oceans, including some of the most iconic life forms of the Mesozoic Era, such as marine reptiles, pterosaurs, and dinosaurs. Nevertheless, this evolutionary party came crashing down with another mass extinction at the end of the Triassic, which took out about half of all species.

But first the good news: Enough organisms survived the devastating end of the Permian for life to start over in the Triassic. In the oceans, a new clade of corals evolved from soft-bodied corals and began forming reefs again. Mollusks, such as clams and snails, came back with a vengeance, occupying nearly every niche on and along ocean bottoms, as well as in freshwater streams and lakes. Echinoderms, such as sea stars and their relatives, were hurt badly by the mass extinction, but did quite well throughout the Triassic; a few even evolved into the first modern-style sea urchins. Sharks, bony fish, and ammonites swam through the water, living in marine ecosystems driven by plankton that rebuilt heavily damaged food webs. At least a few Early Triassic diapsids became aquatic, eventually becoming dolphin-like ichthyosaurs, turtles, and the ancestors of long-necked plesiosaurs. Once fully adapted to marine environments, these reptiles promptly chomped on the already abundant seafood there. Among such marine reptiles were placodonts, which had shell-crushing teeth just right for masticating mollusks. Microbial mats, such as those made by algae and photosynthetic bacteria, made a brief comeback, covering seafloor sediments like thin layers of plastic wrap. These mats persisted until burrowers and grazers evolved in sufficient numbers to disrupt them later in the Triassic.

On land and in freshwater environments, *Lystrosaurus* and its synapsid relatives made it through to the start of the Triassic, as did some seed plants, insects, mollusks, fish, amphibians, and diapsids. The first turtles either evolved on land or in shallow marine environments early on in the Triassic Period, but their descendants later migrated into freshwater environments throughout the rest of the Mesozoic, and at least a few lineages made it back onto terra firma, with some eventually evolving into tortoises. Archosaurs in particular became big and bad on land, as well as in lakes and streams. Among these were 6-meter (20-foot) long carnivorous rauisuchians, 12-meter (40-foot) long crocodile-like phytosaurs, and aetosaurs, large herbivores adorned with bony spikes and body armor that no doubt shielded them against rauisuchians and phytosaurs.

Other new lineages that arrived in the Late Triassic and set the stage for Jurassic and Cretaceous ecosystems were the first pterosaurs, dinosaurs, and mammaliaforms. Pterosaurs, as fully flighted reptiles, represented a completely new strategy for vertebrates, becoming the first to occupy the skies. Dinosaurs (of course) became the most successful land vertebrates of all time, but during the Triassic were relatively smaller and less diverse than their Jurassic descendants. Synapsids were represented by both cynodonts and dicynodonts; some of the latter evolved into multi-ton herbivores. Almost unnoticed among all of the Strum und Drang presented by the evolution of these large vertebrates was a small branch of the cynodonts that developed into mammals, a few of whom either purchased this book, checked it out from their library, or borrowed it from a fellow cynodont.

All in all, life in the last 10 million years of the Triassic was amazingly varied, and most of it was strikingly different from that at the end of the Permian. It was as if evolution had completely rebooted all ecosystems, installing a "Life 2.0" that contained only a few familiar lines of code from the previous version. However, this evolutionary celebration was dampened by another mass extinction at the end of the Triassic, one that took out at least 50% of all species. Admittedly, though, it was not as bad that of the end-Permian; instead of being labeled the "Great Dying," it might be called the "Moderate Dying."

The causes of this mass extinction were similar to those implicated in its predecessor, in that much of the blame can be pinned on plate tectonics. Starting about 200 *mya*, an epic breakup began, prompted by a string of volcanic hot spots that pushed apart North America, South America, and Africa and led to the opening of the ancestral Atlantic Ocean. This rifting released huge volumes of oceanic and continental basalt, which were accompanied by greater amounts of carbon dioxide in the atmosphere. More carbon dioxide in the atmosphere warmed and acidified the oceans, which we now know is harmful for most organisms living in those environments. Some ocean-adapted clades made it into the Jurassic Period, such as ammonites, sharks, bony fish, marine reptiles (ichthyosaurs, plesiosaurs, turtles), but much of the microscopic plankton took a hit. Because planktonic algae formed the base of oceanic food webs, their losses sent ripple effects throughout those ecosystems that resulted in mass extinctions by the end of the Triassic Period at about 200 *mya*.

Land environments also became more arid with global warming, which stressed plants and eventually all of the animals that ate those plants and then the animals that ate those plant-eating animals. With this, another trophic cascade began, toppling nearly every big-bodied animal that lived toward the end of the Triassic. Good-bye, rapacious rauisuchians. So long, dynamic dicynodonts. Farewell, ferocious phytosaurs. Adieu, armored aetosaurs. Survivors of this Late Triassic purge included many land plants and insects, bony fish, amphibians, turtles, dinosaurs, pterosaurs, a few other archosaurs (including the ancestors of modern crocodilians), and mammaliaforms, which were all shrew-sized or smaller at that time.

Although this extinction is mostly attributed to the all too powerful effects of plate tectonics, a few geoscientists have proposed another cause, a *deus ex machina* that only entered our collective scientific consciousness in the past 35 years or so. This would be the delivery of an extraterrestrial bang from above, a meteorite. The evidence for a meteorite impact at the end of the Triassic includes a few chemical signatures and an impact crater in France of the right age, at 200 *mya*. However, the impact crater is a bit too small to be from a major life-killer, such as the meteorite that hit the Earth 66 *mya*. This also does not mean a meteorite impact *caused* the end of the Triassic. After all, many other adverse factors were at play then and many lineages were already weakened by what plate tectonics had wrought.

Did any of the animal lineages that made it past the end-Triassic extinction use burrowing as part of a survival strategy? Of the ones listed previously, the best candidates would have been insects and mammals. With mammals in particular, the only reason they evolved is that their synapsid ancestors were among the few species that saw both the end of the Permian *and* the start of the Triassic. Like *Lystrosaurus*, these mammals may have emerged from a burrow at the end of the Triassic, wondering where nearly everyone else went.

The End of the Cretaceous Was Really Bad, Too

When comparing three of the greatest mass extinctions of all time—the end-Permian, the end-Triassic, and the end-Cretaceous—it might be best to think of their differences. For instance, imagine the end-Permian as a slow-cooked pit barbeque, the end-Triassic as a steak thrown on a hot grill, and the end-Cretaceous as a stick of dynamite thrown onto the grill just before the steak was medium-rare.

As mentioned a few times before, the end-Cretaceous, which happened about 66 *mya*, is probably the most famous of all mass extinctions. This is so because: (1) It is linked with another (and much larger) meteorite colliding with the earth; and (2) all non-avian dinosaurs disappeared with it just before the start of the Paleogene

Period. As a result of these two dramatic events in earth history happening at the same time, and with the meteorite regarded as a dinosaur "murder weapon," the end-Cretaceous extinction is now a part of pop culture. For example, this extinction related to the plot of the 2015 animated film *The Good Dinosaur*. It began with the "What if?" scenario of the meteorite missing the earth, the dinosaurs briefly distracted by its passing, but then continuing to dominate the earth for the ensuing 66 million years. With dinosaurs still around, mammals (which somehow included primitive humans) continued to stay small and relatively harmless.

In reality, though, dinosaurs and many other organisms may have been in trouble anyway, meaning that the "murder by meteorite" of dinosaurs was more like a mercy killing. For instance, in the few million years before the meteorite struck, sea level had dropped with global cooling. This change in climate caused shallow marine habitats—which were widespread throughout the Cretaceous—to diminish. A decrease in habitats in turn adversely affected sea life, including a reduction in plankton, an ecological paucity that would have spread throughout other marine communities.

This cooling was then abruptly interrupted by massive flood basalts, which flowed out of a hot spot between Africa and India during the last few million years of the Cretaceous. These basalts ("traps") now compose the Deccan Plateau of western India. The Deccan Traps, more than 2,000 meters (6,500 feet) thick and spread over more than 500,000 square kilometers (almost 200,000 square miles), comprise one of the most impressive volcanic deposits in the world. The volcanism responsible for the Deccan Traps was very likely similar to that of the Siberian Traps, also releasing huge volumes of carbon dioxide, and initiating rapid global warming in a rapid about-face from cooling. This volcanism also emitted sulfur dioxide, which as we all know from the end-Permian extinction leads to acid rain. The increased acidity was also bad for life, especially land plants and the animals that eat land plants, but also anything living in freshwater environments. Owing to such environmental instability and inhospitable conditions, extinctions rates were probably already on the upswing in the last million years of the Cretaceous.

Then the meteorite hit. The impact from this estimated 10-kilometer (6-mile) wide bolide instantly converted its potential energy into kinetic energy, releasing heat equivalent to about 5 billion atomic bombs. A blast of such enormity and ferocity would have vaporized all life in and near the impact site, which was a shallow marine environment just off the coast of present-day Mexico. Because the impact was in the ocean, it caused a gigantic tsunami, which would have inundated all low-lying coastal areas and buried these under a thick layer of boulders and other sediment. The shockwave of air displaced by the meteorite, combined with heat released by its impact, would have flattened forests throughout nearby North America while also incinerating them.

The impact crater also intersected sulfur-bearing rocks, which added more sulfur dioxide to the atmosphere. Remember what this does? That's right, acid rain, which adversely affects soils and freshwater ecosystems. Also, the aftermath of the fiery impact was rapid global cooling, as dust from both the meteorite and the Earth—along with soot from the forest fires—was tossed into the stratosphere. These conditions were analogous to a "nuclear winter," described in the 1980s by scientists opposing nuclear war. In this situation, suspended particles in the air effectively blocked sunlight for several years, dropping temperatures below freezing in areas that normally enjoyed year-round or seasonal warmth. All in all, the Earth had a really bad day.

This impact was a calamity that no plant or animal lineage could have anticipated or prepared for in its evolutionary history. Accordingly, an estimated 80% of all species died, and quickly. Non-avian dinosaurs vanished, but so did nearly every large-bodied land or marine animal. Little sunlight meant less plankton in the oceans and fewer land plants, which meant less photosynthesis and oxygen. All ecosystems on the Earth's surface would have been undermined, an ecological breakdown that would have hit the big herbivores first, then big carnivores that ate big herbivores.

Scavengers would have had a great time, but only until the bonanza of fresh meat became putrid or scarcer. Of course, such interactions between the deceased and about-to-be-deceased assume that life had any semblance of normality in between the acid-laden rain, sleet, and snow that fell during the longest winter ever experienced by the few surface-dwelling animals that had not been erased by the impact, drowned or buried by its tsunami, or scorched in a forest fire. Whatever species were alive and moving in this post-apocalyptic Earth would have been the walking dead, not knowing they were the last of their kind.

Burrowers: Winners of the Extinction Lottery?

By now, gentle and not-so-gentle readers alike might be wondering what all of these gruesome mass extinctions of the geologic past have to do with burrowing. This is a fair point, which I will now address as I emerge from my underground bunker, nicknamed "The Extinction Preventer."

When paleobiologists began to seriously study mass-extinction events in the 1970s and 1980s, one of them—David Raup from the University of Chicago—coined the rhetorical question, "Extinction: Bad genes or bad luck?" Raup liked this question so much that he even used it as the title of a short book he wrote about mass extinctions of the geologic past. But just to elaborate further on his pithy inquiry, I will attempt to address a two-part problem. One is whether or not the extinction of many species within relatively short amounts of geologic time, such as 50,000 years, 10,000 years, or a day, can be attributed to those species having the "wrong" genes. The other problem is determining if these species had perfectly fine genes developed by millions of years of evolution, but ones that could not have prepared them for worldwide calamities, such as extreme climate change or meteorite impacts.

Unfortunately, questions like this with binary answers tend to discourage creative thinking. For instance, let's consider certain animals that bear the "right" genes for coping with so-called "normal" everyday pressures. Then think how about these genes also program for very lucky adaptations when a once-in-a-hundred-million-years disaster falls on you (sometimes literally). People who study mass extinctions and their causes agree that discerning how only a few plants or animals survive extinctions is far more interesting than figuring out how others did not. This ability to make it past an extinction event is often called *extinction selectivity*, in which natural selection becomes more abrupt with radical changes in ecosystems.

This brings up a very important point about mass extinctions, and extinction in general: Extinction is also part of evolution. New species often evolve when other species are wiped out before them, adapting to changing environmental conditions, whether these are local or global. Mass extinctions in particular have a dramatic effect on evolution, opening ecological niches for those species left behind. The first few million years of the Triassic, Jurassic, and Paleogene were times for surviving organisms to expand into new environments. Furthermore, these species would have been relatively unhindered by competition from species that might have sought the same habitats and resources.

These times of opportunity were also when genetically linked burrowing abilities in certain animals, such as *Lystrosaurus* and other vertebrates at the end of the Permian, mammals at the end of the Triassic, and mammals again at the end of the Cretaceous, helped them survive three of the most egregious extinction events in the history of our planet. Through this span of nearly 200 million years, other burrowing-animal lineages on land and in lakes, rivers, and seas survived the worst that the Earth or cosmos could throw at them. There are essential lessons here, in which the past informs us about the present and the near future as we contemplate how life will deal with the present-day mass extinction.

Animals were not always on land, nor in lakes and rivers, though. In fact, animal life probably began in the oceans more than 600 *mya.* When did these animals depart the sea? How were they able to inhabit the places between the oceans and continents before adapting to the potentially harsh conditions of continental edges and interiors? To answer this question, we must go back deeper in time to more than 400 million years ago, when animals began to dig out homes in these new environments as they sought out an "undiscovered country" that promised continued life.

CHAPTER 6

Terraforming a Planet, One Hole at a Time

Go Landward, Young Trilobite

How did a trilobite get into this river? This question came to mind after I picked up a chunk of sandstone, turned it over, and spotted a perfectly defined trilobite burrow on its bottom. The burrow was originally made as a trilobite dug into a firm muddy bottom with its many legs. This created a hollow that was later filled by sand from above, rendering a natural cast that now projected out of the rock. If a trilobite body had been there when sand flowed into its burrow, its essence had long since departed, dissolved by time. The trace fossil that remained was defined by two long, symmetrical and parallel ridges ending in a bump that looked like a little butt. Ridges leading to this compelling feature denoted where the trilobite's legs dug into and pushed sediment behind it; the rounded bump marked where its body stopped, and the leading edge outlined the position of its head.

Although I was happy to discover this trace fossil, it was a puzzling find. Also, because I was in cell phone–less 1988 and by myself alongside a steep, winding road in north Georgia, there was no easy way to pose my questions about it to anyone else just yet: I had to figure this out by myself. So I did what any good geoscientist should do, which was to repeat the experiment by picking up another nearby piece of sandstone and looking for similar burrows in it. Sure enough, more trilobite-formed "butts" of varying sizes and shapes were on it and other cobbles of sandstone that fallen off the slope next to the road. I had a mystery on my hands.

In my experience, when people are asked to name their favorite fossils, most will say "dinosaurs," but a not so surprising second choice is "trilobites." Trilobites, which were marine arthropods, were among the most successful animals in the history of life. For the latter reason alone, they deserve our attention and adoration, but these animals were also marvelously varied in form and function during their 250 million years of living on, above, and below seafloors. Some were tiny, measuring less than 2 millimeters (0.08 inch) long, whereas others were longer than most domestic cats, exceeding 70 centimeters (2.3 feet) in length. The name "trilobite" comes from the lengthwise divisions of their middles into three lobes, with a central axial lobe flanked by two pleural lobes that look like ribs. At a trilobite's front end was its cephalon (head), and its rear was a pygidium (tail). Some trilobite bodies were perfectly smooth, whereas spikes or other protuberances adorned others, with all sorts of evolutionarily fashioned statements in between.

Nonetheless, a trilobite head was probably its most beguiling part, as it often held a pair of eyes. Trilobites were among the first animals to gaze out into the world's oceans and see what lived there with them. Some trilobite eyes were incredibly complex, composed of a myriad of calcite lenses that were sometimes separate, sometimes connected. However, a few trilobites were born without eyes and hence blind, a condition implying that natural selection had rendered vision superfluous at some point in their evolutionary history. The closest living relatives of trilobites today are modern horseshoe crabs, with each arthropod lineage last sharing a common ancestor more than 445 *mya*. Alas, we have never been close enough in time to see a living trilobite. In our evolutionary lineage, that privilege would have been restricted to our most distant synapsid ancestors during the Permian Period. For instance, a living trilobite washed up on a Late Permian seashore may have exchanged glances with a beach-strolling *Lystrosaurus*. Nonetheless, trilobites were already rare by then and, like most other animals faced with the end of the Permian, the last of their kind vanished with its passage.

So why did I think these particular trilobites—revealed by their trace fossils in north Georgia—were living in a river, and not in their normal oceanic habitats? Because the rocks that hosted them, exposed as strata in the steep slope next to the road, kept screaming "River!" at me. Despite my being a mere geology graduate student in the late 1980s, all of my descriptions of these rocks and comparisons of their geologic traits to known river-made strata kept giving me that same answer. For example, as I strolled along the road and studied the outcrop, I noted cross-sections of thick, white sandstones that were flat on top, thick in the middle, and tapered on either side. These sandstones represented cross sections of river channels, filled with quartz sand as the channels moved laterally from one place to another. The sandstones also contained what geologists call cross-bedding, an internal layering of sand angled ("crossed") from bed top to base. These structures told me that sand was transported downstream on the channel floor, cascading over the tops of ripples formed by the force of flowing water. Thinner, flatter sandstone beds between channel sandstones marked places where river water had breached levees and spilled onto floodplains, taking sand with it. Moreover, finer-grained rocks, such as shales and other mudstones, enclosed these sandstones and were broadly distributed throughout the outcrop. These mudstones were reddish brown, their iron-bearing clays having been oxidized to rusty hues. I was convinced these mudstones were former floodplains, made by repeated flooding of river channels that carried mud, which then settled from waning flows outside of river channels.

But then there were the burrows. Not only were trilobite burrows on the bottoms of sandstone beds, but also vertical, tubular burrows, piercing the tops of other sandstones. Some of these pencil-thin burrows formed straight lines, whereas others were U-shaped, with rounded bottoms joining two tubes. A few of the red mudstones were completely churned by more complex burrows, some filled by white sand from above, composing red-white-pink mottling in these rocks. Despite my failure to find a single bit of fossil plant or animal parts anywhere along this 200-meter (660-foot) long road cut, its abundant trace fossils told me unequivocally the original sedimentary environments that formed these rocks were full of life.

Normally the presence of these burrows would not be a big deal for a geologist looking at strata formed by a freshwater river. After all, burrows made by a variety of insects, worms, crayfish, fish, amphibians, reptiles, and mammals are exceedingly common in the sands and muds of modern rivers. The problem in this instance, though, was that these deposits were far too old for those animals to have dug into them. You see, these rocks originated during the latest part of the Ordovician Period, dating to about 445 *mya*. In my young, naïve, 1980s-era, geology grad student brain, I recalled lessons in which both my textbooks and professors said that animals did not inhabit freshwater and landward environments then.

In that then-reigning scenario, animal settlement of continental environments began about the middle of the Silurian, perhaps 430 *mya*, but was preceded by the first land plants. This would have happened when both plants and animals evolved the means to withstand the relatively harsh conditions of fresh water, which had

THE EVOLUTION UNDERGROUND

lower density and less food, as well as land with its desiccating conditions and lack of buoyancy provided by water. These pioneering organisms, consisting of the first primitive vascular plants and arthropods, then began forming basic freshwater and terrestrial ecosystems, with only a few participants in their food webs. During the Ordovician, worms of any type were not known to have lived in freshwater rivers. Moreover, insects, crayfish, amphibians, reptiles, and mammals did not even exist then, either, let alone burrowing ones. On the other hand, Ordovician seas teemed with animal life: a multitude of worms—including the ancestors of modern earthworms—trilobites, large scorpion-like eurypterids, crinoids ("sea lillies"), brachiopods ("lamp shells"), clams, snails, jawless fish, and much more. It seemed almost unfair that this zoological abundance was restricted to the oceans, with so much new real estate available along its edges and on land.

Nevertheless, the presence of so many burrows sent an important message: Either I and other people were wrong in assuming that Ordovician freshwater rivers were devoid of animal life, or I was wrong in assuming these rocks were formed by freshwater rivers. In either case, I was wrong. For scientists, this sort of noticing is extremely important, especially when self-noticed. After all, it is much less embarrassing than when someone points out your ineptitude in front of hundreds of your peers or when other scientists dissect your once-keen arguments in journal articles. Accordingly, I then set off to figure out why my "ancient river" interpretation was mistaken and took a closer look at these Ordovician rocks.

Little did I know that I had already been subtly primed to solve this riddle, which (spoiler alert) I did. In 1985, I arrived at the University of Georgia and started studying these Ordovician rocks for my PhD dissertation there, work that consisted of a mix of time in the field, lab, and library. In between all of this labor, my PhD adviser, renowned ichnologist Robert (Bob) Frey, regularly placed reprints of his articles in my grad-student mailbox. Weirdly, all of these papers were about his seminal and voluminous work on the ichnology of the modern Georgia coast, which consisted of barrier islands, marshes, and estuaries. At the time, this made no sense to me, seeing that the Ordovician rocks I was studying were more than 440 million years old. In other words, these strata were so far removed in time from the modern Georgia coast that I could not see how they were related. Still, I dutifully read Frey's articles and began absorbing their lessons about the traces and tracemakers of the Georgia coast.

My faithful obedience was fortunate, as this new knowledge later prompted my awareness of what likely happened in north Georgia during the Ordovician Period. This realization came with a second visit to the same outcrop with its trilobite burrow–bearing "river" rocks. The outcrop with these rocks and their trace fossils was a roadcut, where a highway crew had blasted through the rocks with dynamite and otherwise cleared the way for a paved road, decades before my arrival. The road wound up from a valley and over a ridge in a geologic province of the Appalachian Mountains that was called, appropriately enough, the Valley and Ridge Province.

The Valley and Ridge Province in Georgia and elsewhere in the Appalachians is composed of folded sedimentary rocks from the Cambrian through Devonian Periods, spanning about 500–350 *mya*. Tilted Ordovician-Silurian sandstones hold up the ridges, whereas valleys are mostly composed of eroded mudstones and limestones. The road also followed a natural divide on this particular ridge (Rocky Face Mountain), named Dug Gap. In 1864, during the American Civil War, Dug Gap was a strategic pass for Confederate and Union armies, and blood was spilled over these Ordovician rocks. They were even used as weapons of war, as greatly outnumbered Confederate soldiers pushed boulders of Ordovician sandstone down the steep slopes to kill or otherwise slow down the advance of Union troops. (I have often wondered whether any Union soldiers were impacted directly by burrow-bearing boulders, an invertebrate assault by proxy separated by more than 400 million years.) Regardless, I felt little of the human turmoil and history there. Upon arriving at the outcrop, I instead focused on solving the deep-time riddles posed by the burrows of trilobites and other animals that had obviously lived and thrived in these environments.

I am unsure just when a formerly dim lightbulb of geological awareness reached its full luminosity, but at some point in my investigations that day, the trace fossil evidence and physical structures in the strata colluded with Bob Frey's articles about the Georgia coast. Reviews of my field notebooks from then do not provide any clear "Eureka!" moment that some scientists are so fond of telling, if for no other reason but to compare themselves to Archimedes. Rather, this message was more like a conspiratorial whisper replacing the previously shouted word "River!" with another that held both more syllables and layers of complexity: "Estuary."

Estuaries are places where freshwater rivers and the sea meet at the same place and time. These environments differ from deltas by lacking large wedges or fingers of sand and mud dumped into the sea, such as where the Mississippi River comes to its end. Estuaries are more like open bays, or have marshes or lagoons that are sometimes behind a string of barrier islands. Yet estuaries are still connected to rivers, where they receive much fresh water and sediments from landward environments. They also have different parts, or reaches, that change ecologically as one travels from one place to another along their length. For instance, the lower reaches of an estuary are typically in a coastal bay or other seaward area. Here these environments are more dominated by marine processes and organisms, as the salt content (*salinity*) of the water is closer to that of fully marine conditions. In contrast, the upper reaches of an estuary are in the river itself. This overlap happens because saltier waters are also denser, causing these to tightly hug a river bottom while fresh water floats above it. Like an apartment building with tenants on different floors who never see each other, this vertical separation by salinity results in marine animals dwelling in a "river" that may not interact with freshwater animals above them. However, any mixing of water from below is enough to make freshwater animals say "yuck" to this unwanted influx of salt and move upriver, leaving mostly marine animals adapted to lower salinities.

Because of this overlying fresh water, salt-adapted animals may stay near the estuary bottom, where they also might burrow. The in-and-out flux of daily tides also encourages burrowing, as low tides would mean a greater likelihood of fresh water invading the bottom sediments, and burrows temporarily protect against this. Droughts and/or sea level rises also accentuate lateral shifts in the bottom fauna of an estuary, as less fresh water coming from a river causes saltwater wedges to move upstream and further into river systems. These wedges thus bring their ecological posses with them, replacing former freshwater-bottom dwellers with marine fauna. Amazingly, we can observe this process in real time today. For instance, on the Georgia coast, droughts have allowed barnacles and other marine animals to live more than 30 kilometers (19 miles) inland in what, by every other outward appearance to us surface dwellers, is still a river.

Now jump back into the geologic past and think about what an estuary might look like then, and what might have been preserved from it. Because the flow of fresh water in a river would have dominated its processes, many of the sedimentary rocks formed in the upper reaches of an estuary would look very much like those made by rivers. But look closer, and subtle differences emerge, putting salty accents on a bland label. For one, the burrows preserved in those rocks will include at least a few bearing the signatures of marine animals, while also lacking those formed by freshwater animals.

Even in the absence of burrows, sedimentary structures in the rocks left by the movement of water can help to identify marine influence. Think how tides, surging in and out twice a day every day, carry marine mud and sand far upstream on a high tide and then either dump these particles or erode previously deposited sediments on a low tide. Tides thus leave their marks in estuarine rocks. One of these clues consists of thin mudstones on top of rippled sandstones, which formed as mud drapes on sandy ripples left by a falling tide. Another is cross-bedding angled one way, then the opposite way, a pattern called "herringbone" cross-bedding because of how it resembles the rib bones of a fish. This structure would have been made by a high tide moving up the river, shifting the sand landward; this then would have been followed by a reversal as the tide fell, moving the sand more seaward.

When I went back to the outcrop, I reexamined the rocks with such clues in mind. Burrows of marine-only animals, such as trilobites and marine worms? Yup. Any burrows that look like those of freshwater-only animals? Nope. Mud drapes on sandstones? Yup. Herringbone cross-bedding? Yup. These rocks were starting to look less like those of a river, and more like those of an estuary. Sure, it was a river in the sense that the strata were mostly the result of fresh water transporting its sediments from upstream to downstream. Yet it was also a river where salty water from a nearby marine environment moved far up into its channels, bringing in animals that ordinarily would not be there: like trilobites.

I remember well this gathering of evidence that came so incrementally, and how my situational blindness—and perhaps confirmation bias—was overcome by the reality of the trace fossils and other features of these rocks. When I tested this then-radical idea by visiting other outcrops of Ordovician rocks north of this site, they also had plenty of evidence for marine influence. In fact, these strata showed a trend of going farther down river and into the middle and lower reaches of the estuary. This gradual change was defined by alterations in the trace fossils, which were gradually more dominated by those of marine animals, as well as sandstones deposited by offshore tropical storms rather than rivers. Although I did not appreciate it much at the time, these trace fossils and their enclosing rocks were helping me to document the oldest known estuary recorded in the geologic record.

Once I put together a coherent explanation of this estuary, I then went to Bob Frey's office and sat down to have a little chat with him. After listing diagnostic criteria collected in the field at five widely separated outcrops in north Georgia and south Tennessee, I stated, "I think this outcrop has rocks made in the upper reaches of an estuary, and the other outcrops show a transition from up the river to more offshore, but still in an estuary." He listened, nodded his head at much of what I said, and then said (with a smile), "Well, that sounds reasonable." Without having told me directly during the previous years, he suspected these strata were formed in an Ordovician estuary, but he had wanted me to find it out for myself. This was very good mentoring, as I learned how to learn and discover without the imagination-crushing effects of over-direction.

The next year, when all of my results were written into a doctoral dissertation and I had to give an oral defense of it to my graduate committee, there was another moment of Frey's mentorship that stood out, and one that convinced me of the power of ichnology. One of my committee members was being particularly difficult, peppering me with questions related to the sedimentary rocks and their physical structures—such as cross-bedding and ripples—that all expressed skepticism of how I could interpret an estuary from these clues. This committee member was even more doubtful of my interpretation because of the age of the rocks. After all, if I was interpreting the oldest estuary in the geologic record, the conclusion needed solid support. Several other committee members had already grilled me over my statistical methods, probable alternative interpretations, the geologic history of the Ordovician Period, and specific meals eaten by Charles Darwin while on the voyage of the *Beagle*. (That day is now mostly a blur, so the last discussion point may or may not have actually happened.)

Out of politeness, Frey kept quiet throughout this interrogation and waited to go last with his inquiries. Nonetheless, he started his questioning in the best possible way by declaring, "Let's talk about something that really matters, like trace fossils!" Everyone laughed, breaking some of the tension built up over the previous hour or so. He then methodically, logically, and exquisitely led me through a series of rhetorical questions in which my responses neatly laid out the trace-fossil evidence for an Ordovician estuary only a three-hour drive from our room. I explained how the burrows of trilobites and other marine animals in what looked like river deposits meant they were not rivers per se, but must have had marine-flavored waters invading them. I also clarified how the strata showed only a few marine animals, such as trilobites and worms, had burrowed in the "rivers" that I now regarded as an estuary. However, I also noted how the abundance and variety of those burrowing animals increased the farther they were out toward the sea. This is exactly the burrowing trend we see in the modern Georgia coast, so I pulled in that fact to show it was perfectly normal today, and may have been during the Ordovician. Finally, I mentioned how different trilobite burrows were in the upper and lower reaches of the estuary, indicating that different species of trilobites had probably adapted to those environments.

At some point, Frey reached an appropriate end to his questions, and for a few seconds, while all of this information sunk in, everyone was quiet. With a chuckle, he turned to the other committee members, focusing on the one who had catechized me so thoroughly, and said, "So as you can see, the trace fossils alone tell us what environments were there." What he had done in directing my answers into a coherent argument was masterful and devastatingly effective in quashing doubt. With my doctoral defense done, I was dismissed while my committee discussed my fate. About 15 minutes later, Frey emerged from the room of inquisition and greeted me in the hallway of exile. "Congratulations, Dr. Martin," he said with a smile as he reached out to shake my hand. Elated, I thanked him for his help as my adviser. But later, I also made sure to thank the trilobites and other burrowing animals in that estuary from more than 400 million years before for providing me with a most excellent riddle to solve.

Arthropods: They Had Legs, and They Knew How to Use Them

Trilobites were among the first animals to start inhabiting the estuaries, tidal flats, and areas almost on land (but not quite), and many were kept safe by burrows. Based on trace fossils we also know that many other types of arthropods must have plowed onto and into sediments exposed to the air, living along the edges of where the sea met the land. For example, some arthropod trackways are preserved in likely beach deposits from the Early Cambrian Period (more than 500 *mya*), and diverse burrows made by trilobites and other invertebrates are similarly in rocks formed on tidal flats from about 520 *mya*. These traces tell us that animals were seeking out new life and new situations, and boldly going where no invertebrates had gone before then. But the majority of these traces also seem to have been made specifically by arthropods, as opposed to worms or other soft-bodied animals.

Why were arthropods more likely to have burrowed in beaches, tidal flats, estuaries, and other environments on the periphery of the sea? Because arthropods were specially equipped for landward migrations in their evolution, traveling both outward from the ocean and inward to the sediments. One advantage they had was their jointed legs and other body parts, which gave some the right equipment for burrowing. Arthropod legs were not only used for moving about quickly on the seafloor, whether to avoid predators, be predators, or pursue mates, but also could have been used for digging. This excavating ability, in which their legs worked in tandem with their bodies to loosen and push sediments behind them as they moved into it, showed up early in trilobites. For instance, I have seen trilobite burrows from the latest Cambrian and earliest Ordovician Periods (490–480 *mya*) in Spain and Portugal that are long and looping, with beautifully expressed scratches from their legs, looking like thick coils of rope discarded from a rodeo. Sometimes natural casts of these burrows preserve impressions of trilobite legs, enabling paleontologists to count these and, like using fingerprints, narrow down the identity of their makers. Additional help with burrowing came from them using their heads, and literally: Trilobite heads, if shaped like a spade, could have been used for slicing into and otherwise plowing through sediments.

Another adaptation of arthropods for subaerial excursions was their bodies, which were protected by tough yet flexible exoskeletons, made of chitin or similar organic compounds. These bodies differed considerably from those of their wormy or shelly invertebrate cohorts, helping to reserve water and hence delay desiccation in their owners whenever they were out of the water, whether accidentally or on purpose. Such exoskeletons may have originated first as protection against predators and/or support, but this body armor may have later served as an adaptation for going into less saline water or onto land.

Physiologically speaking, going from marine to freshwater environments was challenging enough for arthropods. But how about evolving from a watery environment to land? How did this transition happen? A few replies to such questions come from an unlikely source of information: modern horseshoe crabs. Anyone who has wandered a beach on the eastern coast of the U.S. has probably encountered at least a body part of a horseshoe crab, and occasionally a recently dead or still-alive one. A live horseshoe crab is a sight to behold, pulling one back way before the Mesozoic heyday of the dinosaurs and closer to the start of the Paleozoic Era.

Despite their common names, horseshoe crabs are not true crabs, being only distantly related to crustaceans, but are actually the closest living relatives of trilobites. Classified as chelicerates—the same group containing spiders, scorpions, and other arachnids—these primeval-looking arthropods have a broad head shield bearing a prominent pair of compound eyes, a short rear section, a spiky tail, and ten walking legs. Four species of horseshoe crabs ("limulids") are known, with three in Southeast Asia. However, the most famous lives in the eastern U.S., *Limulus polyphemus*. The largest horseshoe crabs are on the Georgia coast, where they get huge, reaching lengths of 70 centimeters (2.3 feet). This gigantism is probably a function of the wide, shallow marine shelf just offshore there, which allows them plenty of food to eat and places to hide, thus granting the means for longer lives.

Why are horseshoe crabs so significant? They do something no other modern marine invertebrates can do, which is to come up on land to breed. This mass movement of female and male horseshoe crabs happens seasonally, starting in May in the southern U.S., and then in June and July in the northern states. Female limulids leave the relative safety of the sea to lay their eggs on wet sandy beaches, while these lady limulids are hotly pursued by males, who want to be in the right place and time to deposit their sperm directly on top of the female's eggs. This situation sometimes gets a little crazy, as males push one another aside to get immediately behind the female. Unfortunately for the female, who has left the buoyancy provided by the ocean, she was born with a trailer hitch on her rear end. This little notch in the tail section of her body was naturally selected as a trait, eventually allowing male horseshoe crabs to fit the front edges of their head shields onto it. So once a male puts his head on her notch, she pulls him along, too. As soon as she finds the right place to drop her mass of eggs on the beach, the male follows suit by unleashing his sperm. This sex on the beach soon results in thousands of tiny limulids that are not much larger than a period on this page.

Aptly enough, biologists refer to horseshoe crab babies as "trilobites," as a seeming homage to their Paleozoic relatives. These juvenile limulids then grow up in the intertidal sandflats and shallow-water sediments just offshore, burrowing along and in the sand to find food, which mostly consists of tiny invertebrates. While walking along the rippled sandflat of a Georgia barrier island in 2001, I was astonished to find thousands of these combined burrows and trails there, all made by tiny horseshoe crabs. I was further struck by how closely these traces resembled Paleozoic trace fossils that were often attributed to trilobites or marine worms.

All of the preceding is admittedly already pretty sexy, but then (evolutionarily speaking) it gets downright hot. As we now know, horseshoe crabs are the *only* marine invertebrates today that return to the land to lay and fertilize their eggs. This makes them a model organism for what no doubt happened in the Paleozoic Era, when invertebrates began moving out of the water and onto the land. Developing their model status even further, horseshoe crabs can live in estuaries. In fact, the massive ones on the Georgia coast are living in water with some salt in it, but not quite at full marine salinity because of so much freshwater input from nearby rivers. Interestingly, the oldest known body fossils of horseshoe crabs come from Ordovician rocks (about 450 *mya*) that formed in marine environments, but their trackways are in estuarine and even freshwater deposits from the later part of the Paleozoic Era and into the Mesozoic.

How does burrowing benefit modern horseshoe crabs? First of all, burrowing allows them to survive as they grow up. As is true for nearly every animal, a limulid's most vulnerable stage of life is at its start, which is when everything wants to eat it. For instance, migrating shorebirds, such as red knots (*Calidris canutus*), gobble up eggs and newly hatched "trilobites" alike. Although the vast majority of baby limulids are helpless in the face of this mass avian-induced slaughter, at least a few escape fate in a red-knot belly by staying hidden in the sand.

Burrowing also helps prevent young horseshoe crabs from drying out while stuck on land. If you are lucky enough to see a live horseshoe crab crawling along a beach during their mating season, you might notice that after it's done its reproductive duty, it heads back to the sea. However, many get lost along the way and start looping in search of water. If they find a pool of water in a low part of a beach, such as a runnel, they dig into the partly liquefied sand below. Smaller juvenile limulids, however, have a much easier time burying themselves, especially when in saturated sand. On Georgia beaches, I have found little dimples with trackways connected to them; once I scoop my fingers underneath the dimple and wash out the sand, a small horseshoe crab is sitting on my palm. These limulids cause the depressions by rapidly moving their legs, which allows more water in between the sand grains and makes a sort of "quicksand" that collapses around them and aids in their burial.

Limulids are also great survivors, making it past all mass extinctions of the past 500 million years: the end-Ordovician (443 *mya*), the end-Devonian (358 *mya*), the end-Permian (252 *mya*), the end-Triassic (201 *mya*), and the end-Cretaceous (66 *mya*). Assuming that their reproductive mode has stayed basically the same—come up on land, lay and fertilize eggs, and give rise to many madly burrowing juveniles—then this has obviously worked quite well for their lineage.

In the same state of Georgia, I can see these burrows made by trilobite-sized limulids on its coast, then travel to its northwestern corner and see comparable burrows from honest-to-gosh trilobites, but made on Ordovician tidal flats. The trace fossil evidence alone thus shows us that this near-coastal behavior has been happening in invertebrates for more than 400 million years. But were any animals far out of the water and onto land during the Ordovician, burrowing in primeval soils? According to some fossil burrows, yes. These small burrows are in fossil soils (called *paleosols*), preserved in Late Ordovician rocks of Pennsylvania. These trace fossils have been credited to millipede-like animals that used their legs to push and pack sediment behind them in these soils. Given such trace fossils from both tidal areas and soils, invertebrates were clearly pushing the edges of their marine-only adaptations and making their way into estuarine and then freshwater ecosystems, and onto land by at least the end of the Ordovician Period. Hence my graduate-school-era concept of "animals didn't get to land until the last of the Silurian Period" has been shattered, thanks to trace fossils.

It was only a matter of time, then, before some of the freshwater and land invertebrates became as big as their marine counterparts. For instance, by the Silurian, large animals called myriapods (related to modern millipedes and centipedes) occupied freshwater stream and land environments, with some of these animals burrowing. In Silurian rocks of Australia, I have personally seen 30-centimeter (12-inch) wide burrows that were likely made by large myriapods in river deposits. Similar fossil burrows are in rocks of the same age in Scotland, Antarctica, and other parts of the world. Some myria-pods, such as *Arthropleura*, later became the largest land invertebrates that ever lived. *Arthropleura* lived during the Carboniferous Period (about 340–300 *mya*); adults were longer than most people are tall (more than 2 meters or 6.6 feet) and left trackways about 40–50 centimeters (16–20 inches) wide, exceeding tracks left by a kid-powered tricycle. These animals, which surely would have caused more than a little panic if found scuttling across a kitchen floor in the middle of the night, lived at about the same time as the first synapsids and diapsids (around 300 *mya*), but were far larger than those animals. If *Arthropleura* and its close relatives burrowed like modern millipedes, paleontologists may not even recognize their former homes because of their size, which would far exceed the diameter of modern gopher tortoise burrows.

In summary, burrowing was likely one of several strategies that allowed invertebrate animals to make it into estuaries, freshwater rivers, and eventually land during the early part of the Paleozoic Era. Once established in land environments, burrowing inverte-brates were there to stay, and along with plant roots, they soon transformed sediments into soils. This early colonization of the land meant that all of the advantages enjoyed by burrowing ver-tebrates later in the Paleozoic had already been tried and proven true by their invertebrate cohorts long before. But what about verte-brates: When did these backboned critters start invading estuarine and freshwater ecosystems, and how did they survive on land? To answer that question, we look back to the Devonian Period and the first burrowing animals that fulfilled the aphorism of being "fish out of water": lungfishes.

Lungfish: Every Breath You Take, Every Burrow You Make

The first vertebrates to occupy freshwater environments were var-ious types of fishes, which swam up into estuaries and eventually evolved the means to handle less and less salt in the water. Although the fossil evidence for this ecological shift is scanty, it probably began as early as the Ordovician Period, with jawless fishes making their way up rivers during droughts or higher sea levels. "Fish," however, is a vague term, especially once the evolutionary principles of cladistics are applied. For instance, the writer of this sentence and everyone reading it is related to a fish (no offense). This apparently misplaced labeling is entirely appropriate because all four-limbed vertebrates (called *tetrapods*)—amphibians, lizards, pterosaurs, dinosaurs, birds, crocodilians, turtles, mammals, and more—arose from lobe-finned fishes called *sarcopterygians*.

Lobe-finned fish owe their description to four fleshy fins, consisting of a pair each of pectoral ("shoulder") and pelvic ("hip") fins. These fins are supported by a series of bones that correspond directly with limb bones in tetrapods. Natural selection resulted in lobed fins working perfectly fine for moving their owners about in their watery environments, but these parts were later modified in a few lineages to support their bodies on land. The best representative found thus far of a fossil marking this transition from a swimming-to-walking tetrapod is *Tiktaalik*, discovered in 2004 in 375 million-year-old rocks (of the Late Devonian Period) on Ellesmere Island in the Canadian Arctic. The anatomy of this "fishapod" shows that it could hold up its own body weight, nod its head up or down, and otherwise move along water-body bottoms, putting it in close contact with land and air. The body fossil, anatomical, and genetic evidence for the evolutionary transition of lobe-finned fish to tetrapods has been the topic of much research, popular books, documentary TV series, artwork, and songs, with a stage performance (*Tiktaalik!: The Musical*) and screen adaptation (*The Good Tetrapod*) no doubt in the works. Hence I will not review this much-studied and well-publicized subject (and its deserved praise), but instead will talk about something that really matters. Like trace fossils.

The bodily remains of *Tiktaalik*, which consisted of multiple well-preserved specimens, were recovered from freshwater-river deposits. This evidence tells us that *Tiktaalik*'s ancestors had left the oceans behind and adapted to fresh water, perhaps long before some individuals left that not so salty water to pull themselves up onto dry land. However, *Tiktaalik* may not mark the first time and place that the descendants of lobe-finned fishes went onto land. Moreover, these fishes may not have necessarily used freshwater environments as their launching points. For example, in 2009, paleontologists in Poland found a series of depressions in strata that they interpreted as trackways made by primitive tetrapods. This find was remarkable in many ways, but mostly because the tracks preceded *Tiktaalik* by nearly 20 million years, coming from the Middle Devonian (about 395 *mya*). Some of the tracks preserve details of fin anatomy, including more than the usual five or fewer digits we expect in any vertebrate bearing a hand or foot today. Even more thought-provoking was their original environment: The rocks bearing these trace fossils were limestones formed in an intertidal environment, adjacent to an ocean. This means that their makers were coming onto land for a stroll from marine waters, not fresh, considerably changing this fish-out-of-water story.

Other paleontologists have disputed these tracks since then, claiming that the tracks actually are fish nests, hollowed out by fishes that laid and fertilized their eggs in each depression. However, these critics failed to explain why some of these so-called "nests" bear features matching those of lobed fins, such as digits. Paleontologists have found similar track-like patterns in Australia from Siurian-Devonian rocks, but alas, these trace fossils have also been subjected to much criticism and skepticism.

Enter the lungfish. Lungfish are classified as sarcopterygians within the clade Dipnoi, and originated from other bony fish toward the start of the Devonian Period at about 400 *mya*. Lungfishes share a common ancestor with *Tiktaalik* and early tetrapods, meaning those animals did not arise directly from lungfishes, but are closely related. Although lungfishes began their evolutionary history in marine environments, at least a few had shifted into freshwater ecosystems such as rivers by the end of the Devonian. Because modern lungfishes can breathe air and stay out on land for extensive forays, paleontologists assume their ancestors had the same abilities. Their air breathing comes from a swim bladder functioning as a sort of "lung" (hence the name) by taking in air. Indeed, lungfishes rely much more on air than water for their oxygen needs, and only one of the six extant species can use its gills. Lungfishes are also tolerant of low-oxygen conditions in their environments, giving them yet another advantage over other fishes when their watery surroundings are radically altered.

Given such a wide range of habitats and adaptations, lungfishes accordingly became widespread over geologic time, with fossils in both Laurasia and Gondwana. Some also became massive. For example, a single fist-sized Cretaceous lungfishes tooth from the western U.S. suggests its owner may have been as much as 4 meters (13 feet) long, whereas the largest modern lungfish are a mere 2 meters (6.6 feet) long. Lungfishes today are all Gondwanan animals, with four species in Africa (all in the genus *Protopterus*) and one species each in South America (*Lepidosiren paradoxa*) and Australia (*Neoceratodus forsteri*). On the other hand (or fin, whatever), every lungfish in Laurasia went extinct long before humans evolved.

Similar to horseshoe crabs, lungfishes found a way to surpass the worst disasters inflicted on life, surviving the end-Devonian, Permian, Triassic, and Cretaceous mass extinctions. Somehow these fishes adapted to drastic changes, using a relatively simple but effective behavior that enabled them to avoid all of the ecological nastiness visited upon other many other animals, particularly freshwater ones. Gee, I wonder what adaptation could that have been?

Okay, you guessed it: burrowing. Five of the six modern species of lungfishes not only make burrows, but also use these burrows to perform Lazarus-like tricks that allow them to live for years underground while terrible conditions—such as extended droughts—reign on the surface. In my experience, knowing that a fish can burrow is astounding enough by itself, as most people understandably associate burrowing in animals with legs, not fins. This surprise is multiplied if one looks at a photo of a typical lungfish and sees its four small, thin fins, which are indeed ill-equipped for earth moving. Nonetheless, plenty of fish species, in both freshwater and marine environments, make and inhabit their own burrows, once again showing how going down below can save lives and future generations.

Unlike many other animals, when lungfishes burrow, they are not creating a home for popping in and out of at their leisure. Instead, these burrows are made under duress, when the proverbial feces strikes the rotating air-circulation device. Like nearly all fishes, they live in rivers or lakes. But in some places—such as eastern Africa or Australia—their aquatic environments can quickly evaporate during extended times of low rainfall. Once these former water bodies are reduced to mud puddles, millions of years of instinct kicks in and lungfish dig themselves underground bunkers. A lungfish accomplishes this feat by first poking its head into the still-soft mud of a former pond or river and wiggling its body straight down. It then eats mud on its way, pushing this sediment through its gills back and behind it. Their downward movement also compacts the mud immediately around the lungfish, firming the burrow wall and preventing the resulting vertical shaft from collapsing. Once it reaches an optimum depth, a burrowing lungfish flexes its body up into a hairpin turn, positioning its head and mouth up and next to its tail. This doubling of its body width at the burrow end creates a wider, oval chamber, making the overall form of the burrow look like an old-fashioned bulb thermometer.

Once properly ensconced in this burrow, a lungfish secretes mucus from its skin as all-organic, all-natural shrink-wrap. The mucus hardens to form a cocoon around its host, with its mouth as the only opening to the outside world, allowing it to still breathe. The cocoon thus acts as a sealant, preventing moisture loss as the sediment around the lungfish desiccates. Food is not necessary during this time, as a lungfish will reduce its metabolism to a minimum; it also absorbs some of its muscular tail in a form of self-cannibalism. With such protection in place, lungfishes can stay in their burrows for as long as four years, inspiring estivation envy in mud turtles, crocodilians, and many other animals that can only survive a year or less in their burrows. The lungfishes only emerge from their temporary tombs with an influx of water in the surrounding sediment, triggering them to break out of their cocoons and burrows. One BBC documentary film featuring African lungfishes showed this rejuvenation in a dramatic way, as lungfishes buried themselves in mud, and then people mined the mud—some of it containing lungfishes—and formed it into bricks for their houses. A hard rain later transformed their homes into crypts of resurrection as lungfishes writhed out of their walls, a situation that would be hard to explain to your insurance company once repairs were needed.

We know lungfishes have burrowed since near their origin because of lungfish-like burrows from Devonian rocks in Pennsylvania. Granted, lungfish bones or other body parts have not yet been recovered from the Devonian burrows, but these structures are so distinctive that they are very unlikely to have been made by anything else. Lungfish body fossils are also known from the same rocks, so they are implicated in a sort of ichnological "guilt by association." Lungfish burrows containing their skeletons are in Late Permian (about 255 *mya*) rocks of Kansas, having lived and died just before the mass extinction that ended both that period and the Paleozoic Era. I have found lungfish burrows in Early Cretaceous strata (about 105 *mya*) of Australia, where lungfish would have burrowed to avoid polar winters. Furthermore, abundant lungfish burrows are known from Late Cretaceous (about 70 *mya*) rocks of Madagascar, discussed next. Other examples of fossil lungfish burrows are sprinkled through the geologic record, all of which are connected to former freshwater environments.

How paleontologists recognize fossil lungfish burrows is relatively straightforward. Both modern and fossil lungfish burrows have a long, nearly straight vertical shaft ending in an expanded oval chamber. This form translates into circular or oval cross-sections on sedimentary-rock bedding planes, visible because of different sediment fills. However, because their fishy makers had to turn their long bodies around, their burrows double back on themselves, making for "figure eight" patterns of overlapping circles, which also show up in burrow cross sections. This variation in patterns was well documented in a study of Cretaceous lungfish burrows in Madagascar, published in 2012 by Madeline Marshall and Raymond Rogers. Fossil lungfish burrows from there and elsewhere in the geologic record demonstrate that drying or otherwise trying conditions forced their makers to escape into the ground. Granted, a few died in their burrows and thus made themselves into fossils that were much easier for us to find millions of years later. Nonetheless, many more of their ancestors lived long enough to pass on the genes that programmed for burrowing in their descendants, stretching from 400 million years ago to today.

Amphibians Getting Deep and Meaningful

With lobe-finned fishes traveling across land and at least a few burrowing into it, natural selection took a turn toward rewarding better walking abilities in their tetrapod descendants. The very first Devonian tetrapods were amphibians, with some looking superficially like modern salamanders, yet still bearing many fishy traits. Although tetrapods had declared partial independence from aquatic environments by living some of their lives on land, they nonetheless had to use water for their reproduction and early growth. Accordingly, some of these early tetrapods may have been bestowed with burrowing abilities used by their lobe-finned ancestors, or temporarily lost these abilities until natural selection brought them back out of necessity.

Why would early amphibians have needed to burrow? The most obvious evolutionary advantage of burrowing would have been the same as that for lungfishes: When your water supply evaporates, do not dry out with it. Reptiles later mostly solved this dilemma by developing scaled skin, which helped to prevent water loss. Amphibians, though, were a different story altogether, as most need to keep their skins moist year-round.

For example, toads—which are a general category of frogs—evolved a thicker and more moisture-retaining skin than most amphibians. Yet nearly all amphibians, including toads, still need watery environments for laying and fertilizing their eggs. They also need these places for fertilized eggs to go from a fishy stage of body development ("tadpole") to a decidedly four-limbed adult. Reptiles again got around this ecological requirement for their reproduction and development by putting a watery environment inside an already-fertilized egg with an eggshell. Lacking such abilities, though, amphibians instead had to play a waiting game during arid times, in which years might go by before enough rainwater accumulated to form ponds or streams. Burrows thus were the best way to help these animals survive and wait until better conditions arrived or to otherwise dodge the worst dangers offered by the surface world.

Another reason for early tetrapods to burrow is related to temperatures swinging in the opposite direction, going from the ill effects of evaporating heat to freezing cold. As mentioned previously, the vast majority of fish, amphibians, and reptiles are "cold-blooded" ectotherms, relying on their outside environments to provide warmth needed for everyday bodily functions. This means that ectotherms living in aquatic environments that freeze for days, weeks, or months must have some special way to deal with this physiological challenge.

Although paleontologists are not sure exactly when amphibians began living in polar environments, we have a minimum time for this habitat shift during the Early Cretaceous (about 120 *mya*). Some of these polar amphibians were also massive animals. For instance, *Koolasuchus* (named after Australian paleontologist Lesley Kool) lived in polar Australia at this time. This amphibian measured more than 2 meters (6.6 feet) long and, based on its teeth, it was clearly carnivorous. This made *Koolasuchus* the ecological equivalent of an alligator in its freshwater rivers, terrorizing juvenile dinosaurs that might have wandered down to the water's edge for a drink. In order to avoid freezing during polar winters, *Koolasuchus* and other polar-adapted amphibians of the geologic past must have somehow hibernated. (Note that hibernation is similar to but differs from estivation. Hibernation is dormancy more likely to happen during winter so an animal stays warm, whereas estivation is a summer plan for staying cool.)

Plenty of other modern amphibians go down into the ground as a way to stay out of trouble, whether their challenges are environmental (too hot, too cold) or from other predatory species (too tasty). For instance, frogs, toads, and salamanders in more northern latitudes burrow during the winter to hibernate, deftly avoiding freezing temperatures at the surface, then come out in the spring after the snow and ice thaw. New England has its share of amphibians that make it through cold winters by burrowing, but perhaps the most impressive are the amphibians of Siberia and Alaska. The northernmost amphibian is the Siberian wood frog (*Rana amurensis*), which can live as far north as 70 degrees, only 20 degrees from the North Pole. Six species of amphibians are also native to Alaska, consisting of two salamanders, one newt, two frogs, and one toad. Of these, the Alaskan wood frog (*Lithobates sylvaticus*) has natural antifreeze in its tissues that allows it to survive freezing during the winter, but it also protects itself by rotating its body into soil or leaves: more like drilling itself in place, rather than burrowing with its arms. The northwestern salamander (*Ambystoma gracile*) and long-toed (*Ambystoma macrodactylum*) salamander live in burrows year round, coming out of them during warmer times for breeding and feeding.

Speaking of burrowing salamanders, mole salamanders, consisting of more than thirty species of *Ambystoma*, are appropriately named in this respect. At 15–25 centimeters (6–10 inches) long, they are relatively large, and are geographically widespread, occupying a wide range of ecosystems from southern Canada through the continental U.S. to Mexico. Part of their adaptability comes from habitually going underground, whether in other animals' burrows or ones they dig themselves. Mole salamander burrows can be impressive, reaching 50–60 centimeters (20–24 inches) below the surface. Some of these burrows are simple straight to curved tunnels, varying from subhorizontal (25 degrees) to subvertical (75 degrees). However, they can also modify these burrows or otherwise make new ones resembling letters in the alphabet, such as "J," "U," "W," and "Y." Mole salamanders normally stay in their burrows during the day. But at night, small invertebrates and vertebrates must be wary, as these salamanders come out of their burrows looking for prey to consume, such as insects, earthworms, frogs, mice, and other salamanders. Their burrows also come in handy for protecting their occupants from both colder and drier conditions outside year round.

Thanks to burrows, toads and frogs can live in deserts, too, where surface water is either rare or short-lived. One of these is the desert rain frog (*Breviceps macrops*), which lives in a thin area along the southwestern coast of Africa, sandwiched between the sea and vast sand dunes. This spherical little frog with shovel-like feet has adapted to desert conditions by digging 10–20 centimeters (4–8 inches) deep burrows, where it stays during the day. (Incidentally, this species of burrowing frog became a YouTube star in 2013 when a photographer posted a 30-second video of one emitting its squeaky-toy battle cry; the video racked up more than 10 million views.) Desert rain frog burrows are deep enough to reach moist sand, which helps to keep them hydrated in a desert with hot day-time temperatures. The frogs then emerge during the coolness of night to seek out dung, which is where they find their groceries: not the dung, but dung-eating insects. Amazingly, the desert rain frog is completely terrestrial, even in its life cycle. For instance, females lay fertilized eggs in a burrow, where the young skip becoming aquatic tadpoles by developing in the eggs and hatching as juvenile frogs. Similarly dry-adapted species of frogs are in Australia, including the trilling frog (*Neobatrachus centralis*), painted burrowing frog (*Neobatrachus sudelli*), Spencer's burrowing frog (*Opisthodon spenceri*), and ornate burrowing frog (*Platyplectrum ornatum*), to name a few. As one might surmise from their common names, these frogs burrow, and like the people of Coober Pedy in New South Wales, they do it to avoid desiccation. The trilling frog is known to estivate underground for several years, using both its burrow and a secreted cocoon to protect its precious bodily fluids. However, the return of wet conditions triggers these frogs to pop out of their burrows to feed and breed.

Other burrowing amphibians, such as spadefoot toads (*Scaphiopus holbrookii*) in the southeastern U.S., live in areas with seasonal rainfall and cooler, drier winters. These fabulous toads estivate in burrows of their own making during the winter and part of the spring, emerging from these shelters once abundant rain causes standing pools of water. They make their burrows by coordinating their anatomy with behavior. Their common name comes from a hard projection on each rear foot, which serves as the "spade" for digging. They use these feet to loosen sediment, but also dig backward and downward, spinning their bodies as they progress. The resulting burrows are either shallow oval chambers or inclined to vertical shafts with oval chambers at their ends. Interestingly, this backward burrowing means these toads eject very little sediment from the burrow, aiding in success when playing hide-and-seek. Because these toads often appear en masse after heavy rainstorms in the spring or summer, people have nicknamed them "storm toads," which also works as the name of a biker gang. Just as horseshoe crabs have a mating frenzy on shorelines, spadefoot toads congregate when it's time to release their gametes, resulting in orgies that only a toad (or two, or a dozen, or hundreds) would love.

Other amphibians that use burrowing as a normal part of their lifestyles include amphiumas ("congo eels"). Amphiumas are represented by three species of *Amphiuma* that live in burrows like those of lungfishes. These impressive animals, which have tiny limbs but thick bodies that can reach more than a meter (3.3 feet) long, live in wetlands of the southeastern U.S. People who study amphiumas are not sure if they make their own dwellings or occupy other animals' homes—such as those of crayfish—but they probably do a combination of the two. As nocturnal animals, amphiumas use burrows for everyday resting and come out at night to eat a wide variety of animals, such as crayfish, snails, insects, other amphibians (including their own kind), and reptiles. Indeed, their predilection for crayfish may also explain why they sometimes take over crayfish burrows, eating them first and then settling in their homes. They also use burrows for estivation whenever their habitats become parched, and can hole up for months in these refuges. However, females are also known to make nesting burrows, where they lay their fertilized eggs; they then stay in these nests to ensure nothing (including other amphiumas) eats their potential young.

Similar to amphiumas are caecilians, legless amphibians that, depending on their sizes, can look like either worms or snakes, as different species range from 7 centimeters (2.8 inches) to more than 1.5 meters (nearly 5 feet) long. Although tied to aquatic environments because of their reproductive needs, adults of these amphibians live underground in terrestrial environments in South America, Central America, Southeast Asia, and Africa. Despite their lack of digging limbs, caecilians are well adapted for burrowing. Whenever moving through the ground, they use a combination of their heads (which are custom-built for probing into soil) and body musculature. This body movement is similar to that used by earthworms, in which they use hydrostatic expansion and contraction along their lengths to widen the sediment around them. A caecilian widens the burrow further by anchoring with its tail and pushing the rest of its body forward with a series of muscular contractions. However, unlike earthworms, caecilians are vertebrates. A burrowing caecilian must also fold its spinal column while moving through sediment, which sounds painful to you and me, but is perfectly normal for these amphibians. Caecilian burrows have not been described and their trace fossils remain unknown; their traces might be mistaken for those of small to large earthworms. Not coincidentally, caecilians often prey on earthworms, so they use their burrows for hunting these and other underground animals, but of course these burrows also provide personal protection from their predators.

Burrowing amphibians have been around for a long time and may have evolved with the earliest tetrapods during the Devonian, at about 350–400 *mya*. If so, this would have allowed at least a few amphibians to later survive the end-Permian, end-Triassic, and end-Cretaceous extinctions. The oldest interpreted tetrapod burrow, credited to an amphibian, is in a Carboniferous (330 *mya*) sandstone bed in Pennsylvania. The burrow, preserved as a natural cast in which river sediments filled it from above, was an inclined, 15–20-centimeter (6–8-inch) wide tunnel that twisted to the right downward and opened into a 50–60-centimeter (20–24-inch) wide oval chamber at its end. An amphibian is presumed to have made this burrow because of its size and shape, but also because it precedes any known reptiles in the fossil record. Its overall size and shape is like that of a burrow made by a small, lazy, or otherwise unambitious gopher tortoise, but it is still remarkable for its geologic age.

The oldest undoubted amphibian burrows are in Kansas, occurring in rocks formed during the early part of the Permian (about 275 *mya*), with some bearing skeletons of their presumed amphibian makers (*Brachydectes elongatus*). Although this long-bodied and short-limbed amphibian looked like an amphiuma, it was only the size of a mole salamander. Its burrows were also small, measuring only 2–7 centimeters (0.8–2.8 inches) wide and as much as 30 centimeters (12 inches) long, and, like lungfish burrows, were mostly vertical. What was most significant about this burrow-burrower find, though, is what it tells us about their ancient ecology. The burrows were in freshwater ponds that apparently evaporated while the amphibians were living there. This means the amphibians were likely reacting to droughts by burrowing down into still-moist sediments, seeking protection against the loss of water in their habitats: just like lungfishes.

What about reptiles that evolved from burrowing amphibians: Did any retain this as a way of life, or did they stay above ground for their 300-plus million-year existence? We know already from crocodilians and turtles that the latter is not true, and to these two reptilian clades we can add one more: squamates, which include lizards and snakes.

Riders of the Storm

My university students were surprised to see fresh iguana tracks on the beach of the tiny Bahamian cay following our short ocean swim, but I was not. Admittedly, my nonchalance was predicated by prior knowledge that the small island (Cut Cay) had been hosting a transplanted population of the rare San Salvador rock iguana (*Cyclura rileyi rileyi*) for several years. What did surprise me about the iguanas was their abundance and health. Before swimming from the main island of San Salvador, I told the students that we might not see any iguanas while there, but instead were more likely to find a few tracks, burrows, and scat. I was wrong. As we looked up from the tracks on the beach, its maker—a big adult male iguana—looked down on us from a seaside boulder, shaking its head in seeming disapproval. After just a half hour of surveying the cay, we spotted 12 more iguanas, counted and measured 15 active (freshly dug) burrows in sandy areas, and saw more than a few tracks and scat. My students and I were elated by this unexpected success, with some using waterproof cameras to snap photos of these charming lizards. But I also left the island more than a little impressed with the benefits of terrestrial burrows when surrounded by a sea that could rise up and strike the land with hurricane-delivered ferocity.

On October 2, 2015, San Salvador and all of the low-lying cays around it suffered a direct hit from Hurricane Joaquin, a Category 3 cyclone. It was the most damaging storm in recent memory on San Salvador, and when my students and I arrived for a field course there in December 2015, the Bahamian people were still restoring their buildings, infrastructure, and patience. High winds, soaking rain, and storm surges had battered the coastline, leaving behind enormous potholes or thick layers of sand and boulders across the single paved road that hugs the periphery of the island. Even three months later, the island was still a mess, and I felt much sympathy for what residents had endured both during and since the storm.

Keeping in mind that this horrific hurricane had hammered San Salvador made the survival story of the Cut Cay rock iguanas all the more compelling. On January 3, 2016, when my students and I landed on its beach as awkwardly as a herd of Cambrian trilobites, only three months had passed since the storm screamed through that area. Given the low elevation of the cay—which was less than 10 meters (33 feet) on its highest point—and ankle-high vegetation or otherwise sparse cover, all of the iguanas there should have died. After all, these were fully terrestrial iguanas, not the marine-adapted ones of the Galápagos that could be washed over by massive waves and swim around for hours. How did they make it through such a terrible disturbance?

Once again, burrows showed us just how these landlubber lizards could be directly impacted (literally) by the full fury of an oceanic tempest, and then months later casually pose for pictures. For their underground homes, these iguanas and many related species dig relatively simple inclined tunnels that curve to the left or right, and then open into wider chambers at their ends. These burrows then serve as shelters from hot daytime sunshine, cool nights, predators, and yes, hurricanes. In other words, when the weather outside was frightful, the burrow inside was delightful. While 200-plus kilometer-per-hour (125-plus mile-per-hour) winds carried horizontal rain and hurled rocks and tree branches through the air like missiles, the iguanas hunkered down, protected by the earth.

However, considering the amount of rain dropped by a hurricane, one might think that making a basement to prepare for a flood is a really bad idea. Yet these rock iguanas' ancestors had survived far worse storms in the Bahamas and Caribbean by digging "storm cellars" in porous and permeable sand. This meant that water from both rain and storm surges would have washed over and percolated through the sand surrounding each iguana, draining their shelters as they rode out the storm. Although no one had a before-and-after head count of the Cut Cay iguanas, further affirmation that they had a high survival rate following the hurricane came from another inadvertent experiment. At the nearby Gerace Research Centre field station, all six iguanas living there in an educational enclosure made it through the hurricane, with their burrows nearby.

How far back this burrowing behavior goes in iguanas is a good question, as we have no documented examples of fossil iguana burrows, let alone iguana skeletal remains in their burrows. However, we can look at their evolutionary history and other lizards today to get a clue. Iguanas are lizards, and all lizards are diapsids, having shared a common ancestor with archosaurs, which included dinosaurs, pterosaurs, crocodilians, birds, and all of their extinct relatives. Based on fossils and genetics, herpetologists estimate that lizards have been around at least since the Triassic, but possibly since the latest part of the Permian, at about 255 *mya*. If so, this means lizards—today numbering nearly 6,000 species—somehow bested three mass extinctions: the end-Permian, Triassic, and Cretaceous. Whether burrows figured in their surviving these extinction events is unknown, but a few modern burrowing lizards give some clues how.

Among the most noteworthy of modern burrowing lizards are those of the genus *Uromastyx*, represented by 15 species in Africa, the Middle East, and central Asia. (Incidentally, the lizard gracing the cover of this book is one of those species, *U. geyri*.) They are sometimes nicknamed "spiky-tailed lizards" because of their thick tails, which are indeed adorned with predator-discouraging spikes. Most of these lizards live in deserts or otherwise arid environments. Just like lungfishes and amphibians, spiky-tailed lizards use burrows to avoid the worst of heat-related stresses in such settings, controlling both temperature and humidity. Also, burrows are employed to avoid sandstorms instead of hurricanes, after which these lizards then use their excellent digging abilities to unbury themselves. Some species' burrows are admirable for their sheer length and depth compared to body size. For example, *U. aegyptia*, which lives in the inhospitable deserts of the Arabian Peninsula, excavates burrows as long as 5.3 meters (17.4 feet), and with vertical depths of 1.2 meters (4 feet). Such burrowing feats are accomplished while using bodies that are at most 70 centimeters (2.3 feet) long and weigh 2.5 kilograms (5.5 pounds); the latter overlaps with the weight of an overfed Chihuahua. Burrows have only one entrance, which means these also serve as exits, and they build walls around these entrances with displaced soil. Tunnel slopes are relatively steep, at 30 to 45 degrees, and run straight until curving to the right or left, sometimes spiraling. When inside their burrows, these lizards often keep their heads down and block the tunnel entrance with their tails, telling potential harassers that they are at home but will not answer the door.

Although spiky-tailed lizards are commendable diggers, by far the most impressive of burrowing lizards and among the best of all tetrapods is the yellow-spotted monitor (*Varanus panoptes*) in seasonally arid environments of Western Australia. These lizards, which are colloquially called goannas, construct spiraling burrows that are the deepest nests dug by any nonhuman vertebrate, reaching as much as 3.6 meters (11.8 feet) down. For perspective, think of the height of a regulation basketball hoop (3 meters/10 feet), dig down that far, and then a little more. Because these burrows serve as nests, the yellow-spotted monitors that dig these astounding burrows are all females, although both males and females also make plenty of closely spaced dwelling burrows. The inclined upper part of the burrow turns on its way down, descends abruptly into a corkscrew-like form, which ends at an expanded chamber, where the mother goanna lays her eggs. She then packs this part of the tunnel, which helps to protect her eggs from predators, as well as keeping stable humidity and temperatures for the developing embryos. Upon hatching, the goanna babies dig themselves out and crawl up a long, twisting shaft to the surface.

Other than these goannas adapting their nests to environmental extremes, it is worth noticing that this type of nesting coincides with burrowing-bird nests in keeping eggs and hatchlings safe from predators. Among the animals that would be extremely well adapted for easing down a winding tunnel to hunt for both eggs and their freshly hatched occupants are distant relatives of these goannas: snakes.

Sea Serpents or Subsurface Serpents?

Snakes consist of about 3,000 species worldwide, including gopher tortoise roommates such as indigo snakes and rattlesnakes, and are among the most fascinating of all animals. Sadly, though, they are also among the most misunderstood and demonized, primarily because 12% of those species are venomous and hence are perceived (fairly or unfairly) as a threat to humans. Still, snakes represent enduring and powerful myths and symbols in cultures worldwide, many of which connect snakes with the Earth or an "underworld," reflecting how many of these animals live in burrows or other hollows beneath our feet.

In terms of their evolutionary history snakes share a common ancestor with lizards; both groups belong to the clade Squamata. Snakes evolved during the same time as dinosaurs, descending from short-limbed squamates during the Mesozoic, which was probably in the Middle Jurassic, about 170 *mya*. However, exactly when and how snakes evolved from their four-legged ancestors is an evolutionary puzzle, and burrows figure in this mystery. Based on fossil snakes still retaining vestigial limb bones and the original environmental context of their entombing sediments, snakes either gradually lost their legs by adapting to marine environments as swimmers or they went underground.

If the former is true—that snakes evolved in marine environments— then this serves as a great example of how the evolution of lobe-finned fishes into land-dwelling tetrapods during the Devonian eventually led some branches of that tetrapod tree to go back into the water. Indeed, mosasaurs, which were recently made even more famous in the movie *Jurassic World* (2015), were Cretaceous lizards fully adapted to marine conditions. In contrast, in the "grounded" hypothesis, snakes would have evolved as either active burrowers or secondary occupiers of other animals' burrows, and did not adapt to watery environments—whether freshwater or marine—until much later.

The "snakes are marine swimmers" hypothesis certainly received a shot across the bow with the publication of a 2015 article by Hongyu Yi and Mark Norell bearing the unambiguous title of "The Burrowing Origin of Modern Snakes." In this paper, Yi and Norell took many CT (computed tomography) x-ray scans of snake skulls and demonstrated that modern burrowing snakes had spherical vestibules in their inner ears. This anatomical trait corresponds with hearing used to pick up low-frequency underground vibrations, rather than the higher-frequency sounds transmitted by air or water. When Yi and Norell likewise scanned the skull of a Late Cretaceous (90 *mya*) snake, *Dinilysia patagonica* of Argentina, and compared its inner ear anatomy with that of modern snakes, it had a better than 90% fit with modern burrowing snakes. Modern snakes with the greatest number of primitive traits—reflecting their original ancestry—also have underground lifestyles, suggesting that this was likely their primary (and original) behavioral fit. If so, this shows how human-imagined creation myths of snakes coming from the Earth are actually not so far-fetched when viewed from an evolutionary perspective.

Given their low-frequency hearing, long bodies, and tiny limbs, these Mesozoic proto-snakes would have been finely tuned underground hunters, going after small mammals, amphibians, buried eggs, or other underground prey. This evolutionary direction also implies that snake ancestors were descended from predatory burrowing squamates, a supposition affirmed by how all snakes today are carnivorous, whereas iguanas and many other lizards are primarily herbivorous. Regardless, this move to underground environments signaled the start of an arms race between those lizards and other animals that used the underground for safety and those who used it to pursue and kill prey.

The Birth of Burrowing: When Did It Start?

Based on their trace fossils, trilobites began burrowing into seafloors very close to their origin at the beginning of the Cambrian Period and the Paleozoic Era about 545 *mya*, with many other invertebrate animals joining them in this exploitation of the underworld over the ensuing years. Accordingly, trilobites are good candidates for winning the prize of "first burrowing fauna," although some marine worms may have preceded them. The origin of burrowing vertebrates is a bit less defined, as no undoubted vertebrate burrows are known from rocks older than the Devonian. In the meantime, though, invertebrates dug down and deep in the seafloor in the Cambrian, and then in freshwater environments and on land in the Ordovician. Once vertebrates joined in (and especially tetrapods), animals were on their way to terraforming the entire surface of the planet, which they have been doing ever since.

What types of animals preceded burrowing trilobites and other subsurface invertebrates? When did animals start using burrowing to hide from predators? Was there ever a time when no animals burrowed, or were otherwise restricted to two-dimensional surfaces? The answer to the last of these questions is yes. But to know more about the origin story of burrowing, we must hearken back to the alien world of the Ediacaran Period, 630–545 *mya*. It was toward the end of the Ediacaran—more than 300 million years before "The Great Dying"—when burrowing animals evolved for the first time, enabling animals to descend into a world that had been ruled by microbes and algae for nearly four billion years, while also ensuring their survival in upcoming catastrophes.

CHAPTER 7

Playing Hide and Seek for Keeps

Mistaken Point and Foregone Conclusions

Given that the rock in front of me held the oldest known trace fossil made by an animal, I should have been more impressed. Yet if humility could be conferred on a trace fossil, this one just seemed too humble. It was visible only as a straight, narrow strip of light gray cutting across dark splotches on the rock surface. Each end of the strip—half the width of my pinky fingernail—was rounded, but one of those ends had a clearly outlined oval depression. Curved, alternating dark and light lines like nested parentheses were inside its borders, menisci that opened toward the oval. The latter spot was where its maker had stopped moving along the surface, and the curved lines marked where it paused briefly along the surface before pushing forward in an incremental crawl. This trace fossil was clearly from a very small, legless animal left on top of a sedimentary surface and not a burrow; before an overlying layer of sediment buried the trail, all of the animal's motions took place about 565 million years ago. This was quite a long time before human eyes discerned it as a vestige of animal behavior, and about 20 million years before eyes of any kind had evolved. I was now suitably impressed: awestruck, even.

To see this little marvel of a trail, I first had to walk along a much more recently made human path in Mistaken Point Ecological Reserve, located along the southeastern coast of Newfoundland, Canada. Mistaken Point earned its name after one too many ships crashed on its offshore rocks, but is famous for other reasons. It is revered by paleontologists, or rather, paleontologists more interested in the origins of animals, rather than just a few species with backbones and names that include the suffix "saurus." This paleontological eminence is well deserved because it is one of the richest places in the world for body fossils from the Ediacaran Period, a time interval from 630 to 542 *mya*, which was more than 300 million years before dinosaurs. Tens of thousands of Ediacaran fossils are preserved on more than a hundred sedimentary surfaces there, and in marvelous detail. Like precious bas-relief artworks, they project from rock surfaces, displayed on broad, tilted bedding planes next to sheer sea cliffs along the Newfoundland coast.

I was not alone on this little venture but on a field trip organized by Memorial University (St. John's, Newfoundland), accompanied by my wife, Ruth, and a gaggle of nearly twenty ichnologists. Two Canadian park rangers escorted us there, their presence required to ensure that no harm would come to the fossils, accidental or otherwise. Although all of us shared an enthusiastic interest in animal traces, and the field trip leaders—Duncan McIlroy, Jack Matthews, and Liam Herringshaw—promised a few trace fossils would be there, we were also very excited about seeing Mistaken Point's numerous (and renowned) body fossils.

Paleontologists are utterly incapable of describing Ediacaran body fossils without using the words "strange," "weird," "alien," "bizarre," or other descriptors that set these fossils apart from anything else we know about life of the ancient past. Even the word

"Ediacaran" itself is special, as it comes from an indigenous place name in South Australia, the Ediacara Hills, whereas most other English names have Latin, Greek, or European roots. "Ediacara" is a transliteration of the word *idiyakra*, meaning "spring of water," a place name that may have been inspired by a former presence of surface water in this normally parched area of Australia. Much later in human history, though, it became more meaningful because of the Ediacaran fossil assemblage there, comprised of the remains of shallow-marine organisms, which represent the first trickles of animal diversity.

In 1946, while prospecting for mineral deposits in the Ediacara Hills, geologist Reginald C. Sprigg first noticed some of these fossils. Fortunately, he shared his find soon afterward with other geoscientists, who identified most of these bizarre forms as fossils. Preserved as either impressions or popping out slightly from hard sandstone surfaces, these Ediacaran fossils consisted of an astonishing array of shapes and figures: large, flat, segmented "worms," small discs with three "arms" in their centers; jellyfish-like blobs; spindle-like forms; frond-like forms; and other fossils that so defied comparison to any living animals or plants that normally loquacious paleontologists resorted to calling them "things." Perhaps the most intricate and otherworldly of these fossils are the frond-like ones, shaped like cartoon flames but branched in minute detail, their "stems" anchored by a bulbous bottom, termed "rangeomorphs."

To this day, paleontologists still argue over whether some of these Ediacaran fossils are aligned with modern animal groups—such as flatworms, sponges, or sea pens—or whether they reflect a failed evolutionary experiment that left no modern descendants. A few paleontologists have even argued that these fossils are not animals, but gigantic one-celled organisms, and at least one paleontologist proposed the far-flung idea that they were lichens and lived on land. As a result of such uncertainty about the biological affinity of the Ediacaran fossils, some linguistically conservative paleontologists prefer to call them the "Ediacaran biota," rather than the "Ediacaran fauna." With such vagueness they are acknowledging the possibility that at least a few of these fossils, which apparently were all soft-bodied, may not be of animals. Fossil identity crises aside, Sprigg had made a splendid discovery in the Ediacara Hills that ultimately contributed to how we view the history of life, animal or otherwise.

Nonetheless, the combination of oddity and antiquity for these Australian fossils meant most paleontologists initially rejected them in the 1950s and through much of the 1960s. In their minds, either their geologic age was wrong—there was no way these could be from before the Cambrian—or they were not fossils, and certainly not of animals. Something else that did not help was what Australian historian Geoffrey Blainey termed "the tyranny of distance," where the geographical isolation of Australia from the rest of the world impeded most non-Australian paleontologists from visiting and independently investigating the fossils. Fortunately, a few skilled Australian paleontologists, such as Martin Glaessner and Mary Wade, studied them later in the 1960s, with their research papers giving other scientists an inkling of their importance.

Vindication for Sprigg and others in Australia thus came slowly, but happened once other paleontologists in different parts of the world started finding more such fossils in undoubted Precambrian rocks. Starting in the 1950s and through the 1980s, paleontologists and geologists recognized Ediacaran fossils in England, Russia, Namibia, western Canada, and even in the far-off, forbidden land of North Carolina. As of now, Ediacaran fossils have been identified on all continents except Antarctica. Once this biota was recognized as a worldwide phenomenon, and in honor of the original discovery site, a stratigraphic commission decided that a geologic time unit—the Ediacaran Period—would be named after the Ediacara Hills.

As an official geologic time unit, the Ediacaran Period was the last period of the Proterozoic Eon, which lasted about 2 billion years (2.5 *bya* to 542 *mya*) and ended just before the Cambrian Period. The Proterozoic and the preceding Archean Eon collectively represent about 87% of the earth's 4.5-billion-year history; geologists often lump both together into an informal span called the

"Precambrian." The ensuing eon containing the Cambrian at its base, the Phanerozoic, means "visible life" in recognition of its more abundant and hard-bodied fossils. Yet we now know the Ediacaran is when animal life, representing the organization of multiple larger and complex eukaryotic cells into self-producing packages, made its debut in the fossil record.

Nonetheless, there is a twist to this very satisfying story of paleontological redemption. The very first documented discovery of body fossils in Precambrian rocks was actually not in Australia, but in rocks of Newfoundland. Paleontologist Elkanah Billings, who is credited as Canada's "first official paleontologist," described and interpreted these peculiar forms as fossils in 1872, more than 70 years before Sprigg made his discovery Down Under. The fossils, spotted by Scottish geologist Alexander Murray in 1868, were in an outcrop of Precambrian shale in downtown St. John's, and on Duckworth Street. (This also meant Murray was doing urban geology long before it was hip.) Billings not only recognized these entities as evidence of former life, but also gave them a fossil species name, *Aspidella terranovica* ("round shield of the new [found] land").

Aspidella are small (5–10 millimeters/0.2–0.4 inches wide), circular to elliptical disks with concentric rings, many bearing a raised bump ("pimple") in the center. These fossils, which were later found in more Precambrian outcrops of Newfoundland and other parts of the world, can be extremely abundant, sometimes numbering thousands per square meter. Sadly, Billings's peers at the time scorned his interpretation, countering that these fossils must be inorganic structures, perhaps formed by gas bubbles in the original mud. The great age of the rocks prejudiced their views, and the potentially awesome implications of this fossil find were buried, albeit not nearly as long as the fossils.

At least partly because of this setback, more Ediacaran fossils of Mistaken Point and other outcrops along the Avalon Peninsula of Newfoundland took a while longer to get noticed. In 1967, almost exactly a hundred years after Murray noticed *Aspidella* in downtown St. John's, Shiva Balak Misra, an Indian graduate student in geology at Memorial University in St. John's, found the decidedly not so urban Mistaken Point fossils. Misra's momentous discovery was quite serendipitous, as he was not necessarily looking for fossils, but was making a geologic map of this rugged coastal area for his master's thesis. Fortunately, he quickly documented his find in several peer-reviewed articles in 1968 and thereafter. Mistaken Point, with its Ediacaran treasures, was then not only on Misra's map, but also the world map for Ediacaran fossils.

Because the Mistaken Point fossils were so well preserved and abundant, most could be classified into informal categories based on their forms. For instance, Mistaken Point fossils have been described as leaf-like, spindle-like, lobate, dendritic, or radiating. However, like most Ediacaran fossils, they were so different from anything in geologically younger rocks that they could not be easily related to any organisms, modern or extinct. Were the spindles and fronds some sort of sea pens and sea fans, distantly related to modern corals? Were the lobate forms flatworms, or something else? Were the dendritic ones sponges of some sort? Were the radiating fossils akin to jellyfish or related animals? Although some headway has been made since in better connecting these fossils to known organisms, such confusion continues for paleontologists who study Ediacaran fossils.

Taxonomic difficulties aside, once geologists thoroughly studied the sedimentary rock types and their physical sedimentary structures at Mistaken Point, they could interpret the original environment for the fossils. Their conclusions were surprising: These organisms lived in the deep ocean. This environmental setting contrasted with shallow-water settings interpreted for Ediacaran fossil assemblages in South Australia, Russia, Namibia, and, well, almost everywhere else. How deep? Best estimates made by geologists today are that the sea bottom hosting the Ediacaran organisms may have been 1,000 meters (3,300 feet) below the ocean surface. This great depth places them well below the upper part of the ocean that received sunlight (also known as the photic zone), which today, under the best of conditions, extends to only about 100–200 meters (330–660 feet). This also put these life forms out of reach from ocean waves, and they certainly would not have experienced the effects of tides. Whatever these organisms might have been, they were living in the cold, dark, quiet reaches of the ocean. This realization conjures a paleoecological picture both remarkable and counterintuitive, as early animal life was always assumed to have first evolved in warm, well-lit, shallow waters.

Seeing that the Mistaken Point fossils were so far below ocean waves and tides, how did they get buried and rendered so exquisitely? This is where Precambrian plate tectonics lent a helping hand, or rather, big slabs of melting rock. Conservation of these marvelous fossils was facilitated through regular inputs of volcanic ash. The ash erupted from a nearby volcanic island arc, produced by two oceanic plates that collided with each other, similar to the Caribbean today. The collision caused one plate to dip beneath the other into the earth's mantle, partially melting it. This hot mantle material then rose to the surface and expressed itself as erupting volcanoes. Once voluminous clouds of fine particles fell onto the water, these sediments settled and buried the fossils in place, acting like an underwater (and far more ancient) version of Pompeii. The volcanic ash deposits were doubly fortunate for paleontologists and geologists studying the rocks, as they not only conserved the fossils, but also allowed radiometric dating of the rocks at Mistaken Point. Dates revealed ages just on either side of 565 *mya* for most of the layers there. This confirmed that these Ediacaran ecosystems and their biota existed and thrived more than twenty million years before the first trilobites and other more familiar animals of the Phanerozoic.

But enough about Ediacaran body fossils. What about something that really matters, like trace fossils? Tragically, as Ediacaran body fossils became sexier, Ediacaran trace fossils were often mentioned as afterthoughts. This ichnological neglect is not so surprising, though, not just because of disinterest, but also biology. For instance, if you closely examine Ediacaran animals, you will see that most were likely sedentary, locked to the seafloor for their entire lives. Supporting this idea were the numerous spindle-like and frond-like forms we saw at Mistaken Point on our field trip: All had discs on their bottoms, which acted like plant roots, anchoring the organisms to the seafloor. This meant their owners were not going anywhere. Indeed, an inability to let go is why most fossils at Mistaken Point are regarded as in situ, killed and buried exactly where they lived. Hence these organisms would not have been exploring the sea bottom and leaving their traces in or on oceanic sediments. Given this insight, Ediacaran trace fossils should be rare, and are.

Another factor affecting the types and abundance of Ediacaran trace fossils is that no animals had legs then. These appendages were adaptations for movement, which we take for granted in so many animals today—from arthropods to amphibians to reptiles to Rockettes—but apparently did not evolve until just before or at the start of the Cambrian Period. Legs were a huge leap forward in animal mobility, and in fact later greatly aided in leaps, huge, forward, or otherwise. In contrast, animals lacking legs limited their sea-bottom mobility. They could conceivably have been moved about by the water, but at this great depth, there were few ocean currents. And so any such animals moving under their own power instead must have used a creeping sort of motion or muscular contractions of their bodies (peristalsis) to push, pull, or otherwise propel themselves forward. For examples of the latter, just watch modern snails, slugs, and earthworms move across a surface, and then think of the Ediacaran.

Making matters worse for Ediacaran trace fossils is more of a human-perception challenge faced by paleontologists that reduces the number and types of perceived traces. For instance, over the past ten years or so, paleontologists have realized that many forms in Ediacaran rocks identified initially as "trace fossils" are not. Instead, further scrutiny revealed these were either body fossils or strange sedimentary structures that fooled people into thinking they were trace fossils. So my colleagues and I put our ichnological optimism to the test during our field trip to Mistaken Point that day and other spots visited during our three-day field trip.

Alas, we soon found out for ourselves that trace fossils made by mobile animals were indeed rare at Mistaken Point, and did not exactly jump out at us at other outcrops along the Newfoundland coast. Just before seeing those few trace fossils at Mistaken Point, though, we had to hike for about 30–40 minutes to the main fossil site, where we all took off our field boots. The shedding of standard geoscientist footwear was not some odd ichnologist ritual, but required of all visitors at Mistaken Point. (Rest assured, there are ichnologist rituals, and they involve far stranger activities than mere shoe removal.) This rule kept people with hard-soled shoes from breaking or otherwise wearing down the fossils on the bedding plane for as long as humanly possible. Instead of boots, we slipped on soft, bright blue cloth coverings, nicknamed "Bama booties" (I have no idea why), converting us to a flock of blue-footed boobies. Once unleashed so colorfully on the rock surface, we walked along its bedding plane and occasionally fell to our hands and knees to better view the exquisite details of the fossils, while also looking worshipful, which we were.

Make no mistake about Mistaken Point, body fossils are the stars there, and trace fossils are the supporting acts. This means that the paleontological spotlight and red carpet for Ediacaran award ceremonies are reserved for spindles, fronds, discs, and blobs, rather than trails, burrows, and any other indirect signs of animal life. Still, the vestiges there reflecting the oldest evidence of animals in motion also contributed to our understanding of how this movement first happened. And movement is the foundation of being an animal, or "animated": in motion.

For instance, the fossil trail described earlier was ascribed to a soft-bodied animal analogous to a sea anemone. Unlike most corals, some modern sea anemones can move on their own, making trails and even burrows. The researchers studying this trace fossil suspected this and similar fossil trails might be from anemone-like animals, and tested it by cajoling modern anemones to make similar trails in an aquarium. Most impressively, the researchers proposed a potential Ediacaran tracemaker for this trace: *Aspidella*, the very same first-named Ediacaran body fossil, which was also from Newfoundland and in rocks of the same age. Linking a trace fossil with a specific maker is always an impressive feat in ichnology, but is particularly laudable when applied to 565 million-year-old fossils.

Near this trail and on other rock surfaces, we found a few more trace fossils of about the same width as the other one, but shorter. Again using my pinky finger as an international unit of measurement, these were half its width and slightly less than half its length. However, they were a little different from the previous trace fossil, having furrows flanked by levees and lacking menisci in their centers. These traces looked more like their makers had plowed heedlessly through the sediment, rather than just skimmed along its top surface. Were these burrows? Nope. The levees were only about a millimeter high, showing their makers displaced very little sediment as they moved along.

A search for more trace fossils at the site was an exercise in frustration and disappointment, as no others were to be found. However, researchers who looked for many more hours than us at Mistaken Point have so far found about ninety of these small trails in the rocks. Although trace fossils are greatly outnumbered and overshadowed by celebrity body fossils there, they are abundant enough to confirm that animals were indeed moving on the sea bottom at least 565 million years ago.

This brings up what is likely the most important point of all about the Mistaken Point trace fossils: They were shallow. All of us ichnologists noted that these trace fossils were restricted to horizontal planes, their makers having dutifully followed the sedimentary surfaces as if they were driving on an interstate highway across Kansas. In other words, these trace fossils were just trails, not burrows. Granted, the furrowed traces could have been made as burrows just below a surface, in which their makers moved like a mischievous cat crawling under a bed cover. But these animals certainly did not dig down into the sheets, let alone the mattress. Instead, these animals were stuck in two dimensions: length and width, but no depth.

Despite the disappointing paucity of trace fossils and their super-ficial nature, we stayed at the outcrop for a couple of hours, carefully observing the fossils there, sharing insights, and enjoying our usual lively outcrop discussions and debates. Our group of twenty-one ichnologists represented nine countries, and this field trip, like others before it, filled me with gratitude for how our mutual inter-ests and love of life history brought us together every few years as friends. Nonetheless, all good things must come to an end, and our field trip leaders and the rangers informed us we had to leave. We reluctantly shed our Bama booties, laced up our comparatively drab field boots, and began walking away. A fog settled in as we departed, enveloping Mistaken Point in mist, its evidence of long-lost life seemingly returning to great oceanic depths.

The lessons of the Mistaken Point trace fossils and their place in time reinforced those conferred by another seaside outcrop the day before. It was a dramatic exposure, jutting out from the coast as a precipitous peninsula called Fortune Head. The rock sequence consisted of a thick, vertical exposure of layered Edia-caran sandstones and shales that smoothly segued upward into Early Cambrian strata. In fact, the transition was so continuous that Fortune Head is regarded as the international stratigraphic standard for the Precambrian–Cambrian boundary, and every other such boundary in the world is compared to it. Geologists often refer to such international stratigraphic standards as "golden spikes," in which we imagine a gaudy spike hammered into the exact spot between rocks of different ages, acting like a bookmark placed between chapters. In this instance, though, the "spike" separated not just the Ediacaran from the Cambrian, but also the vastness of the Precambrian (almost four billion years) from the rest of the Phanerozoic Eon.

Did the trace fossils at Fortune Head reflect a biological changing of the guard, one in which the organisms shifted in their evolution enough that we mere primate-derived afterthoughts could detect it? For sure. As we scoured the outcrop, our forty-two expertly trained eyes searching for trace fossils, we could only find scarce evidence of horizontal trails below the boundary, and more abundant but millimeter-wide vertical burrows above it. Some of these Cambrian burrows were U-shaped, whereas others spiraled like pigs' tails, but all were quite modest in depth, going only 1–2 centimeters (0.4–0.8 inches) down into the underlying sediment. In contrast, at another outcrop with Early Cambrian rocks from about 540 *mya*, we saw long, curving burrows with short tubes connected to a central shaft. These trace fossils, called *Treptichnus pedum*, provided obvious signs of deeper burrowing and were made by animals that systematically probed the sediment for food. *Treptichnus* is also an index fossil, telling us that we were looking at rocks that were unequivocally Early Cambrian in age. Nothing like this burrow is in the Precambrian, and its presence brazenly announced that animal life had evolved to push aside mud and sand and start plumbing their depths.

What did all of these trace fossils (or lack thereof) tell us about the Ediacaran–Cambrian changing of the ecological guard? Although none of those earliest Cambrian burrowers knew this at the time, the seafloor was about to undergo its biggest and most permanent change, one still with us and likely extending far into the future of life. These profound alterations of ocean sediments—and by extension all marine environments, and then the land—were started by those very same Cambrian burrowing animals that inherited the Earth from a non-burrowing Ediacaran biota.

When All Animals Were Superficial

By now the gentle reader has no doubt gleaned that the life and ecosystems of the Ediacaran Period were radically unlike those of today. Thus the Ediacaran is a superb example of one of many worlds that have shared our planet, only separated by time. Perhaps the most extreme difference between the Ediacaran and today is that nearly all animals of that world lived either on the ocean floor or immediately below it, and nowhere else. As far as we know, no animals were yet floating or swimming in the copious volumes of water above, few animals were probing into sediment below, and certainly no animals were in freshwater environments or on land. It was as if animal life was restricted to an underwater two-story parking garage, with a top floor open to the water and a bottom floor only covered by the deck above. To continue this analogy, a slimy, impermeable layer sealed the top deck, effectively deterring anything from moving downward into the bottom floor. Yes indeed, the Ediacaran was different.

Speaking of slimy, impermeable layers, these covered shallow seafloors during much of the Precambrian. Most of the gooey surfaces were made by colonies of microbial life, such as photosynthetic bacteria (cyanobacteria) and single-celled algae. Such layers were further held together by trapping and binding sediment suspended in the water: Imagine a sticky tape gathering wind-blown sand, or flypaper collecting, well, flies. Geologists who encounter geologic structures formed by these colonies use a fancy term for them, *microbially induced sedimentary structures*, which is often shortened to the memorable (and respectful) acronym MISS. Other geologists who are not so much into multisyllabic terms and trend-setting scientific acronyms simply call their original microbial layers *biofilms* or *biomats*.

Before the advent of mobile, grazing animals, which ate or otherwise tore into biomats, biomats blanketed sedimentary surfaces in shallow-marine environments. Sometimes these films became colonies, growing ferociously upward by collecting sediment and making meters-thick mound-like structures called *stromatolites*. For a long time in the geologic past, stromatolites dominated the shallow seas, with some becoming prominent enough to form the first wave-resistant reefs, more than three billion years before the first corals. Stromatolites, although quite rare today, still exist in the Bahamas, Abu Dhabi, Western Australia, and a few other places.

How did these biomats relate to the first animals and the evolution of burrowing? Think of the biomats as barriers to animal progress, which animals somehow overcame. For instance, biomats were often composed of tough, thick, cohesive layers with sediment mixed in, whereas the first mobile animals had no claws, teeth, or other anatomical attributes that could cut through them. Try to open shrink-wrapped CD cases or vacuum-packed food items, but without scissors, fingernails, or teeth, and you will quickly identify with Ediacaran animals. These films also effectively sealed under-lying sediments from oxygen introduced by overlying seawater. As a result, oxygen-deprived sediments below biomats constituted hostile territory for animals. If animals and biomats were going to live together, then animals had to play by biomat rules: Stay on the surface, or if you want to go under the surface, don't go too deep, or you will die.

What preceded this biomat world, where slimy films and sediment subjugated all animals, was a very cold earth, and one that stayed thermally challenged for a long time. How cold? Geoscientists esti-mate the equator had an average temperature of about –20°C (–4°F), matching modern-day Antarctica. In the later part of the Proterozoic Eon and just before the Ediacaran Period, global temperatures stayed low for more than 200 million years, from about 850 to 620 *mya*. (Remember this next time you grumble about a spring day being too cold for your liking.) Geologists have affectionately nicknamed this interval "Snowball Earth," and it surely had a major impact on the development of all life, not just animals.

Marine sedimentary rocks from that time record much evidence for colder temperatures, but perhaps the most dramatic signs of an icy Earth are big chunks of rock deposited in otherwise fine-grained oceanic sediments. These rocks, which geologists call *dropstones*, are extremely varied in size and composition, which they should be, as they represent a random sample of Precambrian continents. These were rocks plucked from Precambrian landscapes by mas-sive glaciers; once the edges of these glaciers reached the sea, they broke up and calved into icebergs. The icebergs floated out to sea and continued to carry these continental rocks on their bottoms, like someone carrying sand or pebbles on a shoe. Nonetheless, as the icebergs continued to float, some of these errant stones dropped out of the glaciers and plunged to the seafloor. The resulting mix of fine-grained marine sediments and pieces of continental material in a sedimentary rock is a *diamictite*. Whenever geologists recognize diamictites, they know these indicate colder conditions for that time, while also signaling the ghostly presence of long-gone icebergs floating far above.

For a long time, the Earth was frozen—but let it go. Paleontologists consequently assume that animal life was either rare or absent before 630 *mya* because of these brutally prolonged cold conditions in shallow marine environments. Nonetheless, once ocean temperatures got higher and stayed there for a while, animal life went forth and multiplied. When? A partial answer comes from a genetic technique called *molecular clocks*. In this method, geneticists use mutation rates of genes in modern animals and their immediate ancestors to back-calculate divergence times for certain clades. For example, molecular clocks predict that the first tetrapods probably evolved about 400 million years ago, not long before they began leaving a trace fossil and body fossil record. When applied to the origin of animals, molecular clocks suggest a start date as long ago as 800–700 *mya*. This result is both surprising and intriguing, as it suggests that single cells may have organized well enough to form animals during the time of Snowball Earth. If so, environmental stresses may have provided the evolutionary jumpstart that eventually led to sponges, worms, mollusks, arthropods, fish, amphibians, reptiles, and finally reached its zenith with Beyoncé.

Fortunately, we do not just have to rely on geneticists to figure out when animals started, as we can also look at the rocks themselves for clues. For instance, chemical fossils (*biomarkers*), in the form of organic compounds called *steranes*, indicate a presence of animals at a younger date, at about 650 *mya*. Such compounds are also in modern sponges, which happen to be the most primitive animals still around today. The oldest probable animal body fossils, however, are from about 590 *mya*. These microscopic but gorgeously preserved fossils recovered from Proterozoic rocks of China are evidently from dividing cells, mimicking forms we see in the embryonic development of modern invertebrates.

Still, none of these bits of data tell us anything about when animals first moved under their own power, nor when they began burrowing. Based on sheer numbers, once animals began pushing or pulling themselves across the seafloor or just under biomat surfaces to find food, they would have produced telltale signs of their movements. This is why trace fossils are essential for figuring out Ediacaran life.

Still, Ediacaran trace fossils present a problem to paleontologists who want to apply their Sherlock-like powers to intuit who was doing what to whom (or what), how they were doing it, and when. As mentioned before, it turns out that some jots, tittles, and squiggles in Ediacaran rocks initially identified as "trace fossils" are not. Instead, some of these features are physical sedimentary structures or body parts of strange algae or animals that only look like trace fossils. Such misjudgments hint that the number and diversity of Ediacaran trace fossils may have been overestimated, making them more special than we originally thought. This situation in turn creates greater challenges when paleontologists try to discern when animals started transporting themselves.

Paleontological confession time: I unwittingly participated in such an Ediacaran exercise in ichnological doubt and denial. In 2009, I studied enigmatic fossils from Ediacaran rocks of North Carolina with two colleagues from the North Carolina Museum of Natural Sciences, Chris Tacker and Patricia (Trish) Weaver. The rocks were dated from the latest part of the Ediacaran Period, about 545 *mya*, only a few million years before the start of the Cambrian Period. Other undoubted Ediacaran fossils had been reported from the same rocks, so the ones we focused on were part of that same assemblage. Little did we know at the time just how much these fossils would perplex us.

We began our study by assuming that the mystery fossils, which were bundles of millimeter-wide rod-like objects and impressions of these objects, were burrows. Clusters of these fossils looked as if they branched from a central point, which implied that burrowers repeatedly probed into the surrounding mud for food. Indeed, other ichnologists had interpreted these features as feeding burrows and as evidence of sophisticated sensory abilities. If so, these fossils would have represented an advanced behavior in animals during the Ediacaran. Surface trails shared the same surfaces and were denoted by curving furrows and paired levees, looking like tiny meandering streams. Although less exciting and rarer, these trace fossils nonetheless showed that animal life was there, wandering about an Ediacaran ocean bottom.

However, as often happens in science, once we investigated these fossils in detail, we realized that everybody was wrong. The "burrows" were actually bits of an unknown organism, probably an alga with a rod-like growth pattern. Once its bits and pieces were moved and then deposited on the Ediacaran seafloor, bottom currents aligned them into overlapping bundles, like pine needles arranged by a leaf blower. These arrays made them look like feeding burrows made by tiny, wormy animals that methodically poked their heads into the surrounding sediment to eat its organic goodies. Hence when we published our research article in 2010, we stated that these Ediacaran entities were not trace fossils, and people should stop calling them that. Fortunately, our research was not a total loss in an ichnological sense, as the meandering surface trails accompanying the interloping body fossils were real trace fossils. Three years later, Trish Weaver and I published an article describing the trails—which seemed much like the ones I had seen in Newfoundland—and stated how these trace fossils were likely made by mobile, wormy critters on the seafloor.

Since those and other studies, paleontologists scrutinizing Ediacaran fossils have likewise identified other body fossils that were formerly regarded as trace fossils. This discrimination has effectively decreased the number and diversity of real trace fossils, and makes everyone very suspicious indeed whenever someone declares they have discovered "new" Ediacaran trace fossils. The reduction in trace fossils at first might seem like bad news for ichnologists. Still, healthy skepticism has improved the quality of Ediacaran paleontology overall, making it very good for our science.

Given the current state of awareness, paleontologists have put together a short list of Ediacaran trace fossils, some of which have been linked directly to their makers. For instance, a few specimens of *Kimberella*, a coin-sized mollusk-like fossil, have been found at the end of their trails. These trails consist of many overlapping scratches on biomat surfaces, showing how *Kimberella* lived a life of mat scratching, during which it evidently consumed nutrients from this abundant food source. Similarly, a couple of broad, flat, segmented fossils, *Dickinsonia* and *Yorgia*, are connected to a series of overlapping body impressions on biomat surfaces. Such traces, which look like someone took a stamp in the shape of the fossils and pressed it repeatedly into the rock, demonstrate that the animals moved. How they moved, though, is the mystery. Was it under their own power, in which they crept along biomat surfaces? Or did they periodically detach from the surfaces and let bottom currents drift them to the next spot, where they attached themselves anew? Regardless, these beguiling traces are also likely from feeding, in which their makers may have absorbed nutrients from underlying mats.

All animals have to eat. Thus it is not surprising that the oldest trace fossils of animal activity show they were either eating or moving, and the latter may have been motivated by food. (Or sex, depending on Ediacaran priorities. And if so, these would have constituted the very first traces of looking for love in all the wrong places.) However, any list of Ediacaran trace fossils inevitably reveals a big problem when trying to compare them to modern animal traces: Almost none qualify as burrows. Moreover, most Ediacaran burrows identified thus far are simple, horizontal, meandering tunnels. They are also quite shallow, not plumbing depths beyond a centimeter. Their reflected behaviors were quite simple: Seemingly all animals were only going from one place to another, eating, or both. Such superficial attitudes are completely unlike those of burrowing animals today, many of which follow the mottoes of "Deeper is better" and "I want it all." So what was preventing animals from going farther below? Was it a matter of not having the right anatomical tools, and the genes for these traits, or was something else excluding them?

The key suspects are burrow-blocking biomats, which took full advantage of Ediacaran animals' inability to break through. As discussed before, the rocks tell us that biomats were nearly ubiquitous in Ediacaran seafloors, especially in shallow-water areas where sunlight energized their photosynthetic bacteria and algae. These biomats acted like vacuum-packed sealants for sediments below, so unless accidentally ripped by storm waves or bottom currents, they kept animals above. For those few animals that won the biomat lottery and found a way to get underneath torn surfaces, they then faced another challenge, which was not being able to breathe. Just like plastic wraps used to cover leftovers in your fridge, biomats keep out oxygen. These low-oxygen zones made sediments underneath them a paradise for anaerobic bacteria, but a deadly place for animals. Given such restrictions, animals were stuck in a nearly 2-D world, ruled by the whims of microbial slime.

Once animals began burrowing farther down into the seafloor, though, all biogeochemical hell broke loose. Burrowing animals that punched through biomats and went down more than a centimeter abruptly (and quite rudely) introduced oxygenated water from above to sediments that had rarely encountered this element. The storming of this subsurface Bastille meant that anaerobic bacteria, which had ruled the sedimentary underworld for nearly four billion years, screamed, shriveled up, and died. Well, not completely, as these bacteria are still around today, with many of them living happily ever after in our gastrointestinal tracts. But still, anaerobic bacteria became far less common in shallow-marine sediments, restricted to smaller patches of those environments. The resources of the underworld were too rich and tempting to stay out of reach forever, and its colonization was on.

Freed of such barriers, animals switched from being just grazers to homesteaders, and then to miners. The homesteaders, instead of sticking to just above or just below a biomat, became pioneers by picking spots on the seafloor and digging vertical burrows in which they lived as permanent homes. In such burrows their bodies were largely surrounded by sediment, but their mouths stayed near the surface, connected to their food sources. Most of these vertical burrowers were probably suspension feeders, sucking in fine, suspended materials in the water and digesting them. Miners, on the other hand, began plumbing and plundering the organic goodness of the depths, ingesting sediment that had previously been forbidden fruit. Also, all animals poop, meaning that burrowing introduced their bountiful, nutrient-rich feces to the subsurface, radically processing and chemically altering sediments that had never enjoyed the ecological fringe benefits of animal wastes.

This oxygen-laden irrigation of sediments, setting up homes, mass consumption of organic compounds in the water and sediment, pooping, and overall exploitation of increasingly abundant food resources surely drove animal evolution in a myriad of directions toward the end of the Ediacaran and start of the Cambrian Period. Most importantly, though, this collective action changed seafloors throughout the Earth, jump-starting a massive sedimentary and geochemical recycling process that continues today.

Nevertheless, the preceding scenario assumes that animals evolved burrowing down simply because they were hungry. What if some other evolutionary pressure was at play, something that would have caused instant death to those animals that had not found the right genes and behaviors to break on through to the other side of a biomat or other surfaces? What if it was not just a matter of getting a nice home, good food, and adequate mates, but also avoiding harm? This consideration brings up a simple but difficult-to-answer question about the birth of burrows. Which came first: exploiting abundant food resources, or evading predators?

Death from Above? Then Burrow Below

The Ediacaran Period is sometimes nicknamed the "Garden of Edia-cara." This allusion to an Eden-like paradise was applied to an ancient underwater world because of the apparent paucity of predators in the Ediacaran. In other words, this was a time of innocence among animals and their photosynthetic companions. After all, the evolution of snakes was more than 400 million years in the future and oh-so-tempting apples—whether Red or Golden Delicious—took even longer to evolve after that. Nonetheless, toward the end of the Ediacaran, an evolutionary fall from grace took place. Soon Tennyson's "Nature, red in tooth and claw" became a norm for animal relations for the next 550 million years, with no sign of abating any time soon.

The oldest evidence of Precambrian predation comes from minute trace fossils in slightly bigger shelly body fossils. The body fossils, called *Cloudina*, look like a stack of tiny flat-bottomed ice cream cones, ranging from 1–5 millimeters (0.04–0.2 inches) wide and as long as 15 centimeters (6 inches). As in many Ediacaran animals, their biological affinity is uncertain, but they may have been related to marine polychaete worms. *Cloudina* was also among the first animals to secrete a mineralized shell, composed of calcium carbonate. However, a few of these shells are pierced by perfectly circular submillimeter holes, interpreted as drillholes made by an unknown predator. Alternatively, some non-predatory organism might have made these holes after the *Cloudina* animals died. Yet predation is the more likely explanation, as hole diameters increase in proportion with the sizes of drilled *Cloudina*, a mathematical relationship suggesting that their predators were picking on appropriately sized animals. Also, similar drillholes are very common in seashells today, such as those made by predatory moonsnails.

Shells, which we take for granted today in clams, snails, and many other animals, probably did evolve in the latest Ediacaran and were used as protection against anything that might want to eat their owners. However, shells may not have evolved just in response to predators, but also as a way to cope with changing ocean chemistry. For instance, once the oceans warmed after Snowball Earth, their waters contained greater amounts of calcium and bicarbonate in solution. Excess calcium can become toxic or otherwise cause discomfort for animals unused to it: Just ask people with kidney stones. Increased concentrations of these chemicals meant marine organisms somehow had to deal with them or die. Shells may have then evolved through natural selection, in which animals with the right genetic stuff got rid of calcium by precipitating it. These earlier shell makers presumably did this just like modern organisms, by combining calcium with bicarbonate in their tissues and forming calcium carbonate.

This biochemical trick of turning lemons into lemonade is called *biomineralization*. Today, two minerals of calcium carbonate—aragonite and calcite—are exceedingly common in algae and invertebrate animals, but especially corals, mollusks, or anything else one might call a "seashell." Soon after this type of biomineralization evolved, vertebrates then began making calcium phosphate. This mineral then composed internal skeletons consisting of bones, but most importantly from a predatory perspective, it also resulted in teeth. Regardless of how shells first evolved, these accouterments deterred animals that decided their food needed to be freshly squirming, rather than dead and decaying. This in turn meant that predators had to adapt to changing prey and develop their own means for getting past exoskeletons to the juicy stuff inside, which they did through teeth, claws, or other body parts. Skeletons were now not enough for defense. The arms race was on, and is still running.

The coevolution of predators and prey is perhaps best explained by a literary allusion, the "Red Queen Hypothesis." First proposed and named in 1973 by evolutionary biologist Leigh Van Valen, this hypothesis was named in honor of a character in Lewis Carroll's 1871 novel, *Through the Looking Glass, and What Alice Found There*. A close reading of the novel shows that the Red Queen is methodical and persistent, unlike the White Queen, who is odd but mostly innocuous, and unlike the Queen of Hearts, who is just plain crazy evil. (Unfortunately, film directors who adapted Carroll's novel to the big screen, including Tim Burton, conflated the Red Queen and Queen of Hearts, simply depicting them as a single Bond-style super villain.) While playing chess with Alice, the Red Queen assures her that she can start out as a pawn, and then through skillful playing eventually become a queen, neatly fulfilling a Horatio Alger–like fantasy. (Which Alice does, but let's just say this does not end well for the Red Queen, unless her ambition was to become a kitten.) Anyway, while playing chess with the Red Queen, Alice runs as fast as she can, only to stay in the same place.

This is an apt analogy for what happens between predators and their potential prey over the course of geologic time: When natural selection favors a predator's traits—such as speed, stealth, sharp claws or teeth—it also favors the selection of defenses against these traits in prey animals. The biggest difference in the coevolution of predator-prey relations, though, is that if a predator fails, it just goes hungry, whereas if a prey animal fails, it dies. This inequity means that prey have greater selection pressures applied to them than predators, as predators can still pass on their genes while malnourished, but dead prey cannot.

The evolutionary interplay between predators and prey thus connects to how small borings eventually led to big burrows. Prey animals today use a wide variety of defenses against predation, some of which are remarkably sophisticated: spiky bits, pinching parts, bad taste, camouflage, distracting spots and lines, great speed, flocking and herding, poison, and much more. Yet few defenses are quite as simple and effective as hiding. Also, rather than each generation of prey animals waiting for holes to randomly appear in the ocean floor and then slipping into them when trouble arrived, they instead evolved the ability to push aside sediment and conceal themselves in burrows.

Shells, or at least some sort of solid support, also helped with burrowing. Think of a snowplow that tried to use gelatin for its plow rather than steel, or a shovel made of cardboard, or a drill bit made of cookie dough, or, well, you get the point. In other words, soft, squishy parts did not work well for early animals attempting to go deep into ocean sediments. Until animals developed hard parts, they had to rely on peristalsis for moving their bodies through mud or sand. This meant having the proper internal hydraulics to anchor, expand, contract, pull, and repeat those actions. Hard parts broke the rules, as animals then carried their own spades on their heads, sides, and backs. Moving like icebreakers, they could more easily slice through biomats or other sediment surfaces and gain entry to profound (and safe) realms. Some of these formerly all-squishy invertebrates even developed hard parts near their mouths suitable for rasping biomats, or one another. For instance, some modern snails have a radula that scrapes algae off hard surfaces, whereas others use a radula to drill into clams and other snails. When burrowing began, did shells and mineralized exoskeletons in general still provide protection against predators? Sure, but they also paved the way for exploration and exploitation of places that had previously excluded all animals.

The idea that burrowing and biomineralization may both have originated around the Ediacaran–Cambrian transition in response to predation is relatively new, and was given a name: the Verdun Syndrome. This hypothesis, proposed in 2007 by Polish paleontologist Jerzy Dzik, derives its name from Verdun, France, where French and German troops during World War I fought a horrible battle against each other through much of 1916. The main strategies of French soldiers included the use of trenches, tunnels, underground shelters, and fortifications, which they used as their main defenses against German artillery occupying the high ground near Verdun. Like many World War I battles, its scars are still borne by the landscape, enduring traces of ferocious slaughter and mind-searing terror.

This battle thus serves as a metaphor for what happened evolutionarily during the Cambrian Period. In the Cambrian, besieged prey animals would have sought shelter below the seafloor surface, but while also hardening their bodies to withstand assault from above. In other words, the development of burrowing and mineralization may have happened simultaneously, each responding to the instant-selection pressures of predation. When the Red Queen traveled to the fields of Verdun, she also decreed that predatory advances must adapt to and surpass prey defenses. The resulting huge and relatively rapid changes in faunal behaviors and bodies may have then triggered what paleontologists call the "Cambrian Explosion," a twenty-million-year burst of animal diversification during the first part of the Cambrian Period. Burrowing and biomineralization may have even enhanced fossilization, as future fossils with hard parts also buried themselves deeper, greatly increasing their odds of preservation.

Yet another evolutionary innovation that likely drove animals to go underground was the development of eyes. Eyes are deservedly heralded in evolutionary biology as examples of convergent evolution, in which certain traits can emerge in many lineages of unrelated animals that have arrived at the same adaptation. As light-detection tools, eyes take in photons and translate these into shades, shapes, and other stimuli. This ability further led to selection of light-detection tools in animals that could properly translate those shades and shapes into "yummy food," "alluring mate," or "death-dealing predator." Genes for such abilities were accordingly sorted out, with much of this selection driven by predators and prey avoiding them.

Among the very first animals with eyes were trilobites. Trilobites, which had calcite-infused exoskeletons, show up as Early Cambrian fossils in rocks from about 540 *mya*. Trilobite eyes were aided by the development of calcite-crystal lenses and some became wondrously varied and complex. These included the first true compound eyes, present in many insects today, but quite rare early in the history of animal life. But soft-bodied trilobite ancestors with eyes probably preceded mineralized ones in the latest part of the Ediacaran, and other soft-bodied animals likely had eyes, too.

Still, no one has (so far) interpreted eyes in an Ediacaran animal. Perhaps these organs really were absent then, unnecessary in a world where no animal pursued another with the intent to kill it. No predation meant vision was not so necessary, whether for acquiring lunch or not becoming it. However, once vision became standard in most predators, prey animals needed other adaptations, and burrowing was probably one of the most valuable. Long before insects developed elaborate patterns and shapes to disguise themselves, or fish hung out in schools, or octopuses became shapeshifters, sediments would have provided instant camouflage: duck and cover.

Of course, once prey animals started using burrows to protect themselves against predators, their assailants did not respond by evolutionarily shrugging and saying, "Oh well, win a few, lose a few." Predators began burrowing, too, and may have even preceded Ediacaran animals in this behavior. In a 2016 study of Ediacaran body fossils and trace fossils from Namibia done by Simon Darroch and his colleagues, they noted vertical burrows sharing the same sediments with typical Ediacaran animals. They interpreted the burrows as those of anemones, and all anemones today are predators, using stinging cells to stun their prey. So not only were these animals burrowing down into sediments, they were likely eating the floating larvae of animals near them. This combination of a new ecological niche (predation) and habitat partitioning (going down) marked the ecological shift from the Ediacaran to the Cambrian, the start of animals as ecosystem engineers, and the beginning of the end of typical Ediacaran animals.

Once animals starting burrowing below, vertical burrows— rather than just hosting harmless suspension feeders—served as places for ambush predation, a strategy used today by everything from wolf spiders to bobbit worms (*Eunice aphroditois*). Also, as soon as predators developed mineralized tissues, these accessories allowed them to more easily probe substrates of all types, from soft to hard. Mineralized tissues were also used to develop weaponry: Shell-drilling, shell-crushing, and other nasty shell-assaulting behaviors emerged from this extra skeletal support. Indeed, it was not long before animals employed a combination of mineralized bodies and acid to bore into rocks or into one another. The latter especially exemplified the Red Queen Hypothesis, which through borings became more exciting. At the point when animals began drilling into one another—as evidenced by *Cloudina*—prey animals needed more defensive strategies than mere body armor: hence the development of better burrows, which worked in the Cambrian and still do today.

The Revolution in Evolution

The nascent burrowing abilities that evolved in animals toward the end of the Ediacaran Period, whether for finding food or mates, making homes, eluding predators, acquiring prey, or whatever other purposes, ended up having global consequences. As Proterozoic "matgrounds" on shallow-ocean bottoms gave way to burrowed Phanerozoic "softgrounds," big changes happened in both animals' evolutionary lineages and marine environments, changes we live with today.

Once animals developed new real estate in previously pristine sediments below, they launched an unprecedented elemental exchange. Based on Ediacaran trace fossils left by mobile animals, biomats that used to protect sediments from the nefarious beasties above became food. Given the right scraping and digging tools, animals ate or otherwise tore into these mats, turning glass-bottomed boats into window screens. Billions of holes, gouges, and fissures in these mats then allowed in ravenous Visigoths from above. Meanwhile, formerly protected sediments below gave up their organic goodies to these raiders, some of which, like modern earthworms, consumed the sediment as deposit feeders. Holes made by biomat rending and burrows also opened up the floodgates to oxygenated waters from above, introducing an element that had been mostly banished from the subsurface. In essence, this biologically driven stirring and mixing literally overturned the rules for a planet where life knew its place—the ocean floor—for nearly four billion years.

This invasion of the underworld, however, was not a one-way street, but a chemical swap meet with a cast of millions. Formerly anaerobic sediments below biomats, once oxygenated and mixed by animals, became full participants in a conveyor belt that altered elements and compounds as they shifted from one realm to another and back again. With burrowing animals acting as catalysts for these new chemical reactions, major geochemical cycles of carbon, nitrogen, sulfur, and phosphorus in the oceans were no longer separated, but blended. For example, sulfur went from mostly reduced states (no oxygen) to oxidized states (yes oxygen), and began moving in and out of ocean waters above the surface. The evolution of more suspension-feeding animals living in sediments took elements out of the water column and deposited them on and in the seafloor as feces. Then burrowers industriously mixed these neatly packaged and chemically altered wastes back into the sediments. However grand it might seem, burrowing animals were at least partly responsible for the biosphere, geosphere, hydrosphere, and atmosphere becoming part of one unified earth system. In this sense, burrowing acted as a midwife in the birth of Gaia.

Once again, though, I must emphasize that this idea of all-encompassing and permanent change in oceanic environments driven by burrowing is not something I imagined after a Malcolm Gladwell–mandated 10,000 hours of research and pondering. So I must give credit where it is due. In an article published in 1994, German paleontologists Dolf Seilacher and Friedrich Pflüger first proposed this hypothesis and termed it the *agronomic revolution*. Other paleontologists since have also used the less catchy phrase "Cambrian substrate revolution" to describe this concept, but the meaning and import are the same. In this scenario, the development of novel feeding modes—deposit feeding, suspension feeding, and predation—led to a rejection of biomat-imposed hegemony of seafloors everywhere. This in turn was followed by intensive exploitation of previously virgin substrates (in this instance, sediments).

To support the notion of an agronomic revolution marking the end of the Ediacaran and the start of the Cambrian, Seilacher in particular had much authority on his side, as he was much greater than the sum of his parts. Yes, he was a paleontologist, and a fine one at that, but he was also widely acknowledged as the world's most brilliant ichnologist. At the time when Seilacher and Pflüger published their paper, he had studied fossil burrows and other trace fossils for nearly fifty years. Seilacher was also the opposite of a shy, retiring, reclusive professor. Instead, he vociferously promoted his provocative ideas while ferociously challenging those of others. Many of his concepts turned out to be mostly right, though at least a few were flawed. His mixture of genius intellect and occasional error was very good indeed, as Seilacher prompted much good science by daring other paleontologists to prove him wrong. As a result, more than twenty years of testing his hypothesis has fine-tuned it and made it more robust. Mental traces of Seilacher, who died in 2014, are still at the center of debates about Ediacaran life.

As in all revolutions there were casualties. Extreme alterations in the chemistry of the seafloor and oceanic waters caused by burrowing constituted bad news for many organisms maladapted to such changes. These doomed organisms included the life that dominated the Ediacaran Period, such as the spindles, fronds, discs, and other now-compellingly outlandish forms. Soon after the beginning of the Cambrian Period, nearly all Ediacaran-style biota vanished. Apparently none survived for more than a million years into the Phanerozoic Eon, evidence of their lives fading just when all other life became more visible, and visible to one another. The presence of vertical burrows in Ediacaran rocks of Namibia signals the start of the disappearance of the previously dominant life forms, evidence of how the first mass extinction recorded in the fossil record may have taken place. If so, then it is also the first mass extinction connected to burrowing, a biological tragedy that is particularly ironic considering how burrowing has saved animals from problems ever since.

Nonetheless, once the revolution established a new regime early in the Paleozoic Era, smaller revolutions rippled out with each evolutionary shift in sea-bottom faunas. The Cambrian Period was when animals began burrowing deeper, but not nearly as deeply as some around today. For example, a live televised burrowing match between a Cambrian trilobite and a modern burrowing shrimp would be like pitting a kid with a toy shovel against a construction worker with a backhoe. Sure, the trilobite/kid might eventually carve out some artistically intriguing (albeit ingenuous) holes in a sandbox, perhaps earning an ego-coddling award ceremony and a cookie. But none of those holes would approach the size, depth, and complexity of the burrowing shrimp's/backhoe's diggings, nor the sheer burrowing power and efficiency that produced such excavations.

The evolutionary path between the relatively weak burrowers of the Cambrian and present-day earthmovers may have been a gradual descent, but a few researchers have described it more like going down a flight of stairs, with discrete steps. For instance, based on the fossil record of burrows, Cambrian seas had only a few deep diggers, such as trilobites and a few other animals, but the Middle to Late Ordovician (about 470–440 *mya*) had more, including the first burrowing crustaceans. During the Silurian Period, following the recovery of life after the end-Ordovician mass extinction, invertebrates began mixing oceanic sediments on a massive scale. The greater burrowing abilities of marine invertebrates in turn predisposed their descendants to dig in estuarine and freshwater environments and then dry land. Once arthropods and other invertebrates began burrowing in marginal-marine environments and continental environments during the Ordovician, then continued into the Silurian, they were later joined by insects and vertebrates (such as lungfishes) in the Devonian. Accordingly, freshwater ecosystems and terrestrial soils experienced their own drastic changes in chemical cycling, more closely resembling the intensely burrowed landscapes of today.

Nonetheless, probably the biggest step backward in the evolution underground during the past half billion years was the end-Permian mass extinction. Following this extinction, burrowing abilities or other traits saved a few animals from total annihilation.

But for the first 5 million years afterward, during the Early Triassic (252–247 *mya*), burrows were clearly reduced in both number and depth, as were all animals. This even allowed biomats to make a brief comeback. Given far fewer grazers and burrowers around to disrupt them, microbial communities were free to grow and spread, covering shallow-marine and lake bottoms once more. Indeed, some paleontologists have compared Early Triassic environments to those of the pre-predatory and pre-burrowing Ediacaran, perhaps east of Eden.

The survival of a few burrowing animals and at least a few predators among them still guaranteed that subsurface lifestyles would once again rule watery bottoms and soils. Sure enough, Middle Triassic rocks (247–235 *mya*) have much more evidence of burrowing, just preceding the evolution of dinosaurs, mammals, and other important animal clades that produced their own burrowers. Two more mass extinctions—at the end of the Triassic and Cretaceous—would have less of an impact on burrowing animals, which had by then gained a permanent presence below. Burrowing, which brought about the end of the Ediacaran Period and its unique biota, was here to stay.

As a result, the saga of those supporting actors of early animal life continues, in that some of their descendants became directors and producers, altering and controlling ecosystems across the surface of the planet. This encompassing view of underground animals shaping the course of environments and other organisms' evolution is, however, not original at all. Good old Charles Darwin envisioned the pervasive effects of burrowing animals on environments as part of the grand scheme of evolution. And just what animals did he choose to reflect this legacy of evolution being driven from below our feet? Those lowly, legless, constantly burrowing animals that we barely notice in everyday life but which we could barely live without: earthworms. Accompanying earthworms on this journey toward the center of the earth is an all-star team of fantastic invertebrate burrowers who we will now honor for their roles in making the world what it is today.

FIGURE 16. Folk-art rendering inspired by the *Lystrosaurus* saga set during the Permian-Triassic transition (Chapter 5), with a cutaway view of a *Lystrosaurus* burrow. (Artwork by Ruth Schowalter and Anthony J. Martin.)

FIGURE 17. Skeleton of *Lystrosaurus* displaying its stout forelimbs, shovel-like face, and other adaptations for digging. (Photo courtesy of Staatliches Museum für Naturkunde Stuttgart.)

FIGURE 18. Helical burrow of the Early Triassic cynodont *Diictodon*, based on descriptions and figures from Smith (1987) and photos of natural cast of burrow in Iziko South African Museum, Cape Town, Republic of South Africa.

FIGURE 19. Underside of a small trilobite burrow, with tracks (BELOW) and lower body impression (ABOVE), preserved as a natural cast in Late Ordovician sandstone of Georgia; scale in centimeters. (Photo by Anthony J. Martin.)

FIGURE 20. Meandering trilobite burrows standing out in positive relief, originally preserved in Early Ordovician sandstone of Spain. Pictured here is an epoxy resin cast made from the original burrows that was part of Dolf Seilacher's *Fossil Art* display. This one he titled *Trilobite Pirouettes*. (Photo by Anthony J. Martin.)

FIGURE 21. Trackway of a juvenile horseshoe crab (*Limulus polyphemus*) leading to a depression, marking its hiding place during low tide on Sapelo Island (Georgia); scale = 1 centimeter (0.4 inches). (Photo by Anthony J. Martin.)

FIGURE 22. Close-up of juvenile horseshoe crab exhumed from its burrow on Sapelo Island (Georgia); scale in centimeters. (Photo by Anthony J. Martin.)

FIGURE 23. Large invertebrate burrow preserved as a natural cast in Late Silurian sandstones formed by river channels, in Victoria (Australia) and with the author measuring it. (Photo by Ruth Schowalter.)

FIGURE 24. (LEFT) Cross sections of probable lungfish burrows from Late Devonian fluvial (river) deposits in Pennsylvania, in which burrows were filled with sand from above; Swiss Army knife for scale. (Photo by Anthony J. Martin.) FIGURE 25. (BELOW) Lungfish estivation burrows in longitudinal section with lungfish curled in central burrow and mudcracks on the surface. (Figure by Anthony J. Martin.)

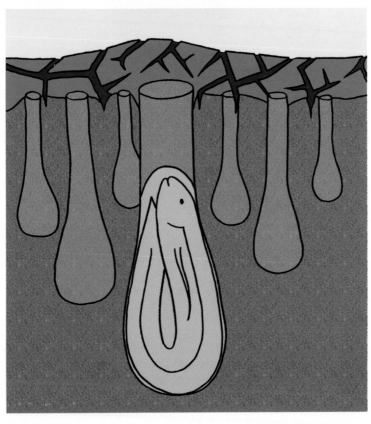

FIGURE 26. (LEFT AND CENTER) Two burrows of tiger salamander (*Ambystoma tigrinum*) with "W" and "Y" shapes expressed by plaster casts; (RIGHT) Early Permian amphibian burrow from Kansas, with shape expressed as natural cast; scale = 5 centimeters (2 inches). (Figure by Anthony J. Martin, but drawn after photos by Dzenowski and Hembree (2014) and Hembree *et al.* (2005).)

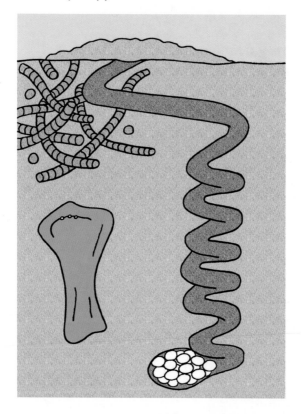

FIGURE 27. Modern vertebrate burrows in Australia: (LEFT) Burrows of marsupial moles (*Notoryctes* sp.), showing back-filled internal structure. (RIGHT) Spiraled (helical) nesting burrow of yellow-spotted monitor lizard (*Varanus panoptes*), the deepest of any known modern non-human vertebrate; Cretaceous sauropod humerus for scale, 1 meter (3.3 feet) long. (Figure by Anthony J. Martin, but marsupial mole burrows based on research by Bensehmesh (2007) and monitor lizard burrow after illustration by Doody *et al.* (2015).)

FIGURE 28. Trail made by a small animal (probably a soft-bodied anemone) on a deep sea bottom 565 *mya*, preserved in sandstone at Mistaken Point, Newfoundland (Canada); Paleontologist Barbie for scale. (Photo by Anthony J. Martin.)

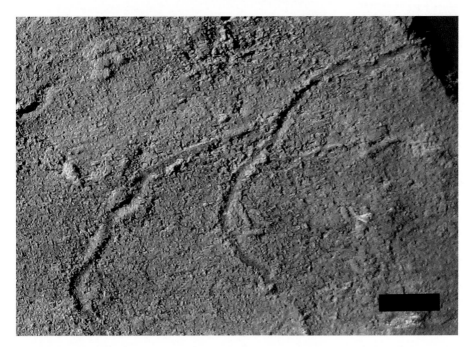

FIGURE 29. Trace fossils made by small, worm-like animals in 545 *mya* Ediacaran mudstone from North Carolina that were originally made as surface trails or shallow burrows that were later eroded; scale = 1 centimeter (0.4 inches). (Photo by Anthony J. Martin.)

FIGURE 30. U-shaped vertical burrow in cross section of Early Cambrian sandstone from 542 *mya*, Fortune Head, Newfoundland (Canada); scale in centimeters. (Photo by Anthony J. Martin.)

FIGURE 31. *Treptichnus pedum*, a complicated feeding burrow made by an animal repeatedly probing into the seafloor, preserved as a natural cast in sandstone, Newfoundland (Canada); scale in centimeters. (Photo by Anthony J. Martin.)

CHAPTER 8

Rulers of the Underworld

Exploring the Earth with Earthworms

"When we behold a wide, turf-covered expanse, we should remember that its smoothness, on which so much of its beauty depends, is mainly due to all the inequalities having been slowly levelled by worms."

—Charles Darwin (1881), *The Formation of Vegetable Mould through the Actions of Worms with Observations on Their Habits*

The "beetle trackway," annoyingly enough, had a live earthworm at its end. My misdiagnosis of the trace and its maker, both spotted on a sandy road near the Georgia coast, exemplifies one of the many reasons why I love doing field-based science. No taxpayer-funded research grants, labs, post-docs, journal articles, persnickety peer review, or other trappings and complications of modern science are needed for me to be wrong. All I require are animals behaving and making traces in their environments, and sometimes helpfully showing exactly who makes which trace. As a result, mistaken identities and findings can be corrected quickly, while also reminding me to be open-minded whenever encountering a phenomenon assumed to be familiar.

This earthworm trail and others like it were definitely novel to me. The traces were located within the Wormsloe Historic Site, a state-managed property south of the picturesque city of Savannah, Georgia. Accompanying me was Craig Barrow, the former owner of Wormsloe, who was also lucky enough to witness my inept trace-maker identification. Wormsloe was originally a plantation dating back to the eighteenth century, maintained during the brutal and inhumane use of slave labor in the American South. However, following the Civil War, Emancipation, and Reconstruction, its agricultural purpose waned, and it became more of a preserve. Now its grounds host much historic and scientific research, including at least one experiment in ecological restoration. During my visit there in October 2014, secondary-growth maritime forests, growing in what used to be plantation soils, were being cleared and replaced by wiregrass and longleaf pine communities. As mentioned previously, these gopher tortoise–driven communities were far more widespread in the southeastern U.S. before colonial times. This restoration effort meant that gopher tortoises and their hundreds of burrow roommates—exiled since the 1700s—would soon return to Wormsloe.

The night before, my wife, Ruth, and I attended a dinner party hosted by Craig and his wife, Diane, and we stayed overnight in a nearby guest cabin. That morning, I got up just before dawn to go tracking and otherwise look for fresh animal signs; meanwhile, Ruth began her daily creative practice of drawing. Given the low light, I only saw a few raccoon and deer tracks, as well as a fresh pile of scat that only recently exited a feral hog's digestive tract.

Low tide had exposed the banks of a nearby salt marsh, its dark mud pockmarked by thousands of holes. Most of these belonged to mud-fiddler crabs, mud shrimp, polychaete worms, and other burrowing invertebrates. A mild breeze brought sulfur-tinged air venting from burrows that penetrated anoxic muds below.

My sojourn ended up at the former plantation house to meet Craig, who was waiting to go tracking. As an outdoor enthusiast, he had happily given Ruth and me personal tours of the woods, fields, and marshes during previous visits to Wormsloe. Because Craig has hunted for most of his life, he is a good tracker, and we have had in-depth discussions on animal tracks and scat. I find such conversations with experienced naturalists refreshing. Academic hierarchies, journal impact factors, and granting agencies go away, and we can instead focus on traces and what they can teach us. After exchanging a few pleasantries, Craig and I hopped onto his golf cart and took off, scouting for tracks and other signs.

Craig had already been out earlier on the sandy roads near his house. In the predawn light he saw three red foxes run in front of him, so he wanted to look for their tracks. Within minutes, we arrived at the spot, got out of the cart, and I quickly confirmed his sighting by pointing to fresh fox tracks in the loose sand. This was when Craig asked me a question that I answered wrong at first, then corrected once I gathered more data—you know, the way every good scientist should. His question—"What's this?"—referred to a thin, shallow, and meandering groove in the sand. "Beetle tracks," I answered instantly, but without looking closely enough. I then squatted to show him the regular sets of tiny tracks that should be on each side of the groove, where I imagined that a beetle dragged its abdomen. Yet there were no tracks. This was a trail made by a legless animal, something more akin to the Ediacaran than today.

"Wait a minute, this isn't from a beetle," I said. "Maybe a worm?" And by "worm," I really meant "earthworm," but my modest experience with identifying earthworm traces prevented me from elaborating any further. After all, I did not want to appear too ignorant about such exceedingly common animals. Fortunately, an earthworm with the common decency to be at the end of one of these trails, moving and otherwise actively demonstrating how its traces were made, saved me from further embarrassment.

With our eyes and brains instantly trained by this coinciding of trace and tracemaker, Craig and I glanced around us and were rewarded with astonishment. The road was crisscrossed by hundreds of earthworm trails for as far as we could see, most of which had living worms at their ends. Intriguingly, some of these trails connected to open, narrow vertical burrows. My second insect-biased mistake of the morning was to think these burrows were the shafts of burrowing bees; like beetles, these insects were also common in the area. Nonetheless, because the earthworm trails repeatedly connected directly to the holes, I changed my mind again. The earthworm burrows showed us exactly where the earthworms exited their subterranean homes before making trails.

What really surprised me, though, was the length and complexity of the trails. These were not simple meandering paths, but complicated records of earthworms making decisions. Many trails branched from multiple probes, and lateral movements produced wider trails that looked much like meander scars left by rivers, telling us exactly where the worm had turned. Wormsloe worms may have been slow, but their traces were not dull. Yet a larger question loomed over these ichnological minutiae: Why were these worms on top of the ground instead of in it? What could have impelled so many to leave their homes and risk the perils of dehydration and avian predation at the surface?

I speculated aloud to Craig that vibrations might have caused this mass stranding of earthworms on the surface. Earthworms associate such tremors with burrowing moles, mortal enemies that can easily eat hundreds of worms per hour. Craig affirmed this idea, recalling how he graded the road the previous day; hence the vibrations from his grader may have sent a false warning and persuaded the earthworms to come up and out. Nonetheless, I later wondered whether another and much broader stimulus had invoked such aversive reactions in the earthworms. What else could have caused these earthworms to flee their underground habitat in such massive numbers?

Then it came to me. A full moon that weekend caused higher tides than normal, ranging from 2.6 to 2.9 meters (8.5–9.6 feet), and the high tide peaked only a few hours before dawn. As we tracked in the maritime forest and along the roads of Wormsloe, it was easy to forget that the extensive salt marshes of the Georgia coast were never more than a few hundred meters away, and that salty groundwater was just below us. Given such high tides, this marine-flavored water would have crept upward into the local water table, going high enough in the soil profile to trigger an adverse and collective reaction in the earthworms. Earthworms, after all, are on strict low-salt diets, and object strenuously to saline water filling their burrows. Assuming this hypothesis was correct, what we saw on that sandy road was a perfect example of a marine environment forcing animals in a terrestrial ecosystem to drastically change their behaviors, while leaving ichnological evidence of this mass movement.

The numerous earthworm trails and burrows on the roads of Wormsloe also served as a reminder of the teeming but hidden abundance of burrowing earthworms below us, living in soils from coastal forests to mountaintops. Consisting of about 7,000 known species (but possibly as many as 30,000), earthworms are broadly classified as annelids, and more specifically as oligochaetes. Other annelids include leeches (Hirudinea), which mostly live in fresh water, and bristleworms (Polychaeta), which primarily dwell in marine settings, like those in the salt marsh next to Wormsloe. Earthworms, which have claimed terrestrial soils as their own, rarely compete ecologically with their annelid cousins.

Oligochaetes are distinguished further from other annelids by a few anatomical traits. All annelids possess obvious segments divided by little rings (*annuli*) and complex internal digestive, nervous, and circulatory systems, that run the length of their bodies. Oligochaetes, however, have short, stiff projections outside of their bodies called *setae*, which provide traction while moving through soils or on the surface. Oligochaetes also tend to have a well-developed internal body cavity (*coelom*), which they use to expand and contract their bodies along their lengths. This helps them push aside sediment and pass through it without actually shoveling. As they move through soils, they line and lubricate the burrow walls with mucus, an endearing sliminess experienced by anyone who has handled a live earthworm. As earthworms move, they can swallow whatever is in front of them, digesting inorganic particles and organic detritus alike, then excreting feces out the other end. Look closely at the ground one or two days after a hard rain where rising groundwater has forced earthworms onto the surface, and you will see tiny grape-like clusters of these feces, calling cards of their short journeys above.

Like many topics in natural history we find fascinating today, earthworms attracted the attention of Charles Darwin, who studied them with his typical mix of passion and meticulousness. Darwin's last book (and one of his most popular) was *The Formation of Vegetable Mould through the Actions of Worms with Observations on Their Habits*. Published in 1881, it summarized what was then known about the biology and behavior of earthworms, but also shared results of long-term experiments Darwin conducted on earthworms in his backyard.

As many biographers have noted, Darwin was a homebody after his formative years of voyaging on the H.M.S. *Beagle*, with home being a spacious two-story dwelling just south of London. In 1842, after moving to Down House (named after the nearby village of Down, now spelled "Downe"), he and his wife, Emma, used it as a place to raise their large family. After his world travels, Darwin never ventured more than 50 kilometers (31 miles) from home and stayed there until his death in 1882. Thus Down House and its grounds became his personal research facility. Here he wrote *On the Origin of Species* (published in 1859) and other important books, papers, and letters, and he also learned about earthworms, a subject that absorbed him for more than half of his life.

Darwin began thinking intensively about earthworms in 1837, when his uncle showed him how layers of lime and cinders sprinkled over his soil fifteen years before were now buried. How? Darwin reckoned earthworms might be responsible, and this idea stuck with him. Once moved into Down House, he immediately began conducting behavioral experiments on worms, while repeating his uncle's inadvertent experiment by depositing chalk and cinders in the backyard. Darwin went a little farther, though, by using flat stones for research equipment.

During a pilgrimage to Down House in 2012, I was thrilled to see one of these stones behind the house, a talisman of a simple but elegant test for assessing the effects of burrowing effects of earthworms. Darwin suspected that the surface just outside Down House and elsewhere was relatively new, having been formed by generations of earthworms overturning the soil. To do this, he used a precise measuring device—invented in 1870 by his engineer son, Horace—and flat, circular rocks set out in the fields behind Down House, nicknamed "wormstones." Darwin then started measuring just how much earth an earthworm could worm. Using Horace's invention, he recorded the height of the wormstone surface above the surrounding soil surface. Any changes in their relative horizons were discerned by fitting Horace's device on three metal slots added to a central hole in the wormstone. Metal rods inserted through the hole were connected to the underlying bedrock, ensuring that these would stay stationary as worms churned the surrounding soil. The rods thus acted as a horizontal datum that could be compared to any changes in the ground surface.

Going into this long-term experiment, Darwin already knew that earthworms burrowed through, consumed, and excreted sediment as fecal casts, which also mixed and chemically altered soils. He further figured that their burrows aerated the soil, allowing passages for oxygen to reach sediments below. Most important, though, Darwin imagined how the burrowing activity of earthworms underneath each stone, as well as fecal castings left at and near the surface, should cause them to "sink" over time: In essence, worms would bury the stones from below. Using his geologically inspired sense of time and rates of processes, Darwin also rightly projected that the daily activities of earthworms, multiplied by millions of worms and enough years, change the very ground underneath our feet in a way so that it, well, evolves. These stones ultimately led him to propose that these so-called "lowly" worms had actually changed the surface of the earth.

Darwin was right. Based on his measurements taken over nearly twenty years, he calculated an approximate "sinking" rate of 2.2 millimeters/year, which was also an indirect measure of soil deposition caused by earthworm feces. Extrapolating these results further, he estimated that earthworms could move 6.8–16.4 metric tons (7.5–18 tons) of soil in a typical acre (0.4 hectares) of land. In short, the land was not static, but in constant motion, reinventing itself via earthworm guts.

As one of the originators of modern evolutionary theory, Darwin has been called many things, but one label he did not live to see was "ichnologist." Yet he was apparently the first scientist to quantitatively measure rates of bioturbation, showing how these rates, when coupled with enormous lengths of geologic time, meant that earthworms could completely alter landscapes. As a geologist (and a fine one at that), he held a long-term view of how small, incremental differences like these, happening year after year, eventually added up to big changes. Or, to put it in Darwin's own words when he responded to (and totally owned) a critic who claimed that earthworms were too insignificant to have a broad effect on their surroundings:

> Here we have an instance of that inability to sum up the effects of a continually recurrent cause, which has often retarded the progress of science, as formerly in the case of geology, and more recently in that of the principle of evolution.

In his book, Darwin also described and showed illustrations of earthworm traces, such as their burrows, estivation chambers, fecal pellets, and turrets. He describes their burrows going as deep as 2–2.5 meters (6.5–8 feet), with many ending in enlarged chambers, where he presumed worms curled up and overwintered. These burrows and chambers also contained seeds, hinting at the powerful role of earthworm in burying these and otherwise helping seed-bearing plants to germinate and propagate. Darwin even noted the orientations and species of leaves earthworms pulled into their burrows to plug them.

Although soft-bodied earthworms have a poor body fossil record, we know through molecular clocks and fossils that annelids likely originated just after the Ediacaran Period, and modern-style earthworms evolved before the end of the Cretaceous. In the Paleozoic Era, oligochaetes descended from a branch of marine polychaetes, and body fossils in Permian rocks (about 260 million years old) give us a minimum time for when such annelids began living in freshwater and terrestrial environments. Earthworm trace fossils also help fill gaps in their history as burrowers. For example, in a 2007 study that surely would have pleased Darwin, Mariano Verde and several other South American paleontologists described fossil examples of fecal pellets and estivation chambers from geologically young (less than 1 million years old) Pleistocene rocks of Uruguay. Estivation chambers point toward how earthworms behave much like some lungfishes and amphibians during droughts. Worms make these life-sustaining spaces by pasting fecal pellets around them, like laying down bricks and mortar. Once sealed in, they slow down their metabolism and go into diapause. Chambers, which are connected to burrows as escape routes, keep their earthworm occupiers moist even as the soil outside dries. As noted by Darwin, these chambers also serve as refuges from cold temperatures, allowing earthworms to live through cold winters, even near the poles.

Did such strategies work for earthworms in the geologic past? Apparently so. In 2013, ichnologist Karen Chin and a few colleagues documented fossil earthworm burrows and feces from just above the end-Cretaceous boundary. This evidence showed that earthworms, by staying below, survived in enough numbers to begin altering the soils of a post-extinction world. Also, if any modern-day robin or mole could speak, it would tell us how much it appreciates the caloric and nutritional value of earthworms. In other words, earthworms that survived the mass extinction at the end of the Cretaceous enabled soils to recover while also supplying smaller, carnivorous vertebrates—such as burrowing mammals and birds—with enough food so that they could live and propagate.

Fast-forward to the twenty-first century, when modern biologists, using tools that Horace Darwin could not have imagined outside of a steampunk fantasy, documented more of the awesome effects of earthworms. In a 2016 study by Anne Zangerlé and her colleagues, they proposed that earthworm burrows and feces were responsible for widespread and mysterious mounds in Colombia. Called the *surales*, these tightly packed mounds, which are as much as 2 meters (6.6 feet) tall and about 5 meters (16 feet) wide, are also numerous and extensive, covering tens of thousands of kilometers along Orinoco River floodplains. The biologists used a combination of satellite imagery, aerial drones, and good, old-fashioned, down-and-dirty field work to determine that local earthworms (*Andiorrhinus*) created the mounds, and by extension, the surales. Members of this genus not only make up more than 90% of the earthworm biomass in the surales, they are also quite large: Individual worms are as long as a meter (3.3 feet) and can make 5-centimeter (2-inch) wide fecal casts. Mounds built upward by these earthworms' fecal casts subtracted from the surrounding land, as the worms transferred massive amounts of soil from the areas between mounds. The result was contrary to the flattening effects assumed by Darwin for English earthworms, yet still exemplifies earthworm terraforming prowess.

So if you go out later today, whether at a place aptly named "Wormsloe" or not, pay attention to the ground beneath you, and think of how it reflects an ichnological landscape, a result of collective traces made by those "lowly" earthworms. Given that more than 130 years have passed since Darwin's death, you might also think of how the descendants of the same earthworms he watched with his appreciative and discerning eyes continue to alter the ground outside Down House. This perspective cultivates an appreciation for a different kind of descent with modification, one that reflects an ichnological worldview appreciated and well articulated by Darwin.

Moreover, these earthworms remind us how they and the small, invertebrate burrowers of today, as well as those of the past 65 million years, are all descended from those that made it past a dinosaur-deleting mass extinction. What other descendants of small invertebrate burrowing animals are still here, which through their burrowing move huge volumes of sediment, dive deeply into the earth, or reshape landscapes and seascapes? The following is a salute to a few of these all-star modern burrowing invertebrates and the considerable effects of their burrowing.

Hail Ants

The preceding subtitle is a reference familiar to fans of the long-aired animated TV show *The Simpsons*, and one that later generated a cultural meme. For those who are not familiar with it or require a synopsis, here it is. In the episode "Deep Space Homer" (1994), Homer, the recurrent hero/buffoon of *The Simpsons*, is recruited and trained as an astronaut to ride into orbit on a U.S. space shuttle. While there, he accidently shatters an onboard ant farm that is being used to study the effects of zero gravity on ants. One of the freed ants passes close to a camera transmitting the mishap, lending a forced perspective and giving the false impression that gigantic (and presumably alien) ants have invaded the space shuttle. This is followed by brief panic on Earth, then a prompt prognosis by a newscaster that the ants would soon enslave everyone. In a sign (literally) of acquiescence, a hastily drawn poster with the phrase "HAIL ANTS" appears behind him, and he declares, "And I, for one, welcome our new insect overlords!" He then pledges his avail-ability to recruit other humans to toil in "underground sugar caves." (This was an apparent reference to burrowing honeypot ants of the genus *Myrmecocystus*, in which some workers store food in their abdomens.) Since this episode, the meme "I, for one, welcome our new [fill in the blank] overlords!" has entered our culture, and is used to signal compliance with a perceived threat.

Although ants in this instance were the source of a recurring joke, we should accept that these insects really are our overlords. Moreover, they are not new, either, but have ruled subsurface and near-surface continental environments for minimally the past 100 million years, rivaling earthworms in their domination of soils. Thanks to the research and enthusiastic public-outreach efforts of entomologist and ecologist Edward O. Wilson, many people know other mind-boggling facts about ants. One is that their biomass at any given moment on Earth is equal to that of our species, with more than one million ants per person. In tropical areas, ants may represent as much as 25% of *all* animal biomass. Represented by more than 12,000 known species, with 20,000 species likely, ants are diverse, occupying and adapting to nearly every available niche in terrestrial environments. Body shapes within each ant species also vary because of how labor is divided through specialized duties. Yet these different bodies can unify instantly, enabling a colony of millions to react and otherwise behave with one mind like a single organism. Thus if an entomologist states that, say, ants are " . . . arguably the greatest success story in the history of terrestrial metazoans [animals]," this is not hyperbole. It is real.

Perhaps the most impressive of ant feats is how some species can form so-called "supercolonies," composed of tens of millions of ants covering vast areas. For example, human-aided dispersal of Argentine ants (*Linepithema humile*) caused this species to make underground nests in all continents outside of Antarctica. Furthermore, Argentine ants maintain big, closely related colonies while widely separated. In a study published in 2009, entomologists reported that mas-sive Argentine-ant colonies in California, the southern coast of

Europe, and Japan were biochemically indistinguishable from one another. Because ants exchange information through biochemicals (pheromones), this meant that ants from Japan could be introduced to ants from Europe or California and still communicate perfectly well with one another. (In contrast, imagine a Japanese software engineer, Spanish flamenco dancer, and California surfer all instant-messaging *and* understanding one another.) The European colony alone spans more than 6,000 kilometers (3,700 miles) along the Mediterranean coast, and colonies in Japan, California, and other parts of the world are likewise extensive. This intercontinental supercolony is the largest known animal society.

Given their domination of places where we live, there is no need to wonder why ants have entered our consciousness in many ways. I remember first becoming intrigued with them when I was six years old, lying on my belly and watching them in the backyard for hours instead of cartoons on TV. I also went to my local public library and checked out many books about ants and other insects, devouring them like a bivouac of army ants. In fact, the preceding metaphor connects to a seared childhood memory of reading the short story "Leiningen Versus the Ants."

Originally written and published by Carl Stephenson in 1938, this story takes place in an Amazon rain forest with army ants as its main characters and antiheroes, in which through cooperative action they outsmart and overwhelm a stubborn plantation owner. (Unfortunately, I did not watch the movie version of this story—*The Naked Jungle* [1954]—because the title said nothing about ants, while also sounding far too forbidden for a prepubescent Catholic boy in the 1960s.) The main theme of this tale, that massive numbers of individuals cooperating as one can defeat us, was updated later when the TV series *Star Trek: The Next Generation* (1987–1994) introduced a new species to its universe: the Borg. These cybernetic humanoids, which assimilated other species and lived in massive cubic spaceships, were clearly modeled after ants. This analogy was further encouraged in the movie *Star Trek: First Contact* (1996), which introduced a Borg queen. In 2015, the Marvel Comics superhero Ant Man also got a movie, imaginatively named *Ant-Man*, whose miniaturized protagonist often enlists ant allies while saving the world. Picnic-ruining aside, ants fascinate us and are part of our lore.

How did ants accomplish their domination of both land and our imaginations? Way back in the Cretaceous Period, just after they descended from their wasp ancestors, ants developed *eusocial behavior*, or *eusociality*. A species is considered as eusocial when it fulfills three main criteria: (1) It lives in large groups that stay together for more than two generations; (2) adults care for young in these groups; and (3) the group divides labor and reproduction, so that only a few individuals breed, whereas most others do not. Ants express the last of these through castes performing different tasks in a colony, such as soldiers ("Defend the nest and queen!"), workers ("Build and maintain the nest!"), fertile males ("Mate with the queen!"), and the queen ("Make more soldiers, workers, and fertile males!"). Owing to form following function, castes often look quite dissimilar, meaning eusocial ant species are polymorphic. Eusociality is rare in most animals, but common in insects. All termites and nearly all ants are eusocial, as are thousands of species of wasps, bees, beetles, and a few other insect clades. Other than insects, only a few species of snapping shrimp and mole rats are eusocial. Eusociality is thus special, and once it became common in ants, there was no stopping them.

As eusocial animals, nearly all ant species have one behavioral trait in common: nesting, in which all other castes ensure that their queen produces more ants. Furthermore, although some ant nests are made above ground, most are below. Ants, however, do not burrow in the traditional sense. The way they make room for their logic-defying nests is by removing one grain of soil after another, creating voids. Worker ants, which are nonreproductive females, normally use their jaws to transport these particles—clay, silt, sand, pebbles, seeds, grass, bits of other animals, or other materials—to the surface, making anthills. Cleared spaces in a nest are sculpted into vertical shafts, horizontal to obliquely oriented tunnels (sometimes called galleries), and chambers. At least some of the vertical shafts go to the surface, where soldiers and workers can take out particles and refuse or bring in food.

In some ant nests, vertical shafts also help to circulate air through the colony. Turrets on the surface built around shafts on the surface direct oxygen-bearing winds down into the nest (think intake vents), while warm, carbon dioxide–rich air rises out of other turrets (think chimneys). Chambers are reserved for the queen(s), housing eggs, larvae, and pupae, storing food, or farming. Yes, some ants can grow their own food. Examples of such ants include leaf-cutter ants, consisting of species of *Atta* and *Acromyrmex*. These ants, which are mostly native to Central and South America, leave the nest to cut chunks of fresh grasses or leaves with their jaws, carry these bits back home, and place them in large, hemispherical chambers. Fungi grow on the plant pieces and are then fed to much of the colony, meaning the ants make gardens that work in complete darkness.

The farming ability of leaf-cutter ant colonies leads to their constructing some of the most astonishing subterranean structures made by any animal. This awareness was increased by a 2004 study, in which several Brazilian entomologists investigated leaf-cutter ant (*Atta laevigata*) nests by pouring cement into them. (Not to worry, ant lovers: They picked abandoned nests, so they did not purposefully smother and entomb millions of ants.) In one nest, decanting required several days, with more than 10 tons of cement flowing down a shaft and out of sight. Once the cement hardened, the field crew used a backhoe, shovels, buckets, and sweat to excavate the casting. The nest covered more than 50 square meters (540 square feet) and went more than 7 meters (23 feet) deep; it was expressed in three dimensions as an intricate network of connected shafts, tunnels, and chambers, resembling a fantastic sculpture. Relatively thin tunnels ran for tens of meters throughout, with many branching off into bulbous chambers, looking like balloons tied to a series of poles. This colony built nearly 2,000 chambers, although other investigated *Atta* nests had more than 7,000. Chambers that could hold as much as 51 liters (13.5 gallons) served as fungus gardens, trash bins, or nurseries.

Based on the volume of cement needed to cast the nest, entomologists estimated the ant colony removed about 36 metric tons (40 tons) of soil, which would require at least three normal-sized dump trucks to move. The leaf-cutter ants, though, redistributed and dispersed most of this soil to the surface. Seeing that this was just one of many leaf-cutter ant colonies in that area of Brazil, the immense impact of these animals on the soils in their ecosystems cannot be underestimated. This research and the considerable human labor required to do it was partially captured in a documentary film, *Ants: Nature's Secret Power* (2006), giving them and this ant nest much-deserved recognition. A three-minute-fifteen-second excerpt from this film, later posted on YouTube in 2010, reached more than 10 million views, rivaling the number of ants that built the nest.

Other entomologists have cast less grand but still deep and complex ant nests, although with a substance not so benign as cement: molten metal. Researchers chose this seemingly unlikely casting medium after experimenting with other materials, such as plaster of Paris and dental plaster. These substances, however, did not go deep enough into the nests, and often were too fragile, breaking up easily once dug out. Metal, however, solved the dual problems of both depth penetration and integrity. Ant biologist Walter Tschinkel perfected this technique by using either melted zinc or aluminum for casting ant nests. To do this, he and his assistants set up a makeshift smelter next to the ant nest, melted the metal, and poured the dangerously hot liquid down the nest entrance. (Yes, this method causes irreversible harm to the ants, but some were much-loathed imported fire ants (*Solenopsis invicta*), which deserved it.) Once the metal cooled, Tschinkel and his field assistants dug out the casts. These silvery casts faithfully rendered nest architectures, produced as delicate and intricate three-dimensional "sculptures." Any one of these beauties could be slipped into a modern art museum and readily accepted by entomologically unaware patrons as the latest creative endeavor of a Chihuly-influenced metalworker.

What Tschinkel and others since found through casting ant nests, whether with plaster or metal, is that colonies express many variations on a main theme. The theme is twofold: Vertical shafts connect its chambers and give a nest its depth, but horizontal chambers give a nest its volume. In some ant species, the shafts are straight-down pipes leading to chambers, whereas in others they spiral clockwise or counterclockwise on their way down. Chambers range from small to big, equal to unequal, flat to bulbous, separate to overlapping, regularly spaced to densely packed, and all combinations of these. Still, the basic plan of each nest can be reduced to shafts and horizontal chambers. For example, nests of Florida harvester ants (*Pogonomyrmex badius*) can reach depths of 4 meters (13 feet) and have four or five spiraling shafts, each of which may be connected to dozens of chambers, creating an exquisitely complicated structure. Yet these nests are basically composed of shafts and chambers. Even enormous leaf-cutter ant nests can be described as a series of shafts (some of which segue into tunnels) and chambers.

Given these insights about modern ant nests, we can predict the forms of their trace fossils and better understand when ants evolved from relatively innocuous eusocial insects to planetary conquerors. Based on molecular clocks and ant body fossils, some of which are beautifully preserved in amber, we know that ants shared the same terrestrial ecosystems with dinosaurs during the Cretaceous. Yet ants somehow survived the end-Cretaceous mass extinction. How? Trace fossils of subsurface insect colonies from Late Cretaceous rocks of Utah provide an intriguing clue. These underground nest structures, attributed to either ants or termites, had galleries linked to horizontal chambers, a complexity consistent with eusocial insect nests. Sadly, the identity of these and other purported Mesozoic ant nests is not yet accepted by most ichnologists, but fossil ant nests from the Paleogene Period (66–23 *mya*), which followed the Cretaceous, closely resemble those of modern ants and become more common.

Entomologists and evolutionary biologists agree that the advent of eusociality in insects, whether it happened first in termites or ants early in the Mesozoic Era, must have required a combination of adequate shelter and food resources to protect a reproductive female (a queen "wannabe"). Furthermore, protection and food also would have had to last long enough for the "queen" to overlap with and ensure the survival of generations of nonreproductive worker females and fertile (but otherwise useless) males serving the prime female. Underground nests neatly fulfill both of these needs, providing both shelter and the capacity to store or grow food. Ant nests also allow these insects to inhabit horrifically hostile environments, from deserts to mountain slopes to mall parking lots in North Dakota. A few ant species have even managed to nest in intertidal zones, such as mangrove swamps. In this respect, these are among the few insects adapted to areas bathed daily by marine waters. Eusociality and living in subterranean nests works for ants, and will likely continue to work no matter what happens at the Earth's surface.

Indeed, some evolutionary biologists view ant nests as an excellent example of an extended phenotype. A phenotype is the outward physical expression of the genetic potential (genotype) of an organism, but one that has also interacted with its environment. This means that genes do not always result in perfectly predictable forms, such as how identical twins become more distinctive from one another throughout their lifetimes. With a "superorganism" like an ant colony, though, its phenotype goes beyond its "body," expanding into and modifying its surrounding environment to better suit the colony. In this sense, an ant nest not only reflects the genes of an ant species, but also shapes its environment, which in turn affects the evolutionary paths of plants, animals, and other organisms sharing the same ecosystem altered by the nest.

Given what we now know, the evolution of eusocial behavior and the takeover of terrestrial environments by ants probably began underground toward the latest part of the Mesozoic Era. If Cretaceous ant colonies living below the earth's surface could have laughed, most would have at least chuckled at end-Cretaceous volcanism, meteorite impact, and mass death above them. Consequently, ants and their extended phenotypes of nests would have expanded considerably during the Paleogene and adapted to new niches left in the aftermath, surviving and thriving. Once adapted, they would have extended their phenotypes into almost all terrestrial ecosystems, changing soils throughout the earth. So here we are today, living in a terrestrial world dominated by our ant overlords, thanks to their going below in the Cretaceous.

Deeply Digging Decapods: Crayfish and Lobsters

Mention crayfish, lobsters, crabs, and shrimp to most people, and the usual response to these prompts are "seafood." In my experience, the word "burrowing" does not enter the conversation, unless I bring it up (which I do). Yet crayfish, lobsters, crabs, and shrimp all include representatives that make magnificent burrows. These invertebrates are classified as decapod crustaceans (where *deca* means ten, *poda* means footed), which bear eight walking legs and two clawed front arms. Among decapods are burrowers that affect their environments in ways analogous to earthworms or ants, driving nutrient cycles while moving enormous volumes of sediment. So decapods are not just menu items, but can be ecosystem engineers.

Let's start with the first course, crawfish étouffée. Crayfish, which have been nicknamed crawfish, crawdads, mudbugs, mud puppies, and yabbies, share an ancestry with lobsters, having diverged from those marine decapods either late in the Permian or early in the Triassic, about 250 *mya*. Although crayfish body fossils are rare in Mesozoic rocks, their trace fossils show how these decapods, once evolved, must have quickly adapted to and spread throughout continental environments. Crayfish today consist of more than 600 species on widely separated landmasses—North and South America, Europe, Australia, New Zealand, and Madagascar—but are most diverse and abundant in the southeastern U.S. and in Australia. Crayfish are also animals that live between two worlds. Because they have gills, they need to stay in contact with freshwater environments, whether these are streams, lakes, or groundwater. Yet many also frequently come onto land, with some species getting much of their food from terrestrial environments.

How do crayfish accomplish this? Burrows, of course. Crayfish burrows are masterworks of adaptation, constructions that allow them to visit the harsh world above and go back below after conducting their business. Although not all crayfish dig, most do, or can, implying that this is a primitive behavior shared with their Mesozoic ancestors. Burrowing crayfish are further divided into three categories: primary, secondary, and tertiary. Primary burrowers make burrows in terrestrial environments and live in these as permanent homes. Secondary burrowers live in water bodies such as lakes and streams, but make burrows on their banks or bottoms. Tertiary burrowers also live in water bodies, but dig only when they must, such as when their lake or stream dries up.

To dig, crayfish use their claws as shovels or sculpting tools, pushing aside and compacting sediments as they move through the ground. Modern crayfish burrows typically have at least one vertical shaft connecting a circular opening to the surface; multiple openings form "Y" or "W" shapes toward the burrow top. Burrow shafts are sometimes marked at the surface by a pyramidal tower ("chimney") of mud balls, rolled up and placed there by the crayfish. The shaft, lined with mud compacted by crayfish moving up and down it, descends to the local water table.

Crayfish burrow depths depend on the water table, so crayfish burrows can go shallow or deep; burrows of the prairie crayfish (*Procambarus hagenianus*) are as deep as 4 meters (16 feet). Below this shaft, burrows branch out into a complex network of horizontal tunnels, galleries, and chambers. This is where crayfish eat, reproduce, fight, and otherwise hang out. Meanwhile, cold winters, hot summers, floods, droughts, fires, volcanic eruptions, meteorite impacts, war, or other horrendous disturbances could be taking place on the surface. Crayfish, ensconced in their burrows, remain oblivious to all such trivial matters.

Thus it comes as no surprise that burrows attributed to crayfish in Late Triassic, Late Jurassic, and Cretaceous rocks show this shelter-building ability was selected for early in their evolutionary history and persists through today. These trace fossils also act as fossil divining rods, telling geologists exactly where the local water table was located, or even how it fluctuated with seasonal changes in water supply, such as during monsoons (burrows went up) or droughts (burrows went down).

Crayfish are becoming more recognized for their role as ecosystem engineers in some ecosystems, altering ecosystems through their burrowing. For example, over the course of a year, Camp Shelby burrowing crayfish (*Fallicambarus gordoni*) of Mississippi can move an estimated 80 metric tons per hectare (180 tons per acre). The shafts, tunnels, and galleries of its burrow systems also cover 30–50 kilometers (19–31 miles) of distance in each hectare. This sort of "piping" affects groundwater flow, which preferentially follows these underground conduits both sideways and down. Where abundant, crayfish towers also concentrate clays at the surface, changing the composition of near-surface soils. Their towers, incidentally, can be extremely numerous. In a 1991 study done in east Texas, crayfish towers numbered more than 63,000 per hectare (157,500 per acre), representing more than 40 metric tons per hectare (90 tons per acre) of soil moved, just by crayfish. Burrows also introduce oxygen to the subsurface, while providing refuges for other animals that can comfortably fit in a burrow. In short, burrowing crayfish can significantly change their environments.

Crayfish adaptability through burrowing is even a factor in how some species successfully invade non-native habitats. For example, the red swamp crayfish (*Procambarus clarkii*), which normally lives in the southeastern U.S., is now scattered throughout the world thanks to humans. In one instance, red swamp crayfish in Lake Naivasha (Kenya) expanded their range by taking advantage of hippopotamus tracks. Hippo tracks on land filled with water became little "ponds" for the crayfish, into which they crawled; from these, they burrowed down to the water table. (Did crayfish of the past likewise occupy and burrow into sauropod dinosaur or mammoth tracks filled with water? The only way to know for sure would be to find these trace fossils directly associated with one another.) The lesson here is that when crayfish are placed in unpredictable environments, burrowing gives them more options for shelter and food. This adaptability gives us a few clues about how their lineages survived mass extinctions at the end of the Triassic and Cretaceous Periods.

Similar to their crayfish brethren, lobsters also slip into hidey-holes of their own making. Lobsters are accomplished diggers, constructing voluminous homes into the seafloor, from near-shore environments to the edges of continental shelves. These decapods, which belong to the clade Nephropodidae, consist of more than fifty species, with most under the genera *Homarus*, *Nephrops*, and *Metanephrops*. The species that shows up most often in North American restaurants is the American lobster (*Homarus americanus*), and in Europe, the Norway lobster (*Nephrops norvegicus*). All lobsters bear two large, pre-buttered claws in front and smaller claws on the ends of three of the four pairs of walking legs. Biologists distinguish these as "true" lobsters, as opposed to other large marine decapods lacking such imposing claws, such as slipper lobsters and spiny lobsters. Although bested by Japanese spider crabs (*Macrocheira kaempferi*) as the largest living crustaceans (and arthropods), lobsters are likely the weightiest, with the record thus far set by a 20-kilogram (44-pound) specimen of American lobster (*Homarus americanus*) captured in 1977. To grow so big, lobsters are long-lived. Those that successfully evade human predation can be more than 70 years old, outliving some Maine lobster fishermen.

Lobster burrows tend to be roomy, with some nearly twice as wide as their makers. This spaciousness is likely an accommodation for their outsized claws and contrasts with the claustrophobically tight burrows formed by crayfish and many other decapods. Making room for claws also explains why their burrows are often wider than tall, resulting in an oval cross section. Owing to their size and strength, lobsters are also capable of burrowing into a broad range of substrates, from clay to cobbles. Like their crayfish cousins, they use a combination of their claws, walking legs, and tails to loosen, push aside, and compact sediment.

Their burrows, though, differ functionally from crayfish burrows because lobsters always live underwater and hence are heedless about reaching a water table. Because most lobsters are omnivores but include a good amount of live prey in their diet, they mostly need their burrows as places to stay out of trouble, molt their exoskeletons, and occasionally ambush prey. As a result, lobster burrows do not have to go very deep, but some have lengthy tunnels and may get complicated. Basic shapes of lobster burrows vary with each species and are influenced by substrate, but can be as simple as J-shaped (one entrance) or U-shaped (two entrances), perhaps with an enlarged chamber somewhere along its length. For instance, Norwegian lobster burrows are typically J-shaped and extend about 30 centimeters (12 inches) below the seafloor, but can be more than a meter (3.3 feet) long. Other burrows, however, can have more than a dozen entrances, leading to a series of interconnected tunnels. Lobsters also construct ventilation shafts, thinner vertical holes punched through the roof of a burrow to allow in oxygenated water.

American lobsters prefer to burrow into sand underneath rocks, the latter lending them solid roofs; their tunnels reach 30–40 centimeters (12–16 inches) deep and more than 70 centimeters (28 inches) long, with multiple entrances attached to the main tunnel. However, where they burrow into loose mud, the same species goes 60–80 centimeters (24–32 inches) down, with sheer, near-vertical shafts leading into the main tunnel. In firmer muds offshore, American lobsters carve out depressions rather than tunnels; these pits are as deep as 1.5 meters (5 feet) and 5 meters (16 feet) wide. Craters this size, far exceeding the depth and width of a typical kiddie wading pool, may even host several lobsters.

Other than providing a home base, another clear advantage of lobster burrows is how these structures help them cope when winter is coming, and when it arrives. For example, species of *Homarus*, which live in cold, northern climates, seal off their burrows and become inactive for weeks or months during the winter. Even in summer, lobsters tend to stay in their burrows throughout most of the day, saving their foraging forays and other adventures for nighttime. The extent of such extra-burrow activities is seasonally affected, as extended summer daylight means less time spent out at night. Furthermore, lower temperatures mean lowered metabolisms, and thus less need for food, causing lobsters to imitate Montreal residents by becoming less active and staying snugly below the surface in winter. Juvenile lobsters also reduce their chances of not becoming dinner for other predators by spending most of their growing-up days in burrows and probably feeding on any animals dumb enough to reside in their burrows.

Like crayfish, lobsters have been digging burrows for a long time, at least since the Early Jurassic (about 200 *mya*), but probably starting in the Triassic Period. Early Cretaceous (125 *mya*) lobster burrows in Portugal actually contain fossil lobsters, leaving little doubt as to their makers. Other Cretaceous lobster burrows I have seen in Portugal were so clearly preserved, I could count the four pairs of walking legs where their bodies pressed into burrow walls. Despite the evolutionary divergence of crayfish and lobsters more than 250 *mya*, their fossil burrows indicate that both clades either retained or developed this burrowing behavior early in their evolutionary history. These trace fossils, along with the prevalence of burrowing in so many modern crayfish and lobster species, suggest their common ancestor was also a burrowing decapod. Did this hypothetical subsurface predecessor originate during the Permian Period, just before the greatest mass extinction in the history of animals, or just before the Late Triassic mass extinction? Regardless, because its descendants also continued their lineage after the end-Triassic and end-Cretaceous extinctions, the readiness of this grandmaster decapod for disaster ensured that we would see them today, whether in their burrows or in our pots.

Around the World with Crab Holes

Unlike crayfish, most burrowing crabs do their handiwork near or in an ocean, with a few overlapping ecologically with lobsters, and some even sharing ecosystems with crayfish. Although all modern crabs descended from marine species, different species may inhabit terrestrial, freshwater, coastal, shallow-marine, and deep-marine environments. This flexibility means that crabs as a clade (Brachyura) are quite diverse, comprising nearly 7,000 species around the world. What crabs still have in common, though, is that they must return to the water to spawn. So for those crabs that live in the margins between ecosystems, with wide variations in temperature, food, salinity, or personal safety, they must burrow and have been for a while. The earliest fossil burrows attributed to crabs are in Middle Jurassic rocks, dating at about 170 *mya*. Because the oldest known fossil crabs are from the Early Jurassic (about 190 *mya*), these trace fossils suggest that crabs, like crayfish and lobsters, may have been burrowing from the start.

Of the burrowing crabs, by far the most abundant worldwide—and among the most comical to watch—are fiddler crabs. Represented by more than a hundred species under the genus *Uca*, the "fiddler" part of their common name comes from male crabs bearing an oversized claw (the "fiddle/violin") and moving the smaller claw near this, giving the appearance of drawing a bow across its strings. In such movements, they are not playing "The Devil Went Down to Georgia," but feeding on detritus, which they scrape off muddy or sandy surfaces near their burrows. At less than 5 centimeters (2 inches) wide, most fiddler crabs are small, but their communal effects on coastal ecosystems, such as salt marshes, mangroves, and tidal flats, are enormous. Mostly because they graze on algae and detritus, as well as produce many closely spaced and deep burrows, fiddler crabs are often ecosystem engineers, cycling nutrients and aerating sediments that normally would be anoxic. Fiddler crabs consequently function as a rough equivalent of earthworms in coastal ecosystems throughout much of the world.

Like many other burrowing animals, fiddler crabs use their burrows to avoid becoming food for other animals, but also to maintain stable temperatures, prevent dehydration, and make more fiddler crabs. Burrows vary widely among species, but have circular cross-sections, are steeply inclined, and can plunge as deep as 75 centimeters (30 inches). Burrow entrances sometimes extend upward as turrets, which either slow water flow into burrows or deter other crabs from popping in uninvited. Once below the surface, burrows have shapes similar to the letters "Y," "J," or "L," but added branches and enlarged chambers complicate these apparent attempts at literacy.

Because of fiddler crab abundance and industriousness, their burrows frequently intersect one another or are taken over and modified by more than one crab. Both home improvements and new homes can be achieved quickly, as fiddler crabs are fantastically efficient burrowers, using their claws, legs, and carapaces to clear out sediment and compact burrow walls. However, they also will take advantage of free holes wherever available. On the Georgia coast, I have even seen where sand fiddler crabs (*Uca pugilator*) burrowed into claw marks of fresh alligator tracks. Within just a few hours, these busy little crabs nearly erased the monstrous footprints with their digging.

Fiddler crab burrows can also be the sites of much entertainment for humans who love to anthropomorphize arthropods. Watch a group of male fiddler crabs outside their burrows for a while, and you soon will be rewarded with a *Magic Mike*-like show of macho braggadocio. To attract lady crabs, male suitor crabs stand by their burrows, bounce up and down like they are doing deep knee bends (only with eight legs), and wave their large claws in the air. With such moves, each crab is clearly communicating that his claw—and only his—is the biggest. If a potential mate is successfully wooed, his next move is to get her into his burrow and to his bachelor pad, or what ichnologists blithely refer to as a copulation chamber.

Mating might also happen in burrows made by female fiddler crabs, although this may not be consensual. In a 2016 study, Christina Painting and several other biologists documented how Australian banana fiddlers (*Uca mjoebergi*) use burrows as "mating traps." Male fiddler crabs, which earn their common name by possessing big, yellow claws, applied what the researchers termed a "ladies first" mating strategy, in which a female crab goes into a burrow, then is followed by a male. The larger male crab then traps the female in this confined space, where mating then becomes her only choice for escape.

Burrow-aided coercion aside, considering how vulnerable crabs (or most other animals, for that matter) can be during mating, burrows later ensure that eggs get fertilized. Baby fiddler crabs later hatch from those eggs, and burrows again help to protect them both before and after hatching. In short, take away their burrows, and fiddler crabs would quickly go extinct, triggering trophic cascades in coastal marshes and mangroves throughout the world.

Closely related to fiddler crabs are ghost crabs. These crabs look similar to their fiddler cousins but are noticeably bigger. They also are camouflaged to match sandy beaches and coastal dunes whenever they travel outside of their numerous and lengthy burrows. Furthermore, both male and female ghost crabs have one large claw, so their sex differences are not so obvious. Most ghost crabs are represented by the genus *Ocypode*, which has twenty-one species living on temperate to tropical beaches throughout the world.

If you are walking along a sandy beach and have a hankering to do some ichnology, do not look for these crabs, which blend so well with their backgrounds. Instead, keep an eye out for their burrow openings, which range from a few millimeters to more than 10 centimeters (4 inches) wide, and with most denoted by fresh tracks and stacked piles of sand balls. When burrowing, ghost crabs use their claws and legs to scratch and pull together sand into balls, which they carry under a couple of walking legs and dump just outside of their burrow entrance. Sometimes they fling the sand outward with a flicking motion, spreading it out in a distinctive stellate pattern around their burrow.

Ghost crab burrows can go deep. Those made by the largest crabs are often more than a meter (3.3 feet) long and steeply inclined, plunging at 30–50-degree angles. Similar to crayfish burrows, depths depend on the local water table, although this level fluctuates daily because of tides, with crabs trying to stay above the saturated zone. Tides also prompt crabs to move up and down in their burrows. The deepest part of a burrow keeps a crab moist during low tides, whereas the highest part keeps it from drowning during high tides. Burrows have either just one opening, serving double duty as an entrance and exit, or two openings next to each other. Burrows with single openings are "J" or "L" shaped, whereas those with paired openings look like a "Y," connecting just below the surface before joining into one shaft. The lowermost end of the burrow is wider, allowing a crab room to turn around or accommodate a guest (if you know what I mean). Burrow widths and depths vary directly with crab size, which also correlate with their distance from the sea: Bigger burrows are farther from the surf, whereas smaller ones are closer. This spatial segregation makes sense not just because bigger crabs dig bigger burrows, but also because they can stay out of their burrows longer than little crabs, which dehydrate quickly away from wet sand.

Like all decapods that burrow, ghost crabs use their homes as safe places from predators, and also for regulating temperature, preventing dehydration, and reproducing. In more temperate climates, crabs use burrows to overwinter by plugging the top and staying below for as long as six months. Burrows also allow large populations of ghost crabs to live in and on beaches frequented by humans. This means that crabs can retain their ecosystem services as the largest invertebrates living along their coasts, where they prey or scavenge from the surf zone to the dunes on a wide variety of animals, from small clams to insects to sea turtle eggs to dead fish. With their burrowing, they also rival (or rather, augment) nesting sea turtles with their mixing and aerating of coastal dune sands, which helps turtle eggs breathe.

Many subtropical and tropical areas with ghost crabs have other burrowing crabs further inland. Generically called "land crabs" or "terrestrial crabs," most of these are true crabs, belonging to the same decapod clade as fiddler crabs, ghost crabs, and swimming crabs that involuntarily contribute body parts to crab legs, crab cakes, and she-crab soup. Most land crabs are terrific burrowers, pockmarking forest floors and other landscapes with their burrows on warm-weather islands or coasts. For example, in the Bahamas and Caribbean islands, only concrete or buildings stop these crabs from poking holes in the ground and leaving piles of soil behind.

The majority of land crabs belong to the clade Gecarcinidae, which includes species under at least six genera. In the Bahamas, the two species' burrows I have seen most often are those of the blue land crab (*Cardisoma guanhumi*) and the purple, red, or black land crab (*Gecarcinus ruricola*). The latter is also intriguingly called the "zombie crab," a nickname related to its nighttime habit of emerging from burrows at night to look for food and seek mates. (Like a teen vampire.) These crabs and their relatives dig big and down. Blue land crab burrows can be nearly 20 centimeters (8 inches) wide and reach 3–4 meters (10–14 feet) long. Burrows typically have more than one entrance, branch below ground, or overlap to create a series of mazes, but also have large, open chambers at their ends. These chambers often partially fill with water, like having an indoor swimming pool to control burrow humidity.

Like ghost crabs, land crabs use their burrows to shield them from the sun and keep moist during the day. However, they differ in how their soil moisture often comes from fresh water, not oceans; some land crab burrows are more than 300 meters (1,000 feet) above sea level, where even tsunamis will not reach them. The only time these otherwise terrestrial crabs leave their burrows to go to the ocean is when they have the urge to merge, prompting a mass exodus.

Residents of Christmas Island in the Indian Ocean are lucky enough to witness such epic migrations of the Christmas Island land crab (*Gecarcoidea natalis*). The island has an estimated crab population in the tens of millions, and seemingly all leave their burrows and go for a stroll across the island when it is "business time." Male crabs try to get to the shore first, where they dig temporary burrows to use as beachfront motel rooms that rent by the hour. When female crabs arrive, at least a few are enticed into the burrows for some gamete mixing and shaking. Once eggs are fertilized, mother crabs go to the nearby shore and drop them into the water by the millions, which hatch into swimming larvae. A few weeks later, at least a few of these larvae survive to become baby crabs, which come back to the island, walk to the interior, and start burrowing.

Sharing Christmas Island with these land crabs and spread throughout the south Pacific is another important burrowing crab, the coconut crab (*Birgus latro*), which is also known as the robber crab. (The latter nickname is because this blue-and-red crab is notorious for pinching people's belongings, not their bottoms.) The coconut crab, however, is not a true crab, but one of the hermit crabs, which has different walking legs and abdomen compared to true crabs; hermits belong to the clade Anomura. Biological classifications aside, the coconut crab is still worth mentioning as a decapod burrower because it is also the largest living land arthropod, with some weighing 3–4 kilograms (6.6–8.8 pounds) and having leg spans as much as 1 meter (3.3 feet).

The coconut crab makes appropriately sized holes in the ground, which it does by using huge and powerful claws to scoop out sandy areas. This immense invertebrate is so well adapted to land that it can climb trees, rip apart coconuts with its claws, and drown if immersed in water too long. Nevertheless, it needs to retain water, which it does by having a tough, thick exoskeleton, but also by using (you guessed it) burrows, which stay humid and relatively cool year round on tropical Pacific islands. Burrows are also good places for bringing and storing food, which consists mostly of coconuts and others fruits, but also includes dead animal parts, as well as live land crabs, which they yank out of their puny and not-so-safe burrows.

Coconut crabs' enormous sizes correlate with great ages. Amazingly, these crabs can live for more than a hundred years, making them among the longest-lived invertebrates. How do they accomplish such longevity? If you said "burrows," you may have been paying attention. Coconut crab burrows extend the lives of their makers by protecting them when most vulnerable. No, not during coitus in this instance, but when they molt. A coconut crab that needs to shed its skin excavates a meter-long and half-meter (1.6 foot) deep burrow with a terminal chamber about twice the volume of the crab, and then closes it, effectively concealing it from anything above. The crab that steps out of its old exoskeleton is a bit bigger, but also softer and weaker. Toughening may take as little as three weeks or as long as four months, during which a crab stays in its underground bunker until its armor is battle ready. To help fortify its skin, it eats the old one, which is rich in calcium and other restorative elements. Burrows are also used for mating, and female crabs afterward dig burrows near the ocean, where they can more readily deposit their fertilized eggs into the water.

Coconut crabs and their burrows were made more famous recently and improbably connected to aeronautical history by being blamed for the disappearance of famed aviator Amelia Earhart. Not for bringing down her plane, but for taking her remains into their burrows. During a 1937 around-the-world flight, Earhart, her navigator, Fred Noonan, and their aircraft went missing in the South Pacific. Subsequent search-and-rescue operations failed to find them, and in 1939, despite the lack of a plane or bodies, the pair was declared officially dead. However, in 1940 a British colonial commissioner, Gerald Gallagher, reported that colonists on Nikumaroro (Kiribati) found part of a woman's skeleton and several other items, including a sextant and a woman's shoe. Nikumaroro is a small coral atoll more than 3,500 kilometers (2,000 miles) southwest of Hawaii, and would have been en route for Earhart's next destination, Howland Island. Unfortunately, Gallagher died of a sudden illness in 1941, and no one seems to know what happened to the bones or items. Despite several expeditions to Nikumaroro since, no definite evidence of Earhart's former presence was found. However, the people associated with those expeditions suggested that the absence of Earhart's and Noonan's remains was because of coconut crabs, which lived on Nikumaroro then, and still do, in great numbers. Given their robust claws and hefty sizes, they easily could have sliced off and carried human bits and pieces down into their burrows. Could this grisly scenario have actually happened? Maybe. Did it happen? Not likely. Still, the burrowing coconut crabs of Nikumaroro helped to inspire this imaginative tale, in which they hid the last bodily remnants of the legendary explorer and her copilot in their lairs.

Ghost Shrimp Busters

Other decapods that deserve fame for their remarkable burrowing abilities are shrimp. Understandably, we normally associate the word "shrimp" with cocktails and other festive mealtime treats. But those shrimp and all others we consume are swimming shrimp, living most of their shrimp-net-shortened lives above the ocean floor. Real shrimp dig. These would be thalassinidean shrimp, with most of them nicknamed "ghost shrimp," presumably because they are rarely seen, but also because most are whitish and somewhat translucent. Like burrowing crayfish and crabs, thalassinidean shrimp are ecosystem engineers, modifying their environments to better suit them, which in turn affects other species in those habitats. However, when considering the great volumes of sediment they churn and process in shallow-marine environments, ghost shrimp are actually more comparable to earthworms, ants, and fiddler crabs in their impact. Based on fossil burrows in Brazil closely resembling those made by modern shrimp, thalassinidean shrimp have been affecting marine environments since at least the Early Permian (about 280 *mya*).

Which environments do these shrimp alter? They live in muds and sands of temperate to tropical shallow-marine ecosystems through much of the world, such as mangroves, lagoons, seashores, and shelves. These shrimp are seen far less often than ghost crabs, preferring to live their entire lives underneath marine sediments. This proclivity for remaining hidden from our exterior world is understandable when you consider their teeming abundance and the marvelous constructions of their burrows.

Take a stroll along sandy beaches of most of the southeastern U.S. during low tide, and you will likely see 5–15-centimeter (2–6-inch) wide, low-profile mounds dotting otherwise smooth or rippled surfaces on the lower part of the beach. Look a little closer, and you will notice tiny, dark cylinders collected on and around these mounds, looking much like chocolate sprinkles adorning a cupcake. Resist the urge to taste these enticements, though, as they are ghost shrimp fecal pellets, freshly flushed from their burrows. Watch these little "volcanoes" for a few minutes, and you may even see one "erupt" with a mixture of water, sand, and pellets emanating from a central millimeters-thin hole.

The hole is the very top of a thin chimney—or rather, vertical sewer—just below the wider, main part of the burrow system. The active cleansing you may witness reveals that a ghost shrimp is in its burrow and immediately beneath its chimney. The upper portion of its burrow is shaped more like a wine bottle, with the thin opening representing the former position of the cork (or screw top, depending on your tastes and budget), whereas the wider part is where the shrimp actually lives. Shrimp form these chimneys by pumping water in and out of the burrow, which they do by beating little leg-like parts (*pleopods*) under their tails. The pleopods flutter in one direction to direct flow into the burrow, carrying clays and suspended organic detritus to feed the shrimp, while also pulling in oxygen-rich water. Pleopods then flutter in the opposite direction to reverse the flow and eject feces. Although these chimneys are frequently eroded by waves and tides, they go down 15–20 centimeters (6–8 inches) before connecting to the main burrow, which is where the feeding and house-cleaning shrimp lives.

To get a sense of how many ghost shrimp occupy a beach, count the burrow openings in a given area and start multiplying. For example, on beaches in the southeastern U.S., these openings can exceed 400 per square meter (40 per square foot), all made by just two species, the Carolina ghost shrimp (*Callichirus major*) and Georgia ghost shrimp (*Biffarius biformis*). This means as many as 40,000 ghost shrimp could be in a hectare (100 square meters) of beach, or 100,000 in an acre. That is a lot of shrimp. Yet if not for these openings, you would never know a decapod metropolis was just below your feet. (Incidentally, because knowledge is power and smart is sexy, be sure to share this insight about ghost shrimp burrows with your partner during your next long, romantic walk on a beach. Or not.)

In contrast to its "bottleneck," the main part of a ghost shrimp burrow can descend to 3–4 meters (10–13 feet); fossil examples are as long as 5–6 meters (16–20 feet). Viewed from a human scale, this is deeper than most people's basements. But if a 10–15-centimeter (4–6-inch) long burrowing shrimp were human sized and it dug out a proportionally deep shaft, it would go 50–70 meters (164–230 feet) down, a well from which little Timmy could never escape, no matter what Lassie does. In some instances, shrimp burrows do not go straight up and down, but twist or spiral, adding even greater lengths to each vertical segment. The shaft may also widen along the way, swellings that serve as little roundabouts where a shrimp can change direction.

Vertical shafts, however, may only represent a small part of a ghost shrimp burrow system. For one, many species fortify burrow exteriors with rolled-up balls of muddy sand cemented with all-organic and all-natural shrimp spit (mucus, actually). These walls presumably slow erosion from waves while also deftly deterring hungry predators that might fancy a shrimp snack. The bumpy outer surface contrasts with a smooth muddy lining inside, pipes through which shrimp occupants slide up and down when feeding, irrigating, or tidying their homes. This inner tube is a tight fit for the shrimp, so its diameter can be used to judge the approximate size of its maker.

Go farther down to where the sand is fully saturated by the ocean's water table, and the burrow is still walled, but goes oblique, horizontal, and every which way but loose. Here the burrow segues into a complex, branching, interconnected system of tunnels, shafts, galleries, and chambers, a jungle gym of geometric complexity. At nodes where branches meet, the burrow widens, again providing enough room for a shrimp to turn around and move from one tunnel to another. Such systems are apparently where ghost shrimp stop being "rugged individualists" and get communal, as their burrows join one another, which abets mating. Living burrowers also may intersect with or reuse previously abandoned burrows, mixing newer and older systems. Such burrow systems thus resemble Cappadocian underground cities, connecting with one another far below ground.

Ichnologists began investigating ghost shrimp burrows and those of other coastal invertebrates in earnest during the 1960s, with much of that work done by Bob Frey and his graduate students on the Georgia coast. Because many of these burrows went so far down, and those made in saturated muds or sands below the low-tide zone often collapsed as soon as excavated, these researchers had to come up with creative solutions for studying them. Like entomologists who later poured molten metal into ant nests, they decided to cast these burrows. Rather than zinc or aluminum, though, these ichnologists used far less risky epoxy resin to make the casts. Once the resin hardened, the researchers simply dug out and washed off the casts. Still, some of the burrows were so deep that the resin did not reach the base of the burrow system, resulting in incomplete casts and understanding. Many other ichnologists and biologists have since cast many burrows from many thalassinidean species, a low-tech and inexpensive method that still provides a nearly complete snapshot of these remarkable formations.

Thalassinidean burrow casts, at first glance, resemble those of ant nests. Recall that ant nests have a basic plan: a vertical shaft with short horizontal to oblique tunnels connecting to chambers. However, shrimp burrows are far less predictable, varying not only among species but also among individuals within species. Environments with different substrates also affect burrowing behaviors, meaning that the same shrimp may make wildly diverging burrow shapes in, say, a firm mud versus loose sand. Fortunately, these challenges did not stop ichnologists and biologists, who have tried to classify thalassinidean shrimp burrows by combining architectural elements (how they look) with ecological purpose (how they function). For example, shrimp that ingest sediment (deposit feeders) are more likely to make lots of galleries. In contrast, shrimp that suck in suspended food out of the water (filter feeders) are more likely to form deep, branching, and reticulate systems. Making this diagnosis less simple are shrimp that imitate submarine leaf-cutter ants by filling burrow chambers with pieces of sea grasses. In a form of underwater gardening, shrimp storing these grass clippings likely encourage microbes to grow and nutritionally enrich these, like adding dressing to their salad.

Probably the most useful architectural classification scheme for thalassinidean burrows (that is, the simplest) came from Richard Bromley, one of the world's best ichnologists and a great admirer of burrowing decapods. Bromley suggested that we categorize their burrows in three ways: (1) mazes and boxworks, consisting of complicated branching and interconnected tunnels, shafts, and galleries that look like scaffolding; (2) spirals, in which the burrow systems become helical downward, although with many side branches emanating from the main tunnel; and (3) dendritic, in which the burrow system branches downward in a fractal-like way, looking like an upside-down tree. Does this method account for every variation of thalassinidean burrow out there? Of course not. But it does provide a basic framework for describing the bewildering myriad of burrow forms designed by these shrimp.

Aside from how their burrows look, the overall ecological impact of so many ghost shrimp doing so much feeding, burrowing, and pooping in temperate to tropical shallow-marine environments worldwide is currently incalculable, but can be summarized with one word: huge. For instance, deposit-feeding shrimp, which swallow sediment wholesale and digest the good stuff from it, pump out major volumes of feces and otherwise greatly disturb sediment, forming impressive conical mounds of sand around the tops of their burrows. In a study of a Bahamian burrowing shrimp, *Glypturus acanthochirus*, ichnologist Al Curran estimated that if given eight years of time to burrow, one *Glypturus* could potentially mix a cubic meter (more than 35 cubic feet) of sediment. Mounds formed by *Glypturus* attest to their substrate-altering power, with each spanning more than a meter (3.3 feet) wide and about a half meter (20 inches) tall. Burrow mounds number by the thousands in lagoons or bays of the Bahamas and other tropical areas of the world. In a biannual field course I teach on San Salvador Island, Bahamas (where hurricane-surviving iguanas dwell), my students and I have often snorkeled over these burrow mounds, where we marvel at their sizes and abundance. The awe-filled appreciation of shrimp-burrow ubiquity is accentuated at low tide, which reveals a lagoonal bottom entirely composed of white, coalescing sediment mounds, a wavy moonscape stretching to the horizon.

Owing to the potentially great volume and complexity of thalassinidean burrows, it is then easy to imagine how such systems—like gopher tortoise burrows—might also host numerous other species that decide to call these burrows home. This is indeed the case, as many different species of crustaceans, polychaete worms, small fish, and other animals are often found living in thalassinidean burrows. Some species are specially adapted to these safe havens, having coevolved with the shrimp over many generations of burrows. Thalassinidean burrows even affect microbial communities in oceanic sediments, with greater bacterial diversity within burrows than in the surrounding sediments. Shrimp burrows are a key factor in promoting the biodiversity of shallow-marine environments, from microbes to animals.

In summary, whenever you encounter a muddy area in a salt marsh, lagoon, sandy shoreline, or other coastal environment, do not think of how the clay particles in the mud settled out of the water like a gentle rain from heaven above. Instead, think of feces that gushed out from burrows below. All thalassinidean shrimp—regardless of whether they are filter feeding, deposit feeding, or farming—poop. What filter-feeding shrimp and other filter-feeding invertebrates do with their poop, however, is even more ecologically dramatic. By staying in their burrows and sucking in water with suspended clays, filter feeders concentrate these clays into little shrink-wrapped packages (feces bound with mucus), which get dumped on the seafloor as sand-sized particles. (Remember the "chocolate sprinkles" mentioned before?) This is how most mud in coastal environments gets deposited. Take away these shrimp and other filter-feeding burrowers, and say good-bye to mud as a part of that ecosystem. This in turn impacts all bottom-dwelling deposit-feeding animals that depend on mud for concealing and nutrition. Water quality would also suffer, as clays would stay suspended indefinitely in the water, causing permanent murkiness. So the next time you swim, snorkel, or scuba-dive in beautiful, crystal-clear tropical waters, think of the burrowing shrimp and other invertebrates that help make it that way.

Lessons from the Rulers of the Underworld

In 1987, Edward O. Wilson wrote a succinct essay with a provocative title, "The Little Things that Run the World (The Importance of Invertebrates in Conservation)." Despite Wilson's penchant for directing his essays to general (non-scientist) readers, he published this one in the scientific journal *Conservation Biology*. The article had a purpose of persuading his fellow biologists to devote more of their conservation-minded efforts to smaller invertebrate animals, rather than just "charismatic megafauna," such as mammals and birds. Granted, a few of these vertebrates, such as pandas and penguins, are endearing and far more likely to inspire plush toys and popular cartoons than, say, coconut crabs. Still, his points that invertebrates make up the vast majority of biomass and biodiversity in any given ecosystem, and hence have far greater influence over how that ecosystem functions than vertebrates, are undeniable. In this essay, he mentioned leaf-cutter ants and their nests as specific examples of how invertebrates shape their ecosystems, affirming an invertebrate hegemony that has reigned for nearly 600 million years, thriving and diversifying long before vertebrates played any significant role in the history of life.

To Wilson's views, I will add that much of this invertebrate success can be credited to their evolving the means to burrow. Furthermore, given that the Earth's surface has been ruled by burrowing invertebrates since the end of the Ediacaran Period, then supplemented later by burrowing vertebrates, it is perfectly reasonable to think of how burrowing in general shaped the surface of the Earth for millions of years. But that was then, and this is now. What lessons can we draw from this long evolutionary history of burrowing animals, and particularly our fellow mammals, as we look toward an uncertain future of climate change?

CHAPTER 9

Viva La Evolución:
Change Comes from Within

Small Burrows, Big Impacts

The pocket gopher did not know it then (and never would), but her problems did not start specifically on May 18, 1980. Instead, they could be traced to the Jurassic Period (about 150 *mya*), when the offshore Farallon plate collided with the western edge of the North American plate. In this battle of the plates, the Farallon lost. Its veneer of burrowed oceanic sediments and basaltic bedrock began to be subducted beneath the North American plate and melted in the asthenosphere, a consumption that eventually left only a small bit, called the Juan de Fuca plate. The magma made by this melting mixed elements of oceanic and continental plates, producing basalts, andesites, and a variety of other igneous rocks. The birth of each volcanic vent uplifted the ground surface in that part of North America, eventually making a volcanic mountain range in which the pocket gopher and her kin lived. Much later, some non-native, bipedal mammals in North America called this terrain "the Cascade Mountains," or simply "the Cascades." Some volcanoes erupted during the past few millions of years; a few became dormant, while others remained active. The long history and build-up caused a certain volcano next to this specific pocket gopher to experience earthquakes, a breach explosion, superhot and superfast volcanic-ash-laden gases screaming overland, massive mudflows, and more ash falling from above over the course of nine days. Fortunately, though, the pocket gopher slept through it all.

From the geologically brief perspective of the mammals, birds, amphibians, reptiles, insects, crustaceans, and other animals in the area, the volcano—recently renamed "Mount St. Helens" by the bipeds—had been simmering for only a few months. It was ready to blow and unleash an experience far more horrific than anything they had witnessed during their lifetimes. The pocket gopher's ancestors had lived in this area for quite a while and survived far worse eruptions than the one about to happen. Still, there was absolutely nothing they could do. Even the technologically advanced immigrants on the surface, who were carefully monitoring the volcano, had no control over what happened next.

The eruption began at 8:32 A.M., initiating a rapid series of events. The precipitating incident happened when the north side of the volcano, bulging for weeks from a buildup of steam and other hot gases, was shaken by a 5.1 Richter-scale earthquake. The seismic jolt unleashed a debris avalanche of rock and soil, the largest witnessed in human history. This mass movement took pressure off the top of the mountain so that groundwater heated by the 1,000°C (about 1,800°F) magma inside the volcano turned into steam and burst out. The void left by the avalanche allowed for a rapid release of hot vapor, carrying millions of tons of fine, silica-rich pyroclastic material (volcanic ash) with it. The breach explosion and collapse of the volcano top dropped the height of

Mount St. Helens by more than 400 meters (1,300 feet), and was like taking a heavy lid off a boiling pot of water. As a result, the finely crystallized magma flowed upward and outward, spewing a second vertical plume of ash to join the one punching out the mountainside. Within just 20 minutes or so, the column of ash rose more than 25 kilometers (15-plus miles) up into the stratosphere, 125 times the height of Trump Tower in New York City, making it look very, very small. Once there, the finest ash particles dispersed, drifted, and stayed suspended in the atmosphere for weeks.

The blast of superheated ash quickly outran the debris avalanche that unleashed it, moving downslope at speeds approaching 1,000 kilometers per hour (more than 620 miles per hour). This *nuée ardente* ("glowing cloud") extinguished all surface life in a 600-square-kilometer (230-square-mile) area north and east of the volcano, including whatever might have been in the North Fork Toutle River and Spirit Lake nearby. Surrounding old-growth forests—composed of Douglas and Pacific fir, mountain hemlock, and other trees—were destroyed. Those trees that were not instantly incinerated closest to the blast were knocked down like dominoes and buried under ash, while others left standing on the periphery were scorched from the heat and also died. To further enhance a sense of inescapable doom, this primary explosion was followed by more pyroclastic flows, as well as thick, fast-moving mudflows called *lahars*. These were caused by snow, ice, and alpine glaciers on the mountain instantly liquefying in the eruption, the water mixing with ash, rocks, and soil. These fatal slurries smothered any living thing downslope still clinging to a naïve hope of making it out alive.

Just before the volcanic ash cloud reached out and demolished all that it touched, the debris avalanche imparted some damage of its own. Consisting of 2.5–3 cubic kilometers (0.6–0.7 cubic miles) of earth material moving at 175–250 kilometers per hour (110–155 miles per hour), the avalanche dumped its load into the North Fork Toutle River valley and Spirit Lake. The abrupt filling of Spirit Lake by the landslide was analogous to a sumo wrestler performing a cannonball dive into a Jacuzzi, but scaled up a bit. The displaced water flowed uphill into the forested slopes more than 200 meters (670-plus feet) above the lake. Then, once gravity took over, the downward rush of water uprooted and pulled thousands of trees and their topsoil back down into the lake basin. Just like that, the forest was gone.

The devastation wrought by these multiple disasters rapidly spread over more than 400 square kilometers (more than 150 square miles), an area larger than metropolitan Atlanta. Once the breach explosion commenced at 8:32 A.M., David Johnston, a U.S. Geological Survey geologist monitoring the volcano from about 10 kilometers (6 miles) away, called it in on his radio while watching it happen. "Vancouver [Washington], Vancouver, this is it!" he shouted. And that was it for Johnston and fifty-six other people who were in the area, killed by the extreme heat of the *nuée ardente*, suffocated by the ash, or buried by the avalanche and lahars. Most of their bodies were never recovered.

Meanwhile, the northern pocket gopher and her same-species neighbors (*Thomomys talpoides*) were still deep in their respective burrow systems, mostly oblivious to the chaos above. The few gophers unfortunate enough to have ventured out of their burrows early that morning died instantly. Still, most survived, perhaps having only felt vibrations from earthquakes leading up to and coinciding with the eruption. Part of their tardiness in getting out and about that morning was because of the late spring on and around the mountain that year. Until that day, snow and ice still covered much of the area, and mornings were cool enough to keep most gophers inside, including this one, whom we will call Loowit.

Pocket gopher burrows serve as excellent winter homes, and Loowit's was no exception. It had a main tunnel 5–8 centimeters (2–3 inches) wide and about 30 centimeters (12 inches) below the soil surface, a depth that doubled or tripled whenever the soil was covered by snow. The main tunnel branched frequently along its length, making a burrow system about 150 meters (about 500 feet) long. The upper story of this system held several short tunnels that ended in latrines, little toilets that effectively concealed her feces from aboveground predators, but also kept her home as tidy as a dirt-walled burrow could be. Deeper parts of Loowit's burrow system were 1.5 meters (5 feet) down and included food-storage chambers. These well-stocked larders held seeds and other plant parts that helped her get through the winter, a diet she augmented with fresh foods by burrowing to and gnawing on plant roots in the soil. Short, inclined lateral tunnels branched off the main tunnel, providing routes for pushing excavated soil to the surface. However, like conscientiously closing a door, these openings were all plugged to keep intruders out and warmth and humidity in.

Winter had not stopped Loowit's foraging, and she often popped out of the soil and into the bottommost part of the snowpack, moving soil behind her. When the snow melted, natural casts of her tunnels stood out on the surface, looking like sedimentary renderings of elk antlers. As soil temperatures rose toward the end of May, she was a bit more active, preparing to reap a spring and summer harvest from freshly sprouted and tender plants above. Spring was also when she passed on and shuffled her genes with those of another pocket gopher; safety and food would not be the only items on her biologically driven agenda whenever journeying from the burrow. When the time and male gopher were right for her, she would soon hollow out a nesting chamber at the same depth as the food-storage chambers.

The day after the eruption, Loowit felt motivated to go above. She ran along one of the many tunnels in her burrow system, her whiskers guiding her in the complete darkness. After taking a right at a burrow junction, a left, up, then a right, followed by another right, and up again, she got closer to the main tunnel. Once on this avenue, she climbed up a ramp toward the surface, pausing briefly to snack on exposed roots along the way. Although her evolution-arily imparted behavior did not fill her with expectations, spring normally meant thinner to absent layers of snow on the ground surface above, meaning less effort for her to access blue skies and sunshine overhead. Encouraged by warmth sensed on the way up, she popped through the plug at the top of the tunnel.

If she could register surprise, she would have upon encountering a layer of sediment that behaved like snow, yet did not melt at her touch. It was volcanic ash. Using her hands, feet, and prominent front teeth, she started digging a shaft at a steep angle upward through the ash, pulling it underneath and past her. The ash was compacted but otherwise loose enough to move out of the way. Elsewhere in the area, other pocket gophers performed similar tasks in their burrows, all trying to make their way toward the surface, where food and possible mates awaited.

At some point in Loowit's upward progress, the digging got tougher and filled with obstacles. A tree trunk blocked her way, so she diverted her tunnel to go around it. A chunk of andesite three times wider than she was likewise proved immovable, so around that she went, too. The sediment changed upward from a moist blend of ash and debris to dry ash; then, once close to the surface, she encountered a thin, rigid crust. Fortunately for her, she and her lineage had dealt with much tougher substrates in the past. Pushing up and out with her head and front paws, she broke through the crust, her exit depositing a mound of loose sediment to one side of the opening.

Loowit was always at her most alert whenever breaching the surface, which took her right into the danger zone. She sniffed the air, her head swiveling right to left, left to right, and above, as she scanned for movements of nearby predators on the ground or in the air. Daylight bathed the area, but otherwise everything was gray. Smoke drifted overhead, bringing scents of burnt fir, as well as burnt fur. The latter was from an elk body only a few meters away, its rack still on its head but its torso blackened and torn asunder. Loowit's little round ears heard only the wind, which had suspended some of the finer ash near the ground and caused her to blink more rapidly.

After a few minutes of watching and listening, her naturally inherited paranoia eased somewhat once she realized she was alone. No other animals—not even insects—were within sight, an assumption verified by an absence of morning birdsong. Ordinarily Loowit ventured only a meter or two from any given opening of her burrow system, but the unusual dearth of bad vibes encouraged her to boldly go where no gopher had gone before and search for greenery. Sadly, after several meters of crawling, only a few shades of gray were in her limited vision. The landscape, now devoid of vegetation and predation, was reminiscent of Ediacaran conditions, its smooth, hummocky top surface a clean slate waiting to be marked by life.

At some point in this fruitless (and seedless) journey outside her burrow, Loowit sensed minute tremors. She stopped. Such vibrations typically correlated with footsteps of an approaching big mammal, such as an elk, which was fine, but sometimes it was a puma, coyote, or bobcat, which was not so fine. In any case, because she was out in the open and now more than five meters from her home, her best defense was to stay motionless. A few seconds later, the source of the tremors made itself apparent, as another spot on the volcanic crust bulged and cracked in front of her. It was another gopher. Facing away from her, he rapidly performed the same nervous glancing, sniffing, and listening she had minutes before. On his second turn to the left, his peripheral vision discerned her, and he saw her standing there. Somewhat interested, she looked back at him, their eyes met, and he promptly vanished. Not quite knowing what to do, she slowly crawled back to her burrow, got halfway in, and turned around to face the opening of his man cave. This was when he popped back up, having changed his look.

Loowit was instantly attracted to his bulging cheeks. This meant he had food, and lots of it. While below, he raided one of his food caches, stuffed the "pockets" in his cheeks with seeds and other vegetative bits, and brought up some to share with her. As he approached, she stood her ground, or at least stood in her hole in the ground. As he got to within a half meter of her burrow, she turned around, flicked her tail, and went down the tunnel.

Considering himself invited to bring in dinner and get cozy, he followed her, but not before turning around and stopping up the entrance to ensure their privacy. Within minutes, they were doing what they liked, and doing it naturally.

While Loowit and her momentary companion were undercover, other gophers began punching holes through the gray terrain. These gophers likewise emerged in a land without predators, allowing them to roam freely between one another's burrows to mate and share food supplies. As a result, many scenes like that enacted by Loowit and her mate took place below ground over the next several weeks.

As for Loowit, she became pregnant a few days after briefly sharing her burrow with her gene donor. By the middle of June 1980, she birthed five pups, three female and two male, all of which she suckled and raised in the newly dug nesting chamber; she was sustained by food from her storage chambers. Not all offspring survived the next few weeks, but the ones that did—two females and a male—dispersed from their birthplace burrow within two months and made their own homes underground.

Although Loowit did not give birth again and only lived to be two years old, her daughters lived long enough to have their own litters, and her son sired two additional litters. Hawks, the first gopher predators to move back into the ravaged area, picked off a few of their offspring, but earthbound hunters were slow to return. In the meantime, the gophers had virtually unrestricted access to the sparse (but growing) plants reseeding the surface. Given their food caches, access to plant roots buried in the soil, and little competition from other herbivores, they had more than enough sustenance to get by. Within just two years, any dip in the pocket gopher population caused by the eruption of Mount St. Helens was erased and followed by a boom.

Given their increased numbers, these little ecosystem engineers began terraforming the previously desolate landscape, first by helping plants take root and grow. Each individual pocket gopher was capable of overturning more than a ton of soil each year, so Loowit's progeny and other gophers readily plowed the land in just a few years. Through their burrowing, they pulled up darker, organics-rich soils from below the volcanic deposits and mixed the old with the new, creating soils more capable of hosting plant growth. Soils from below also contained more clays than volcanic sediments, which better retained water. Burrow mounds became prime spots for windblown seeds to settle and germinate, joining those dropped in fertilizing feces by birds flying overhead. Tunnel casts made under snow during the winters, as well as burrow mounds outside of plugged entrances, also contained seeds brought up from below. Activated by sunlight near the surface, they sprouted and grew.

Gophers even grew plants from beyond the grave, as seeds, nuts, and other plant parts in food-storage chambers of old, abandoned burrows also sprouted. Gophers further helped plant communities by reacquainting them with their fungal cohorts. Most plants have a form of symbiosis with fungi that wrap around their roots, a mutually beneficial relationship in which fungi extract phosphorus and nitrogen from the soil, and plant roots provide carbohydrates for the fungi. By burrowing, pocket gophers introduced fungal spores from soil below to burrow mounds above, where these fungi then were much more likely to attach to plant roots. In a nutshell, the plant communities around Mount St. Helens had no better friends than pocket gophers.

Pocket gopher burrows even changed the hydrology of the area, and again for the better of the ecosystems. Their numerous holes perforating the hard, impermeable volcanic ash crust caused surface water, whether from rain or snowmelt, to flow down into burrow systems. The burrows thus increased soil moisture necessary for plant growth and animals, while also slowing surface-water runoff, which prevented soil erosion. Also, though Spirit Lake was filled with dead trees from the surrounding forest and most of its aquatic life had been extinguished, the hummocky land surface held many more depressions than before the eruption. With just a little precipitation and runoff reserved by burrows, these depressions turned into more than a hundred new ponds.

Of the 55 mammal species in the area of Mount St. Helens in May 1980, only 14 survived the volcanic eruption and its collateral damage. Surface-dwelling elk, deer, black bears, coyotes, mountain goats, pumas, bobcats, hares, and all other large- to medium-size mammals perished. On the other hand, nearly all of the small mammals that lived were burrowing rodents, including yellow-pine chipmunks (*Tamias amoenus*), deer mice (*Peromyscus maniculatus*), Pacific jumping mice (*Zapus trinotatus*), and golden-mantled ground squirrels (*Callospermophilus lateralis*). One of the few non-rodent survivors was the tiny Trowbridge's shrew (*Sorex trowbridgii*), which (not coincidentally) is also a burrower. Pocket gophers are active year round, but many other small-mammal species were both underground and still hibernating when the eruption took place. The fortuitous timing of this disaster at the transition between winter and spring thus greatly enhanced the chances of these minutest of mammals to emerge and thrive. Of the rodents that had already come out of hibernation, nocturnal species were doubly lucky to have already retired for the day in their burrows when the blast occurred. Had the volcano erupted at night, many more would have died.

Defying expectations, 12 of 15 native species of amphibians, such as the northern red-legged frog (*Rana aurora*), western toad (*Anaxyrus boreas*), and northwestern salamander (*Ambystoma gracile*), also showed up after the explosion. Most presumably survived by staying buried in mud banks and under ice in water bodies as part of their overwintering. The northwestern salamander is also a well-known squatter in small mammals' burrows, suggesting that some found shelter in these. Paradoxically, the high number of amphibians afterward was attributed to the eruption itself. As mentioned previously, the reshaped surface held more than a hundred new ponds, which served as host habitats for amphibian life cycles. However, adult frogs, toads, and salamanders emerging from these ponds normally would have found protection in the nooks, crannies, and shade of surrounding forests. No forest? No problem, thanks to pocket gopher burrows, which in the absence of trees provided plenty of shady, moist refuges. Likewise, forest-dwelling reptiles, such as the northern alligator lizard (*Elgaria coerulea*) and common garter snake (*Thamnophis sirtalis*), may have ducked and covered in mammal burrows, which they also used as hiding places from predators.

Elk herds that returned to the area a few years later also interacted with gopher burrows in ways that aided amphibian and plant populations. Elk lent a helping hoof by walking across shallow tunnels of older gopher burrows, opening these abandoned homes as humid and otherwise protective places for frogs, toads, and other amphibians caught between wetlands. Elk hooves also broke up tunnel casts and mounds so that seeds inside these were more easily dispersed, whether by sticking to elk hooves or washing away from places disturbed by them. Traveling elk herds made their own contributions to the plant population by depositing seed-laden feces, at least a few of which landed on burrow mounds.

Ants (of course), such as *Formica pacifica*, also made it through the disaster, their nest holes nudging through volcanically placed sediments only months afterward. Survival rates were apparently determined by ash thickness; those buried under less than 20 centimeters (8 inches) of ash were more likely to find their way up and out, ensuring the continuation of their nests. From 1981 through 1987, researchers sampled other insects and arachnids in the blast zone, finding beetles, crickets, millipedes, centipedes, and spiders. Nearly all such arthropods had some part of their life cycles underground, below the snow and ice, in downed trees, or a combination of these. These ground-dwellers were later joined by flying and otherwise aerially dispersed insects, which as pioneers helped to rebuild arthropod roles in food webs. Among these arthropods were aquatic species, such as the signal crayfish (*Pacifastacus leniusculus*), which overwinters in burrows as deep as 2 meters (6.6 feet). Hence its survival strategy during the eruption successfully mimicked that of the small hibernating mammals on land, which was staying below the chaos. Although the ecological effects of ants and other burrowing arthropods were not as well documented as those of the pocket gophers, they no doubt assisted in stirring pre-eruption soils with the newly arrived overlying sediments. By 2008, at least ten species of ants were thriving in ecosystems around Mount St. Helens, nesting in its post-eruption sediments.

In October 2009, I went on a field trip to Mount St. Helens and saw for myself some of the ecological handiwork of burrowing pocket gophers and other animals since the fateful eruption twenty-nine years hence. The field trip preceded the annual Geological Society of America meeting, which was in Portland, Oregon, that year, so we left from there. An hour and a half later, we arrived at Mount St. Helens National Volcanic Monument in Washington. Although this was a geological trip, and hence attended primarily by geologists, it was co-led by an ecologist (Charlie Crisafulli) and a geologist (Jon Major). This partnership helped communicate how Mount St. Helens was a superb example of geology and ecology as sister sciences, especially when assessing how ecosystems recover after geologically inflicted disasters.

The day was gloomy and drizzly, but as field-oriented scientists, we accept that when in the field together, we get wet together. Part of our time, though, was spent indoors at the visitors' center at Johnston Ridge Observatory, named in honor of David Johnston, who died very close to that spot. In the center, we watched a brief film about the 1980 eruption, followed by the "screen" sliding apart to reveal the vista below of the area affected by the avalanche, ash flow, and lahars. Later, on our too-brief hikes near the observatory, fog obscured the top of Mount St. Helens, but we could see many of the still-healing scars on the landscape. Nonetheless, much seemed green and alive. As geologists, we of course were struck by what happened there as a result of geological forces, but also came away impressed with the resilience of ecological forces.

For the pocket gopher populations that survived the eruption of Mount St. Helens in 1980, their collective actions were the key to turning a desolate, monochromatic landscape back into a vibrant and verdant one. From a geological perspective, their effects were astoundingly quick, with partial ecological restoration apparent within just five years of the eruption. Consequently, pocket gophers and other burrowing animals that lived beyond May 18, 1980, send a powerful message about the benefits of burrows for surviving such an ecologically traumatic event, as well as for their role in restoring an ecosystem after it is nearly destroyed.

What happened at Mount St. Helens may provide a microcosmic example of how ecosystems recover from other horrific events in earth history, such as the end-Permian greenhouse effect from hell, the end-Triassic ecological crises, or the end-Cretaceous meteorite. Considering how pocket gophers and all other mammals—including us—are likely descended from burrowing synapsids that survived all three mass extinctions, we can be thankful for these predecessors and what they may teach us about coping with current and upcoming environmental crises. Severe events like the Mount St. Helens eruption and its outcome tell us that this was survival of the luckiest, not the fittest. Luck favored the burrowers, as well as plants and animals that could take advantage of the ecological services supplied by burrowers.

Small Burrowing Mammals, Naked and Otherwise

Mammals, numbered at about 5,000 species, are only half as diverse as birds. Yet a startling fact about mammal diversity is how four out of any ten species are rodents. The fossil record of rodents goes back to the Paleocene Epoch, with the oldest rodent fossils discovered thus far dating from around 58 *mya*, but molecular clocks hint at their origins at 62 *mya*, within 4 million years of the end-Cretaceous extinction. Rodents live on all continents other than Antarctica and have adapted to terrestrial and aquatic environments ranging from coastal dunes to mountains. This adaptability coupled with great abundance means that rodents often have disproportionate effects on their ecosystems, especially because many (if not most) are fossorial. Moreover, the vast majority of subsurface rodents make their own burrows, although others might reside in another animal's burrow, such as mice in gopher tortoise burrows.

Still, many rodents reveal their genetically held penchant to burrowing (and hence their ancestry) by digging their own burrows if circumstances demand it. For example, I recall seeing the handiwork of opportunistic rodent burrowing in downtown Philadelphia while walking by a raised, streetside and cement-encased flowerbed. There in its soil were perfectly round 10-centimeter (4-inch) wide holes. Normally I would identify such cavities as chipmunk burrows, but was dissuaded by their size (too big), their location (too urban), and tracks outside their entrances (too ratty). Yes, common rats, whether brown rats (*Rattus norvegicus*) or black rats (*Rattus rattus*), are excellent burrowers, and often call on this innate ability as a survival strategy in the concrete jungles of big cities. Mice are especially good at going underground, with many simply obeying their genes when they do so. In a fascinating 2013 study, researchers linked burrow architecture with DNA in oldfield mice (*Peromyscus polionotus*) and deer mice (*P. maniculatus*), showing that oldfield mice burrows were more complex than those of deer mice owing to small genetic differences between them. This provided a superb example of how the extended phenotypes of mammal burrows can reflect their genotypes.

Other rodents that are obligate burrowers, such as pocket gophers, are so diverse that it is impractical to list them all here. So I will name just a few that most people will likely recognize, such as mice, chipmunks, ground squirrels, voles, prairie dogs, hamsters, gerbils, kangaroo rats, lemmings, muskrats, and groundhogs. Other burrowing rodents include: marmots (*Marmota*) in Europe; Old World porcupines, consisting of various species in Europe, Africa, and Asia; jerboas, represented by more than 30 species in northern Africa and Asia; giant pouched rats (*Cricetomys*) in sub-Saharan Africa; agoutis (*Dasyprocta*), nutrias, also known as coypus (*Myocastor coypus*), tuco-tucos (*Ctenomys*), and chinchillas (*Chinchilla*) in South America; and bamboo rats (*Rhizomys* and *Cannomys*) and zokors (*Myospalax* and *Eospalax*) in Asia. Many other rodents also construct burrows if the situation calls for it. Even beavers (*Castor canadensis*) dig burrows on the banks of lakes or streams if vegetation is insufficient for building dams. Beavers' burrowing behavior also goes back quite a while, as paleontologists have found beaver skeletons in deep, corkscrew-shaped burrows from the Oligocene Epoch (about 25 *mya*).

Owing to the abundance and diversity of Asian rodents, biologists currently think that rodents originated there and later dispersed to other continents. Unfortunately, humans later aided in the spread of mice and rats, particularly when Europeans colonized the Americas, but also as Polynesians went seafaring to islands throughout the Pacific Ocean, including Hawaii and New Zealand. Whether they are native or non-native, burrowing rodents are extremely important parts of their ecosystems in most parts of the world.

One unusual species of burrowing rodent is the naked mole-rat (*Heterocephalus glaber*). Granted, the platypus (*Ornithorhynchus anatinus*) of Australia, with its duckbill, webbed feet, venomous leg spurs, and egg laying, has an unfair advantage in any contest for strangest mammal. Nevertheless, naked mole-rats always deserve to be in the top five of such competitions. These rodents, which live in eastern Africa (Ethiopia, Kenya, and Somalia), are almost always described as "bizarre" or "weird," so I will break slightly with this tradition by simply saying they are a bit peculiar.

What is the main noteworthy trait of naked mole-rats, besides their startling hair-challenged appearance, wrinkled pink skin, prominent incisors (earning them the nickname "saber-toothed sausages"), tiny ears, and squinty little eyes? They are one of only two species of eusocial mammals; the other is the less known and markedly more hirsute Damaraland mole-rat (*Fukomys damarensis*) of southern Africa. Intriguingly, a group of naked mole-rats acts less like a herd of elephants, a pack of wolves, or a pod of killer whales, and more like an ant colony. They are also the longest-lived rodents, with some in captivity exceeding thirty years (fifteen times longer than pocket gophers), and are extraordinarily resistant to both pain and cancer. These and other bragging points of mole-rats are directly attributed to their evolving entirely underground, along with their development of burrow systems that act as an extended phenotype for each colony. Their uniqueness is what makes naked mole-rats such excellent survivors under challenging environmental conditions, as well as shapers of their environments.

Naked mole-rats live in tropical savannas and grasslands at relatively high elevations of 1,000–3,000 meters (3,300–9,850 feet), which have patchy distributions of food resources. Their ancestors evidently coped with such scarcities by evolving a caste system. For instance, if their food (roots and tubers) was too widely dispersed for any one individual mole-rat to gather, then it made evolutionary sense to divide labor throughout a colony. A social network of naked mole-rats thus mimics that of ants or termites by having a queen responsible for producing most of the offspring in a colony, a few breeding males, and the rest of the colony (as many as 200, but usually around 80) performing duties as either workers or soldiers.

To assist in her matriarchy, a breeding queen picks one to four subservient males, and if any of them adequately deploy their special purpose, she becomes pregnant for about seventy days. Naked mole-rat litters typically consist of ten to eleven pups, but can be larger, with most (but not all) surviving to adulthood. Because a queen can give birth to five litters per year, she may then add anywhere from thirty to fifty new mole rats annually to a colony. Queen mole-rats do not go through menopause, meaning that they can keep giving birth throughout their entire lifespans. Female pups reach maturity sooner than males; moreover, they are genetically predisposed to perpetuating female dominance, a terrifying concept for men's-rights activists everywhere.

Because the queen is busy making and nursing babies in a nesting chamber, she is assisted by a bevy of handmaidens, nannies, butlers, and other servants in the colony that help with her young, effectively showing how it takes an underground village to raise a mole-rat child. However, another female in the colony may decide she's had it with serving the aristocracy and instead wants to be "Queen Rat." If so, she must fight the reigning queen, in which they wickedly wield their incisors like staple removers. If a rebel female wins and becomes a new queen in this death match, her vertebrae loosen and she elongates to supermodel proportions. With such physical assurances accompanying her new status, she then starts picking out her breeding stock: The queen is dead, long live the queen, and time to get busy.

How do such "House of Holes" dramas relate to burrowing? Because food resources are limited and scattered in their environments, naked mole-rats must dig great distances to reach roots and tubers, and they must do it cooperatively. Imagine if every time you had to shop for food, you had to dig a tunnel from your home to the grocery store. Making matters worse, the store might be in a different place each time, and perhaps miles away. If you had to do this yourself, you might briefly get extremely muscular, then exhausted, and then starve. Alternatively, naked mole-rats work together and are specially equipped for such challenges.

Mole-rats scratch through soils with their front teeth, rather than depending solely on their limbs for digging. Their teeth are positioned in front of their lips to avoid swallowing dirt, and their limbs are used for passing sediment underneath and behind them. Their loose, wrinkled skin also allows them to press through small openings or perform 180-degree turns in their tunnels. The few hairs they have are used to feel their way through dark tunnels. Their tight and compact ears are another part of their sensory toolkit, as colony members employ a variety of calls, such as soft chirps, to communicate throughout the burrow system. Digging is likewise collaborative, as workers form conga lines to excavate tunnels, with the lead mole-rat scraping with his or her teeth and sweeping the dislodged dirt back to others who move it further behind. Once tunnels are dug and open, mole-rats can also run forward or backward with equal ease through them. Given all of these adaptations and colonies working jointly, naked mole-rats find plenty of food for everyone while making new tunnels.

Not surprisingly, then, naked mole-rat burrow systems are among the most incredible of any animal, mammal or otherwise. Unlike many other rodents, they can easily spend their entire (and long) lives underground, having little to no reason to connect to the surface. Most burrow systems are relatively deep for such small rodents, about 2 meters (6.6 feet). However, workers often approach the surface to push up excavated sediment into molehills, producing hundreds of these: A single colony may displace more than 4 metric tons (4.4 tons) of soil. Tunnel widths are slightly more than the largest mole-rat in the colony, about 10 centimeters (4 inches) wide; living and foraging tunnels in a mole-rat system can add up to 3–5 kilometers (2–3 miles) in total length. When compared to an interstate highway system, this may not sound like much to us humans, but keep in mind how the animals making these burrow systems are normally only 10–15 centimeters (4–6 inches) long and weigh around 30 grams (about 1 ounce). (For perspective, this is 6% and 0.04% of my height and weight, respectively. However, I am admittedly a petite scientist.) Granted, queen mole-rats are fuller-bodied, approaching 80 grams (2.8 ounces), but smaller worker rats do all the tunneling. If workers do find a large tuber, they treat it as an everlasting gift by chowing down on the inside first, letting the rest keep growing. Also, however incongruous it might seem to use the words "naked mole-rat soldiers," larger workers defend the nest against intruders. One of these interlopers is their sworn mortal enemy, the rufous beaked snake (*Rhamphiophis oxyrhynchus*), which also makes its own burrows but slips into mole-rat tunnels for toothed-sausage snacks.

As mentioned before, naked mole-rat nests are more like ant or termite nests in how their burrows act like another body part. Indeed, take away its burrows and other members of its colony, and a single naked mole-rat would die very quickly. This fate is partly because of their unusual physiology, in that they can flip back and forth between poikilothermic (varying its body temperature according to its environment) and homeothermic (maintaining a steady temperature). For instance, in order to compensate for low oxygen and high carbon dioxide levels in deeper parts of their burrows, mole-rats can slow down their metabolism, like turning down a dimmer switch. However, if they get too warm, they can also go down into the lower, cooler parts of their burrows. To elevate body temperatures, they either get higher up, or they pile on top of one another in large spaces used as communal bedrooms.

A mole-rat colony uses at least two other types of chambers, a food-storage (pantry) chamber and latrine. Interestingly, mole-rats decrease wastes in their burrow system by recycling one another's feces. Although this may not sound appetizing for most animals, poo-eating (more elegantly known as coprophagy) provides both predigested food for younger mole-rats, while also bestowing gut bacteria needed for digesting cellulose-rich foods.

Admittedly, naked mole-rat burrow systems are more extreme in size and complexity for rodents in part because these are a product of eusocial behavior. The bewildering variety and number of other burrow systems by subsurface rodents, whether solitary or colonial, similarly make describing their burrows a challenge. Fortunately, pocket gopher burrows provide a succinct summary of what one might expect in a typical rodent burrow, whether made by an individual or colony:

- Many foraging tunnels just under the ground surface.
- A main, central tunnel that connects to the foraging tunnels but is deeper than them.
- At least one nesting chamber.
- At least one food-storage chamber.
- At least one toilet chamber (latrine).
- Shorter tunnels that end abruptly (so-called "blind" tunnels).

Burrowing rodents can then serve as a comparative model for other mammals that make and use burrows, such as badgers, moles, meerkats, foxes, hedgehogs, rabbits, pangolins, warthogs, and many more.

However, the preceding mammals are all placental. What about marsupials (pouch-bearing mammals) and monotremes (egg-laying mammals)? In Australia, where marsupials have dominated their environments for the past 60 million years or so, burrowing has also been a key strategy for coping with its arid interior, a lesson learned well by the opal miners of Coober Pedy. So upon hearing the word "marsupial," do not restrict your imagination to just aboveground kangaroos, wallabies, and possums. Instead think of burrowing wombats, bandicoots, Tasmanian devils, numbats, bilbies, quolls, and marsupial moles. Marsupial moles (*Notoryctes*) in particular represent a remarkable example of convergent evolution, neatly matching the ecological niche and behaviors of American and African moles as finely tuned invertebrate-eating instruments of subterranean ease.

It is also not surprising to know that, of the few extant monotremes, all are accomplished diggers. Monotremes include the platypus of Australia, as well as four species of spiny, ant-eating echidnas, such as the short-beaked echidna (*Tachyglossus aculeatus*) of Australia and New Guinea; the other three species live in New Guinea. Platypuses, which swim and hunt in lakes and streams, do double duty with their strong swimming muscles to pry crayfish out of their burrows. Platypuses also relax or nest in bank burrows of their own making; female platypuses can hollow out nesting burrows as long as 20 meters (66 feet). Echidnas are similarly accomplished diggers, owing to their steady diet of soil-dwelling insects. However, they are also adept at using the ground for disappearing acts. I once watched a short-beaked echidna bury itself vertically by rapidly loosening the soil beneath it, leaving only the tips of its back spines exposed at the surface. Mother echidnas also carve out dens for their young, made after their eggs hatch and echidna puggles (babies) grow up enough to leave a mother's pouch.

Given the great number and variety of mammals that burrow and live at least part of their lives underground, one might wonder if such abilities and lifestyles are connected to the evolutionary origins of mammals. Maybe, maybe not, but it is an idea certainly worth exploring.

Night Moves and Tunnel Vision

Because all mammals—placental, marsupial, and monotreme—are synapsids, they (and we) can trace their (and our) ancestry back to those synapsids that survived the greatest of all mass extinctions at the end of the Permian Period, about 250 *mya*. Moreover, we know at least a few successors of synapsids that became mammals also made it past the end-Triassic and end-Cretaceous mass extinctions. How many post-Permian synapsids—including mammals—burrowed? Although we cannot say for sure, trace fossils of complex burrow networks from Triassic rocks of Argentina, South Africa, Poland, and elsewhere point to synapsid makers. Moreover, some Jurassic fossil burrows match sizes and architectures of modern-day small-mammal burrows. For those sad people who do not trust trace fossils but instead look to bones for paleontological salvation, Middle and Late Jurassic fossil mammals from China and Portugal are also interpreted as fossorial based on their anatomy.

This evidence for burrowing synapsids and their mammaliaform descendants before and after the Triassic extinction shows not only how these animals would have lived through major ecological calamities, but also how they avoided direct competition with dinosaurs on the surface. Does your world have predators with nasty, big, pointed teeth? Then go underground. Does your world also have animals large enough that, when they step on you, instantly turn you into a furry crepe? Then go underground. Are you competing for the same food resources on the surface as these aforementioned dinosaurs? Then go underground. Are forest fires, volcanic eruptions, storms, extreme heat or cold, or other disasters making life a little more difficult? Then, well, you get it. Granted, living in burrows was not the best tactic for every dangerous situation in the past. For instance, paleontologists have described Late Cretaceous dinosaur trace fossils they think were made by clawed theropods digging into mammal burrows. Nevertheless, burrows still offered far more safety for smaller non-mammal and mammalian synapsids than taking their chances out on the surface.

Yet another way Mesozoic mammals removed themselves from direct competition with dinosaurs and other animals was by having an active nightlife, which may have been facilitated by burrowing lifestyles. In a 2016 analysis of genes linked to eyesight in mammals, geneticists concluded that low-light vision was an original ancestral trait in their lineage. That is, the proportion of rods and cones in early mammal eyes favored seeing in the dark. This led the scientists to posit that an early adaptation to nocturnal behaviors was a way for mammals to avoid competition with daytime dinosaurs. However, I hereby propose that the low-light conditions of burrows might have been the original incubator for this inheritable trait, which later was easily adapted to nighttime activities.

So which came first, a burrowing synapsid using "tunnel vision" during the Permian Period, or a nocturnal mammaliaform with "night vision" during the Triassic Period? Like most evolutionary puzzles, the origin of mammalian sight probably has no easy answer, but it also might have been coupled with adaptations for hearing. For instance, *Morganucodon*, a proto-mammal cynodont from the Late Triassic (205 *mya*), retained two jaw bones (quadrate and articular) that later evolved into middle-ear bones (incus and malleus, respectively) in "true" mammals. *Morganucodon*, which was about the size of a naked mole-rat, is often interpreted as nocturnal, burrowing, or both. This assumption is based on its species' descent from burrowing cynodonts, as well as how nearly all small modern mammals are nocturnal. In nocturnal mammals or their immediate ancestors, eyesight would have been naturally selected in dark conditions, as would good hearing, whether for burrow or nightscape. For one, improved sound reception would have aided communicating with their own kind, which would have been especially useful if early mammals were gregarious. If these nascent mammals ate terrestrial invertebrates, hearing helped them to find food at night, especially if that food included noisy insects. Hearing also would have provided an early-warning system for detecting dangers, such as approaching predators, thunderstorms, or floodwaters.

As mammals, then, we might ask ourselves this: Did the fundamental senses of sight and hearing, which most of us take for granted every day, originally evolve because our mammalian ancestors were in burrows during the early part of the Mesozoic Era? If so, this would be an example of how we not only have an evolutionary history represented as "your inner fish" (*sensu* Neil Shubin), but also as "your inner burrow."

So now let's imagine the end of the Cretaceous as an instant event, like what happened at Mount St. Helens, but with global consequences. Also imagine an extinction of large animals that was neither local nor temporary. This disaster would have quickly nixed apex predators such as tyrannosaurs and other theropods, as well as big herbivorous dinosaurs, such as hadrosaurs and ceratopsians. With non-avian dinosaurs out of the way, small burrowing mammals, amphibians, lizards, crocodilians, and fish that were fortunate enough to be in burrows or other forms of protection emerged to a clean slate. In the next 66 million years, birds certainly rebooted dinosaurian history, but mammals were the biggest success story among vertebrates in their domination of world ecosystems.

Starting with small, burrowing mammals, none of which were much bigger than a modern Virginia opossum (*Didelphis virginiana*), mammals expanded into every terrestrial environment and ecological niche owned previously by dinosaurs. Some stayed mouse-sized, but others got really big, such as the rhinoceros relative *Paraceratherium* from the Oligocene Epoch (34–23 *mya*), which may have weighed as much as 18 metric tons (20 tons). Like sauropod dinosaurs, mammals this grand did not need to burrow, but many of their distant relatives meanwhile continued to happily pass on their genes below their enormous, stomping feet.

Given such a history, modern burrowing mammals offer us important lessons, especially for those living in environments where many other animals risk extinction. Again, like pocket gophers, it pays to stay small and to use burrows as ways to forward their genes to future generations, all while avoiding being eaten, drying out, starving, or otherwise meeting an abrupt end before reproducing. As climate continues to change too quickly for most animals to adapt, small burrowing mammals—and especially burrowing rodents—are the ones most likely to make it past the worst of times once more. For this reason alone, we should study and learn from our little cousins living beneath our feet.

Nonetheless, there was a time not long ago when burrowing mammals became much larger than their Mesozoic ancestors and returned to burrowing as a way of life. In fact, some of these mammals tested the upper limits of body size for subterranean animals, far surpassing any invertebrate or vertebrate from the preceding more than 500-million-year history of burrowing animals. How did they cope once they encountered a combination of environmental change, habitat alteration, and extreme predation pressures? Spoiler alert: Burrowing alone is not enough to save you, especially if you are large. In fact, burrows or no burrows, doomsday scenarios become much more likely once a certain special invasive species of mammal arrives in a new environment, one acting like the biological equivalent of a meteorite strike or volcanic eruption. To identify this species, find a mirror and take a long look into the past, present, and future of the planet.

Giant Underground Sloths: Megafaunal Mammals Making Megaburrows

Despite much contentiousness in pop culture, and especially in cinema, virtually everyone agrees that the movie *Tremors* (1990) was the best ichnology-horror-comedy film of all time. This movie also starred Kevin Bacon, meaning that ichnology as a science scores a Bacon Number of 1 on the Six Degrees of Kevin Bacon scale. However, if you lived in caves or similar underground dwellings for the past few decades and somehow missed this movie, here is a synopsis. The plot focuses on the exploits of its heroes, 2-meter (6.6-foot) wide and 9-meter (30-foot) long carnivorous burrowing worm-like animals called "graboids." Unfortunately, the graboids are invaded by amoral and unappreciative surface-dwelling humans, who do everything they can to hasten their extinction. Two of the antagonists in the movie, played by actual humans Reba McEntire and Michael Gross, are preppers, bonded by a shared interest in stockpiling food, weapons, and other gear in a bunker below their house. True to their hardcore survivalist aspirations, they successfully deploy overwhelming firepower to kill a graboid that burrows into their bunker, affectionately called the "rec room." It is a senseless slaughter of a magnificent and misunderstood burrowing creature simply seeking food and comfort in its own underground environment.

Not to spoil the film for anyone who has not seen it, the humans (including Kevin Bacon) similarly dispatch other graboids throughout the film, leading to a so-called "happy" ending in which the last one dies. Still, one can take solace knowing that until then the graboids managed to consume and evade many of their surface-dwelling oppressors, thus neatly demonstrating the advantage of staying below ground when fighting against a technologically more advanced alien species. Moreover, although all of the graboids in *Tremors* were killed, their huge tunnels must have had a superb chance of being preserved in the geologic record. This further implies that graboid trace fossils would indicate their prehuman existence, upholding a glorious tradition since the Cambrian of predatory worms using burrows for ambush predation.

Did any real animals ever construct graboid-sized tunnels, perhaps large enough that a person could stroll into them? Yes, and their makers even overlapped temporally with humans, having lived during the latter part of the Pleistocene Epoch. Although the animals that made these huge burrows were not fictional people-eating worms, their identities were nonetheless almost as surprising: They were mammals. Despite the long geologic record of burrowing animals, including dinosaurs, there is poignancy in realizing that the record for largest non-human burrows was by recently defunct furry critters. Even more intriguing, some of our ancestors might have even witnessed these big, burrowing mammals shortly before their extinction, leaving only their bones and trace fossils behind.

Owing to their great size, these mammal burrows at first defied ichnological imaginations. Geologists in Argentina and Brazil noted the massive structures in the 1920s and 1930s as features cutting into a variety of soft and deeply weathered igneous, metamorphic, and sedimentary bedrock. Some were preserved as open tunnels, while others were partially or wholly filled with sediment, visible in outcrops as either oval cross-sections or cylinders in lengthwise views. In the 1970s and 1980s, archaeologists studied some of the tunnels, assuming that they had been dug out by indigenous peoples and used as cave-like shelters. Considering their proportions and geological setting, this was a perfectly reasonable hypothesis, as they superficially resembled human-made tunnels and chambers in Cappadocia and elsewhere. The smallest, however, are only about 60 centimeters (24 inches) wide, 50 centimeters (20 inches) tall, and only a few meters long. Such cavities might have made a temporary place for a child to hide from her or his parents, but certainly were not suitable dwellings for adults. Nonetheless, the largest tunnels are much more voluminous, with some as wide as 4 meters (13 feet), 2 meters (6.6 feet) tall, and more than 100 meters (330 feet) long. Moreover, some tunnels connect with one another or join larger, sub-spherical chambers to make more complicated networks. Once put together, some of these spaces feasibly could have served as underground homes for families or small communities. A few even contained petroglyphs, showing that pre-Columbian people entered at least some of them.

At some point, though, the archaeologists realized that the tunnels were not of human origin. For one, the paucity of human artifacts and bones in these "caves" implied that people might have only visited them, rather than lived there. Secondly, indigenous peoples in the southeastern part of South America did not have the right materials for making rock-carving tools, and no such tools were in the tunnels or chambers. Third, petroglyphs were rare, giving the impression that folks were not inspired enough to hang out in these places and make art. So starting in the 1980s, paleontologists and geologists took over investigating the tunnels, with intense research in early twenty-first century. So far, hundreds of these burrows have been discovered, mostly in Rio Grande do Sul (Brazil), and with many more likely awaiting discovery.

Given all of this attention and scrutinizing of the tunnels, the paleontologists subsequently noticed enough details that eventually revealed their creators. Their conclusions were based on: the size of the largest tunnels and their architecture; other trace fossils on tunnel walls; and fossil bones of the potential tunnel-builders in the same parts of Argentina and Brazil. As a result, paleontologists concluded that giant ground sloths were the likely makers of the largest burrows, and giant armadillos were credited with the smaller ones. Both types of mammals shared Pleistocene landscapes with many other big mammals in the Americas, which were collectively referred to as the "Pleistocene megafauna."

Now, normally upon hearing or reading the word "sloth," most people imagine a relatively small (think "cuddly"), lethargic, long-limbed, hairy, slow-moving, cute, tree-climbing mammal cosplaying as Chewbacca. Yet if pressed, you may also recall museum displays or artistic depictions of ground-dwelling sloths, and then remember that the word "giant" often precedes the term "ground sloth." How giant? Adults of the South American ground sloth *Scelidotherium* from the Late Pleistocene were about a meter (3.3 feet) wide, 2.7 meters (8.9 feet) long, and ranged from 600 to 1,000 kilograms (1,320 to 2,200 pounds) in weight, but more likely were about 850 kilograms (1,870 pounds). For the sake of comparison, this sloth was about the length and weight of a Smart Fortwo car. Impressive? Sure, but its Pleistocene ground sloth contemporary in South America, *Glossotherium*, was even bigger: 1.2 meters (4 feet) wide, 3.2 meters (10.5 feet) long, and weighed 1,200 kilograms (2,600 pounds), capable of taking up the parking space for a midsize car.

Once Argentine and Brazilian paleontologists looked closer at the burrows, their megafaunal origins became more obvious. First, parallel grooves on burrow walls acted as "fingerprints" for identifying their perpetrators, matching the number (usually two) of claw-bearing fingers and claw widths on ground sloth hands. Secondly, many of the Brazilian tunnels were not smooth cylindrical tubes, but more like a series of semi-elliptical chambers, linked like rosary beads. Floors were flat in each chamber, but walls and ceilings were bowed outward (concave inward) and separated by ridges, corrugating each tunnel along its length. Chambers were regularly sized, ranging from 0.5 to 2.0 meters (1.6 to 6.6 feet) long. Such spacing probably reflects digging cycles, in which a ground sloth dug a chamber just short of its body length, stopped, and then came back later to dig another. In a few of the more complicated structures, short tunnels connected to expansive chambers as much as 1.5 meters (5 feet) high and 7 meters (23 feet) wide. Upper portions of tunnels and chambers are also smooth in spots, attributed to sloths rubbing against and hence buffing them with their fur. Judging from their sizes and complexity, paleontologists conjectured that the tunnels and chambers might have been constructed and occupied over hundreds of years by many generations of sloths.

Giant ground sloth anatomies, especially in *Scelidotherium* and *Glossotherium*, also confirm these mammals were well equipped to scratch out the gigantic burrows. Their claw-bearing hands had thick, flat, straight, and closely spaced fingers, serving as natural shovels when applied against soft rock. Furthermore, their hands were powered by brawny muscles attached to thick bones in their forelimbs and shoulders; calculated forces and stresses generated by these forelimbs equal or exceed those of galloping mammals. Giant ground sloths, however, did not gallop: They dug. Sloth anatomy further affirms this form following function, as the center of mass (or "center of gravity") was more toward their rear, rather than in the middle. This freed up their hands for burrowing, especially when making horizontal tunnels.

Joining the ground sloths in digging big burrows were giant armadillos. South America is where armadillos originated, and is still home to twenty species, ranging in size from the impossibly adorable pink fairy armadillo (*Chlamyphorus truncatus*) to the modern giant armadillo (*Priodontes maximus*), all excellent burrowers. (The latter is often credited as largest burrowing mammal, but is not, explained later.) Giant armadillos are typically 1–1.5 meters (3.3–5 feet) long and weigh 20–30 kilograms (44–66 pounds), but can weigh as much as 50 kilograms (120 pounds). This species, however, would have been dwarfed by some of its Pleistocene relatives, reminding us how being a "giant" is often a matter of timing. For instance, two genera of Pleistocene armadillos, *Pampatherium* and *Holmesina*, weighed at least twice as much as the modern giant armadillo (*Priodontes*). The Pleistocene armadillos *Eutatus* and *Propraopus*, however, were closer in size to *Priodontes*, which makes comparing their burrows easier. Present-day giant armadillos dig dens that are slightly taller (45 centimeters/18 inches) than wide (35 centimeters/14 inches) to better accommodate their armored bodies. In contrast, the Pleistocene burrows are 50 centimeters (1.6 feet) to a meter (3.3 feet) wide. Other signs of armadillo origin for the smaller burrows are parallel scratches on the walls made by three digging claws, which is exactly what armadillos had on their hands.

One important question hangs over all of these revelations that a few members of the mammal megafauna of South America went underground: Why? For armadillos, building long tunnels and chambers made them a little safer from carnivorous contemporaries of the megafauna. These threats included saber-toothed cats (*Smilodon*) and short-faced bears (*Arctotherium*) that might have fancied having an armadillo on the half shell. However, the wider tunnels built by ground sloths would not have excluded these predators.

But then, the sloths were much larger than saber-toothed cats and bears, and thus may not have needed to hide.

So once again, we have to look at burrows as multipurpose. For instance, many of the tunnels in Brazil are located in the sides of hills near stream valleys with running water, providing plenty of bankside vegetation and drinking water. From a longer perspective, having your home near these life-sustaining resources would have been especially handy during extended droughts and drier climates in general.

In fact, the paleontologists who are exploring, mapping, describing, and otherwise studying the Brazilian tunnels proposed just that: Armadillos and ground sloths dug tunnels and chambers to better cope with climates much drier than today's, as tunnels would have maintained more humid conditions than outside. Nowadays these same researchers have to crawl through constantly wet or submerged tunnels, demonstrating their unsuitability for year-round living under current conditions. These homemade "caves" also would maintain the area's average annual temperature, regardless of whether Pleistocene climates were warmer or cooler. Lacking cave-forming limestones, local bedrock geology also gave incentives for these mammals to burrow, impelling them to embrace a "DIY" (do it yourself) mentality back in the Pleistocene.

Life Underground and the Sixth Extinction

The realization that the largest burrowing animals in the history of the earth were not only mammals, but also that they overlapped with human existence, is startling for several reasons. But the main reason why is that before the detailed studies of these gigantic burrows by South American paleontologists and geologists, zoologists had assumed that body sizes restricted burrowing, preventing big animals from living below the surface. Such a mistake is perfectly understandable because, after all, we have only been living in a minuscule interval of earth history and the fossil record is vast. A long-term perspective helps us understand how extinctions of the past filtered out potential exceptions to the patterns we infer about burrowing abilities based solely on modern examples.

So just who is the biggest burrowing animal today? Surprisingly, it is the grizzly bear (*Ursus arctos*), with the Kodiak bear (*Ursus arctos middendorffi*) as its largest subspecies, exceeding 600 kilograms (1,300 pounds). Grizzlies, which live in northern latitudes of North America from the Rocky Mountains to Alaska, apply their digging skills to unearth insects for food (moth larvae and ants are among their favorites) and also to make roomy dens in soil toward the end of autumn. Grizzlies dig dens with their robust claws and powerful forelimbs, sometimes displacing more than a metric ton (1.1 tons) of soil. Burrows have a grizzly-wide tunnel leading to a subspherical chamber large enough to accommodate its occupant. If the grizzly is an expectant mother, dens are also places where they give birth during the winter and suckle their young until springtime. Once cozy, grizzly bears can hibernate for as long as seven months. This means that grizzly bears, which we normally visualize as active, wide-ranging terrorizers of surface life, actually live nearly half of their lifespans underground, one burrow at a time.

If you count snow as a substrate for burrowing (which I do), the runner-up for largest modern burrower is not the modern giant armadillo, but another bear, and more specifically its females. Female polar bears (*Ursus maritimus*), which are about half the size of males, can still approach 350 kilograms (770 pounds) as adults. Like grizzlies, mother polar bears dig maternity dens large enough to hold them over the winter, but unlike grizzlies, they usually do this in snow. (Male polar bears are normally active year round, but make temporary dens in uncommonly cold weather.) They first make a narrow tunnel, which leads to a main chamber about 1.5–2 meters (5–6.6 feet) wide and a meter high; however, an expectant bear could also dig more than one chamber. The denning chamber is placed a bit higher than the tunnel, which keeps warm air in. Polar bear mothers seal themselves in, lower their metabolism, bear their cubs, nurse them through the winter, and break out for fresh food in the spring.

Anyone who has watched a documentary film about the Arctic in the past decade cannot help but notice that polar bears have become a symbol of awareness about climate change and its repercussions, including their possible extinction. Their peril is due to decreasing volumes of sea ice over the past few decades. This situation is forcing polar bears to swim longer distances when hunting for prey, causing them to drown or starve. However, yet another factor is how decreased amounts of snow also lead to less maternity denning during the winter: no dens, no overwintering and giving birth to cubs, and far fewer bears each spring. And if you want to make an animal go extinct, just stop it from reproducing.

Luckily, polar bear dens are not limited to snow and ice. If these frozen substrates are unavailable, mother bears will go onto land and burrow into soil. In a long-term study of polar-bear behaviors published in 2016, Todd Atwood, Elizabeth Peacock, Melissa McKinney, and other researchers showed that polar bears have already shifted their behaviors in accordance with shrinking volumes of sea ice since the 1990s. Less ice has forced bears to spend more time on land than before, which includes digging more maternity dens on land. So if polar bears survive as a species in the face of climate change, it will be partly attributed to the behavioral plasticity of burrowing bestowed upon their ancestors and so many other mammal lineages.

The topic of extinctions brings us back to thinking about the most recent one, in which we are implicated. "The Sixth Extinction," following the big five of the end-Ordovician, Devonian, Permian, Triassic, and Cretaceous, began about 50,000 years ago in Australia, when its terrestrial megafauna declined rapidly and vanished within just a few thousand years. This scenario was repeated toward the end of the Pleistocene in broad swaths of Europe, North America, and South America starting about 12,000 years ago, taking out mammoths, ground sloths, saber-toothed cats, and many others. Large islands, such as Madagascar and New Zealand, also experienced rapid extinctions of their big land-dwelling animals, each of which happened just 2,000 and 800 years ago, respectively. In these extinctions, most victims were mammals, but some were birds and reptiles. Although many ecological factors were likely involved, including climate change, what they all had in common was the arrival of *Homo sapiens*. Upright bipeds using ichnology (tracking) and cooperative hunting, while also carrying sharp sticks and fire, dealt a lethal wild card to the hand of the megafauna, and not just through overhunting, but also via habitat alteration.

Still, one would think that with all of the protective qualities of burrows, at least the underground members of the different megafauna would have survived these new ecological crises. Alas, they did not. Part of this failure may have happened because very few of these animals were burrowers in the first place. However, it also might have occurred because humans could easily crawl or walk into tunnels or dens built by these animals to kill or evict them, leaving survivors homeless in environments already undergoing ecological crises because of the exotic invaders. Rest assured that small burrowing mammals, such as pocket gophers and prairie dogs in North America, were fine in the face of this human-led blitzkrieg, but most larger animals did not make it.

This scenario is roughly analogous to what likely happened with mammals at the end of the Cretaceous Period, 66 *mya*. In a 2016 study, paleontologists Nicholas Longrich and others examined end-Cretaceous mammal extinctions in North America to see which ones made it into the brave new world of the Cenozoic Era. Of fifty-nine mammal species living toward the end of the Cretaceous, fifty-five were extinct at its end and only four survived. The remaining four apparently rebounded quickly, as populations increased going into the Paleocene Epoch. What did these mammals possess that gave them permission to live through the horrific effects of volcanically altered climates and a meteorite impact? They were small, widespread, and common, just like many modern resilient rodents today. But were they also burrowers? That we do not know yet, but it is a prediction we can test by looking for trace fossils (burrows) made by these mammals, anatomical clues pointing toward burrowing abilities, or (best of all) skeletons of these mammals in appropriately sized burrows.

In 2004, more than ten years before this study focusing on mammals, Douglas Robertson and his coauthors published a speculative-science article titled "Survival in the First Hours of the Cenozoic." In this article they estimated the probable effects of a meteorite impact on continental life, particularly vertebrates. This required assessing resultant damages and ecological stresses, how long those effects would last (minutes, hours, days, months), and ways that animals could survive them. For means of survival, the solution they offered for every ecologically awful consequence—extreme heat, fires, atmospheric dust, abnormal soils, decreased plant production (and hence food shortages)—could all be summarized with the following sentence:

> We argue that sheltering underground, within natural cavities, or in water was the fundamental means to survival during the first few hours of the Cenozoic.

Sound familiar? Nonetheless, these researchers also stated an important caveat to the "Burrows save!" message I have repeated throughout this book, and one with which I agree: Sometimes burrows are not enough. This is especially true when an entirely unexpected environmental stress is added to an already precarious ecological situation. Indeed, in a recent (2016) study focusing on the South American megafaunal extinctions from about 12,000 years ago, researchers proposed that the arrival of humans weakened megafaunal populations, which then were hit hard by climatic warming, a one-two punch that reduced their numbers to a point of no return. In short, we were the meteorite, and climate change following our impact was the last straw for the megafauna, including the burrowing ground sloths and giant armadillos.

Our species has come a long way technologically since the end of the Pleistocene, and these advances have accelerated environmental changes sufficiently that it may be leaving a definite mark in Earth history. Geologists are even debating whether to name this new time the Anthropocene Epoch in recognition of how humans have become a geological and ecological force of nature. Part of the evidence for this powerful effect is the addition of worldwide extinctions from the past few hundred years to those of the Pleistocene.

Another marker of the Anthropocene is the rapidly rising seas associated with the increased melting of polar ice. Heightened sea level means that marine sediments will soon bury terrestrial sediments and the traces of their organisms. For example, as I write this, citizens of the island nation of Kiribati in the Pacific Ocean are making plans to abandon their islands as the seas overtake the land. This means the land crab and coconut crab burrows on these islands will become part of the geologic record later this century, crosscut by the burrows of ghost shrimp and other marine animals. On the eastern coast of the U.S., fiddler crab burrows, ghost crab burrows, sea turtle nests, and ant nests are likewise shifting landward, and their former coastal environments will be occupied by ghost shrimp and other burrowing marine animals.

Meanwhile, on land, species that already live in burrows or use these as an important part of their life cycles—such as burying eggs, underground development and growth, and denning—will not only survive, but will also enable many other species to live by offering their burrows as havens from climate change. In a 2013 research article, ecologists David Pike and John Mitchell proposed that burrowing ecosystem engineers, such as gopher tortoises, pocket gophers, wombats, and seabirds, provide "thermal refugia" for other species. Even as global temperatures go up, burrow commensals will be living in air-conditioned comfort. As a result, these researchers recommended that conservation efforts should make sure to include subterranean animals—including alligators and other burrowing animals, great and small—so these species can provide underground bunkers for other animals.

Living on Burrowed Time

In a sense, we all live in a burrow. Not literally, of course, but we are living in a world that has been and will continue to be shaped by burrows. One facet of this ichnological perspective is manifest in the everyday mixing of the Earth's "skin" by burrowing animals, from deep-sea abyssal plains to the Tibetan Plateau. The variety and abundance of fossil burrows in the geologic record—ranging from the Ediacaran Period to the Pleistocene Epoch, and far outnumbering the bodily remains of their makers—reflect another aspect. A third and perhaps most important aspect is how the evolutionary paths taken by most modern animals, whether these are crocodilians, turtles, birds, lungfish, amphibians, earthworms, insects, crustaceans, or mammals, are connected to their burrowing ancestors. This subterranean heritage is especially important to keep in mind for lineages that survived mass extinctions by ducking and covering. Omit burrowing from the past 550 million years of animal life, and most ecosystems today would be more like those before the end of the Ediacaran Period, never resulting in a species that could interpret and document the history of life, let alone predict its future.

But is this holistic knowledge of our intrinsic connection to burrows and burrowing animals through time just an interesting intellectual exercise? Or does it contain a value that is to our benefit? I will argue that learning more about burrows is both interesting and useful, as interdependent on one another as the different parts of a leaf-cutter ant nest. For instance, curiosity-driven research about modern and fossil burrows led to more accurate interpretations of ancient environments. This in turn allowed geologists to more easily explore for economically valuable deposits of fossil fuels, such as oil and natural gas. Curiosity-driven research about the lasting effects of burrows on the properties of rocks bearing fossil fuels then led to better recovery and production of them. Perhaps most important for now and our near future, curiosity-driven research about changes in fossil-burrow assemblages aided geologists in more accurately interpreting when sea level went up or down in the geologic past.

Ironically, then, ancient burrows and their important role in our acquisition and exploitation of fossil fuels indirectly led to the current crisis we face with rapid climate change. Thanks in part to fossil burrows, we were all too successful in finding fossil fuels. Consequently, we have liberated immense volumes of previously locked-up carbon dioxide and methane, both "greenhouse gases" that contribute to global warming. As I write this, 2015 was the warmest year on record, with eight of the previous ten years' global temperatures also among the top ten warmest; July and August 2016 tied for the warmest months documented thus far. Based partly on assessments by the Intergovernmental Panel on Climate Change (IPCC) and peer-reviewed research by climatologists, the U.S. Department of Defense (DOD) identified climate change as a global security threat. The latest DOD report on this topic in 2015 said, ". . . the department must consider the effects of climate change—such as sea level rise, shifting climate zones and more frequent and intense severe weather events—and how these effects could impact national security." The Pentagon and other U.S. governmental entities may have stopped building nuke-proof underground bunkers, but they are looking ahead at the latest and perhaps more lasting of human-made threats.

How do burrows of the past relate to our future, and perhaps hold the key to better predicting how we and the rest of life will adapt to a warming world? This is again where curiosity-driven research about how burrowing animals reacted to and survived mass extinctions in the geologic past might apply. For one, I predict that research on fossil burrows used originally to find fossil fuels will have its utility flipped. Instead, we will use these traces of past lives to discern climatic cycles, which in turn should help us better predict and prepare for the grimmest of climate change scenarios in our near future. Knowing more about modern burrows and burrowing animals will also improve planning for ecosystem restoration, as well as conservation of endangered species, helping to slow the "The Sixth Extinction." We can even look to mammal burrows specifically to inform us of our own evolutionary legacy, and what lessons we might draw from this awareness as ecosystems alter, such as ensuring that more urban development does not go up, but down. One thing is certain: Our future world will continue to be affected by burrowing animals, but in the face of accelerated climate change, we know those that burrow—including us—are also more likely to survive.

In summary, our curiosity about the natural history of burrows and burrowing animals matters in a way that affects our survival and those species sharing the earth with us now. Considering the environmental challenges we face in the near and far future, we cannot afford to remain ignorant of our subterranean contemporaries, the vestiges of their ancestors, or the lessons imparted by populations of our own species who wisely sought the mollifying protections offered by the underworld. So before the Red Queen of environmental degradation runs faster than us, let's go down the rabbit hole with all of the inquisitiveness of our humanity, learning what we can about the grand history of evolution underground.

APPENDIX

Genera and Species Mentioned in *The Evolution Underground*

Note: Most of the genera and species listed here are animals and burrowers, and most of them are still alive (extant). However, some species are extinct, and not all of those are confirmed as burrowers. Names are arranged alphabetically by genus.

Acrolophus pholeter—gopher tortoise acrolophus moth; extant in southeastern U.S.

Acromyrmex—leaf-cutter ants, about thirty species; extant in Central and South America.

Alligator mississippiensis—American alligator; extant in southeastern U.S.

Alligator sinensis—Chinese alligator; extant (but endangered) in eastern China.

Alloblackburneus troglodytes—gopher tortoise scarab beetle; extant in southeastern U.S.

Ambystoma gracile—northwestern salamander; extant in northwestern North America.

Ambystoma macrodactylum—long-toed salamander; extant in northwestern North America.

Amphiuma—congo eel; extant in southeastern U.S.

Anaxyrus boreas—western toad; extant in northwestern North America.

Andiorrhinus—earthworm, at least two species (*A. kuru* and *A. motto*); extant in northern South America (Venezuela and Colombia).

Aptenodytes forsteri—emperor penguin; extant in Antarctica.

Apteryx—kiwis, five species; extant in New Zealand.

Archaeothyris—synapsid; extinct, lived during Late Carboniferous Period, fossils in Nova Scotia (Canada).

Arctotherium—short-faced bears, five species; extinct, lived during Pleistocene Epoch, fossils in South America.

Aristida stricta—pineland threeawn (plant); extant in southeastern U.S.

Arthropleura—myriapod, five species; extinct, lived during Late Carboniferous Period, fossils in North America and U.K.

Aspidella terranovica—uncertain affinity, but probably a soft-bodied animal; extinct, lived during Ediacaran Period, fossils in Newfoundland (Canada).

Athene cunicularia—burrowing owl, with many subspecies; extant in North, Central, and South America.

Atta—leaf-cutter ants, 15–20 species; extant in Central and South America.

Biffarius biformis—Georgia ghost shrimp; extant in southeastern U.S.

Birgus latro—coconut crab; extant on many South Pacific islands.

Brachydectes elongatus—amphibian; extinct, lived during Early Permian Period, fossils in western U.S.

Breviceps macrops—desert rain frog; extant in southwestern Africa.

Broomistega—amphibian; extinct, lived during Early Triassic Period, fossils in South Africa.

Calidris canutus—red knot; extant, migrates from North America, Europe, and Asia to South America, Africa, and Australia (respectively), then back again.

Callichirus major—Carolina ghost shrimp; extant in southeastern U.S.

Callospermophilus lateralis—golden-mantled ground squirrel; extant in western North America.

Cannomys—lesser bamboo rat, may be only one species (*C. badius*); extant in eastern and southeastern Asia.

Cardisoma guanhumi—blue land crab; extant in southern coasts of North America, Caribbean, Bahamas, and parts of South America.

Castor canadensis—American beaver; extant throughout much of North America.

Ceratophaga vicinella—tortoiseshell moth; extant in southeastern U.S.

Ceuthophilus latibuli—camel cricket; extant in southeastern U.S.

Chelyoxenus xerobatis—gopher tortoise hister beetle; extant in southeastern U.S.

Chinchilla—chinchillas, two species (*C. lanigera* and *C. chinchilla*); extant in western South America.

Chlamyphorus truncatus—pink fairy armadillo; extant in Argentina.

Clepsydrops—synapsid, four species; extinct, lived during Late Carboniferous Period, fossils in Nova Scotia (Canada).

Copris gopheris—gopher tortoise copris beetle; extant in southeastern U.S.

Cricetomys—giant pouched rats, four species; extant in sub-Saharan Africa.

Crotalus adamanteus—eastern diamondback rattlesnake; extant in southeastern U.S.

Ctenomys—tuco-tucos, 50–60 species; extant in southern South America.

Cyanoliseus patagonus—Patagonian conure; mostly in Argentina.

Cyclura rileyi rileyi—San Salvador rock iguana; extant in various cays and mainland of San Salvador Island, Bahamas.

Cynognathus—cynodont therapsid, only one known species (*C. crateronotus*); extinct, lived during Early-Middle Triassic Period, fossils in Africa, Antarctica, Asia, and South America.

Cynomys—prairie dog, five species; extant in western North America.

Dasyprocta—agoutis, eleven species; extant in Central and South America.

Dermochelys coriacea—leatherback turtle; extant in Atlantic, Pacific, and Indian Oceans, females nest on tropical-subtropical shorelines.

Desmatochelys padillai—sea turtle; extinct, lived during Early Cretaceous, fossils in Colombia.

Dickinsonia—animal of uncertain affinity, nine species; extinct, lived during Ediacaran Period, fossils in Australia, Russia, and Ukraine.

Didelphis virginiana—Virginia opossum; extant in southeastern U.S.

Dimetrodon—synapsid, thirteen species; extinct, lived during Early Permian Period, fossils mostly in western U.S.

Dinilysia patagonica—snake; extinct, lived during Late Cretaceous Period, fossils in Argentina.

Drymarchon couperi—eastern indigo snake; extant in southeastern U.S.

Echinerpeton—synapsid; extinct, lived during Late Carboniferous Period, fossils in Nova Scotia (Canada).

Elgaria coerulea—northern alligator lizard; extant in northwestern North America.

Eospalax—zokors (rodents), three species; extant in China.

Eruca—flax (flowering plant), one to five species; extant in Mediterranean area.

Eudyptula minor—fairy penguin; extant in Australia and New Zealand.

Eunice aphroditois—bobbit worm; extant in Atlantic and Pacific Oceans.

Eunotosaurus—diapsid reptile ("turtle ancestor"); extinct, lived during Middle Permian, fossils in South Africa.

Fallicambarus gordoni—Camp Shelly burrowing crayfish; extant in Mississippi.

Formica pacifica—ant; extant in northwestern North America.

Fratercula arctica—Atlantic puffins; extant in northeastern North America, Greenland, Iceland, Norway, and U.K.

Fukomys damarensis—Damaraland mole-rat; extant in southern Africa.

Gecarcoidea natalis—Christmas Island land crab; extant mostly on Christmas Island.

Geomysaprinus floridae—equal-clawed gopher tortoise hister beetle; extant in southeastern U.S.

Giganotosaurus—non-avian theropod dinosaur; extinct, lived during Late Cretaceous Period in South America.

Glossopteris—seed fern (plant), more than seventy species; extinct, lived during the Late Permian Period, fossils in Africa, Antarctica, Australia, India, and South America.

Glossotherium—giant ground sloth, two species (*G. robustum* and *G. chapadmalense*); extinct, lived during Pleistocene Epoch, fossils in South America.

Glyptemys insculpta—wood turtle; extant in north-central and northeastern North America.

Glypturus acanthochirus—Bahamian ghost shrimp; extant in Bahamas.

Gopherus agassizii—Mojave desert tortoise; extant in southwestern North America.

Gopherus berlandieri—Texas tortoise; extant in Texas and Mexico.

Gopherus flavomarginatus—bolson tortoise; extant in Mexico.

Gopherus morafkai—Sonoran desert tortoise; extant in southwestern North America.

Gopherus polyphemus—gopher tortoise; extant in southeastern U.S.

Heterocephalus glaber—naked mole-rat; extant in eastern Africa.

Holmesina—pampathere (related to armadillos), six species; extinct, lived during Pleistocene, fossils in South and North America.

Homarus—lobsters, two species (*H. americanus* and *H. gammarus*); extant in northern Atlantic Ocean.

Homo sapiens—modern humans; extant (so far) with worldwide distribution.

Ictidosuchoides—synapsid (therapsid); extinct, lived during Late Permian through Early Triassic Periods, fossils in South Africa.

Kimberella—mollusk-like animal; extinct, lived during Ediacaran Period, fossils in Australia and Russia.

Koolasuchus—amphibian, one species (*K. cleelandi*); extinct, lived during Early Cretaceous Period, fossils in Australia.

Lepidosiren paradoxa—South American lungfish, extant in South America.

Limulus polyphemus—horseshoe crab; extant in southeastern U.S.

Linepithema humile—Argentine ants; extant and native to Argentina but an invasive species throughout much of the world.

Linum—flax (flowering plant), about 200 species; extant throughout subtropical to temperate areas of the world.

Lithobates sylvaticus—Alaskan wood frog; extant from mid-continent of North America through Alaska.

Lystrosaurus—dicynodont therapsid, seven species; extinct, lived during the Late Permian-Early Triassic Periods, fossils in Africa, India, Russia, and South America.

Machimus polyphemi—gopher tortoise robber fly; extant in southeastern U.S.

Macrocheira kaempferi—Japanese spider crabs; extant in Pacific Ocean.

Marmota—marmots, fifteen species; extant in North America, Europe, and central Asia.

Megaceryle alcyon—belted kingfishers; extant in North and Central America.

Megalochelys atlas—tortoise; extinct, lived during Miocene to Pleistocene Epochs in central and southeastern Asia.

Meles meles—European badger; extant in Europe and Eurasia.

Merops—bee-eaters, twenty-four species; extant in Europe, Asia, and Australia.

Mesosaurus—reptile; extinct, lived during Late Permian Period, fossils in South America and Africa.

Metanephrops—lobsters, eighteen species; extant in Atlantic, Indian, and Pacific Oceans.

Morganucodon—mammiliform ("mammal-like" animal), five species; extinct, lived during Late Triassic Period, fóssils in North America, Asia, and Europe.

Moschorhinus—therapsid, one species (*M. kitchingi*); extinct, lived during Late Permian through Early Triassic Periods, fossils in South Africa.

Myocastor coypus—nutria (coypu); extant in South America but invasive species in Asia, Africa, Europe, and North America.

Myospalax—zokors (burrowing rodents), three species; extant in China, Mongolia, and Russia.

Myrmecocystus—honeypot ants, about thirty species; extant in North America.

Neobatrachus centralis—trilling frog; extant in central Australia.

Neobatrachus sudelli—painted burrowing frog; extant in southeastern Australia.

Neoceratodus forsteri—Australian lungfish; extant in Queensland, Australia.

Nephrops—lobsters, one species (*N. norvegicus*); extant in northeastern Atlantic Ocean and Mediterranean Sea.

Notoryctes—marsupial moles, two species (*N. typhlopa* and *N. caurinus*); extant in central Australia.

Ocypode—ghost crabs, twenty-one species; extant on tropical-subtropical beaches throughout much of the world.

Odontochelys semitestacea—turtle; extinct, lived during Late Triassic, fossils in China.

Onthophagus polyphemi polyphemi—punctuate beetle; extant in southeastern U.S.

Onthophagus polyphemi sparsisetosus—smooth gopher tortoise beetle; extant in southeastern U.S.

Opisthodon spenceri—Spencer's burrowing frog; extant in southeastern U.S.

Ornithorhynchus anatinus—platypus; extant in Australia.

Orodromeus—ornithopod dinosaur; extinct, lived during Late Cretaceous Period, fossils in Montana.

Oryctodromeus cubicularis—ornithopod dinosaur; extinct, lived during Late Cretaceous Period, fossils in Montana and Idaho.

Pampatherium—pampathere ("armadillo-like" mammal), three species; extinct, lived during Pleistocene Epoch in Central and South America.

Pappochelys—diapsid reptile ("turtle ancestor"), one species (*P. rosinae*); extinct, lived during Middle Triassic, fossils in Germany.

Paraceratherium—rhinoceros; extinct, lived during Oligocene, fossils in Eurasia.

Peromyscus floridanus—Florida mouse; extant in southeastern U.S.

Peromyscus maniculatus—deer mouse; extant in much of North America.

Peromyscus polionotus—oldfield mice; extant in southeastern U.S.

Philonthus gopheri—gopher tortoise rove beetle; extant in southeastern U.S.

Philonthus testudo—western gopher tortoise rove beetle; extant in southeastern U.S.

Pinus palustris—longleaf pine (tree); extant in southeastern U.S.

Platyplectrum ornatum—ornate burrowing frog; extant in northern Australia.

Pleuromeia—spore plant, five species; extinct, lived during Early Triassic Period, fossils in Asia, Europe, and Australia.

Pogonomyrmex badius—Florida harvester ants; extant in southeastern U.S.

Priodontes maximus—giant armadillo; extant in northern through central South America.

Procambarus clarkii—red swamp crayfish; extant in southeastern U.S., but also an invasive species throughout much of the world.

Procambarus hagenianus—prairie crayfish; extant in southeastern U.S.

Procolophon—procolophodonid (reptile), three species; extinct, lived during Early Triassic Period, fossils in South Africa, South America, and Antarctica.

Proganochelys—turtle, one species (*P. quenstedti*); extinct, lived during Late Triassic, fossils in Germany and Thailand.

Proterosuchus—archosauriform (reptile), four species; extinct, lived during Early Triassic Period, fossils in South Africa and China.

Protopterus—African lungfishes, four species; extant in central and southern Africa.

Psittacosaurus—ceratopsian dinosaur; extinct, lived during Early Cretaceous Period, fossils in Asia.

Puffinus puffinus—Manx shearwater; extant in north Atlantic Ocean on islands in Europe and North America.

Rana amurensis—Siberian wood frog; extant in northern Asia.

Rana aurora—northern red-legged frog; extant in northwestern North America.

Rattus norvegicus—brown rat (Norway rat); extant, native to Asia but an invasive species spread throughout the world by humans.

Rattus rattus—black rat; extant, native to Asia but an invasive species spread throughout the world by humans.

Rhamphiophis oxyrhynchus—rufous-beaked snake; extant in eastern Africa.

Rhizomys—bamboo rats, three species; extant in southeast Asia.

Riparia riparia—bank swallows; extant in all continents except Antarctica and Australia.

Scaphiopus holbrookii—spadefoot toads; extant in eastern North America.

Scelidotherium—giant ground sloth, two species (*S. leptocephalum* and *S. parodii*); extinct, lived during Pleistocene Epoch, fossils in South America.

Smilodon—saber-toothed cats, three species; extinct, lived during Pleistocene Epoch, fossils in North America and South America.

Sorex trowbridgii—Trowbridge's shrew; extant in northwestern North America.

Spermophilus—ground squirrel, about forty species; extant in North America and Eurasia.

Spinosaurus—non-avian theropod dinosaur, one species (*S. aegyptiacus*); extinct, lived during Early Cretaceous Period, fossils in northern Africa.

Stelgidopteryx serripennis—rough-winged swallows; extant in North and Central America.

Tachyglossus aculeatus—short-beaked echidna; extant in Australia and Papua New Guinea.

Tamias amoenus—yellow-pine chipmunk; extant in western North America.

Tetracynodon—therapsid, two species (*T. darti* and *T. tenius*); extinct, lived from Late Permian to Early Triassic, fossils in South Africa.

Thamnophis sirtalis—common garter snake; extant in much of North America.

Thomomys talpoides—northern pocket gopher; extant in much of North America.

Thrinaxodon—cynodont therapsid, one species (*T. liorhinus*); extinct, lived during Late Permian to Early Triassic Periods, fossils in South Africa and Antarctica.

Tiktaalik—sarcopterygian ("lobe-finned") fish; extinct, lived during Late Devonian Period, fossils in Ellesmere Island, Canada.

Treptichnus pedum—trace fossil (burrow); tracemaker extinct, lived during Early Cambrian with a widespread distribution then.

Tyrannosaurus—non-avian theropod dinosaur, one species (*T. rex*); extinct, lived during Late Cretaceous Period, fossils in western North America.

Uca—fiddler crabs, about a hundred species; extant in tropical-subtropical coastal areas throughout the world.

Uromastyx—spiky-tailed lizards, thirteen species; extant in northern Africa and the Middle East.

Ursus arctos—brown bear or grizzly bear; extant in northern North America, Asia, and Europe.

Ursus arctos middendorffi—Kodiak bear; extant in northwestern North America (southwest Alaska).

Ursus maritimus—polar bear; extant in northern part of North America and Siberia.

Varanus panoptes—yellow-spotted monitor; extant in Australia and Papua New Guinea.

Velociraptor—non-avian theropod dinosaur, two species (*V. mongoliensis*, *V. osmolskae*); extinct, lived during Late Cretaceous Period, fossils in Mongolia and China.

Yorgia—animal of uncertain affinity; extinct, lived during Ediacaran Period, fossils in Russia.

Zapus trinotatus—Pacific jumping mice; extant in northwestern North America.

Notes

CHAPTER 1: THE WONDROUS WORLD OF BURROWS

p. 4   "This idea stems from knowing how alligators descended from a lineage of crocodilians and their kin that were alive more than 100 million years ago . . ." M. Bronzati *et al.*, "Diversification events and the effects of mass extinctions on Crocodyliformes evolutionary history," *Royal Society Open Science* 2 (2015): 140385; http://dx.doi.org/10.1098/rsos.140385.

p. 4   ". . . this disaster caused a devastating worldwide crisis for life everywhere, whether in the oceans or on land . . ." (1) D. S. Robertson *et al.*, "Survival in the first hours of the Cenozoic," *Geological Society of America Bulletin* 116 (2004): 760–768; (2) P. Schulte *et al.*, "The Chicxulub asteroid impact and mass extinction at the Cretaceous-Paleogene boundary," *Science* 327(2010): 1214–1218; (3) P. R. Renne *et al.*, "Time scales of critical events around the Cretaceous-Paleogene boundary," *Science* 339 (2013): 684–687. Although now a bit dated, a good popular science book on the end-Cretaceous extinction (and with one of the best book titles ever) is: W. Alvarez, *T. Rex and the Crater of Doom* (Princeton, N.J.: Princeton University Press, 1997).

p. 5   "The first birds descended from theropod dinosaurs about 160 mya . . ." (1) P. Godefroit *et al.*, "A Jurassic avialan dinosaur from China resolves the early phylogenetic history of birds," *Nature* 498 (2013): 359–362; (2) M.Y.S. Lee *et al.*, "Sustained miniaturization

and anatomical innovation in the dinosaurian ancestors of birds,"
Science 345 (2014): 562–566. Several popular science books on the
transition of feathered "non-bird" dinosaurs to birds are available
now, but you might start with the most recent one: J. Pickerell,
Flying Dinosaurs: How Fearsome Reptiles Became Birds (New York:
Columbia University Press, 2014).

p. 6   "Hence these birds and alligators may have coevolved their
respective behaviors . . ." A. N. Nell *et al.*, "Presence of breeding
birds improves body condition for a crocodilian nest protector,"
PLoS One 11(3) (2016): e0149572; 10.1371/journal.pone.0149572.

p. 8   ". . . they will use these as places with plenty of fresh water (which
baby alligators need) . . ." (1) D. J. Lauren, "The effect of chronic
saline exposure on the electrolyte balance, nitrogen metabolism, and
corticosterone titer in the American alligator, *Alligator mississippiensis*,"
Comparative Biochemical Physiology 81A (1985), 217–223; (2) G. Grigg,
and D. Kirshner, *Biology and Evolution of Crocodylians*, (Clayton,
Victoria, Australia: CSIRO Publishing, 2015).

p. 9   "For alligators, the ideal is about 27–32°C (80–90°F); any higher
or lower than this range, they get torpid and die . . ." (1) I. L. Brisbin
et al., "Body temperatures and behavior of American alligators
during cold winter weather," *American Midland Naturalist* 107 (1982):
209–218; (2) F. Seebacher *et al.*, "Body temperature null distributions
in reptiles with nonzero heat capacity: seasonal thermoregulation in
the American alligator (*Alligator mississippiensis*)," *Physiological and
Biochemical Zoology: Ecological and Evolutionary Approaches* 76 (2003):
348–359; (3) V. A. Lance, "Alligator physiology and life history: the
importance of temperature," *Experimental Gerontology* 38 (2003):
801–805.

p. 9   "But in North America, these big reptiles can live as far north
as North Carolina . . ." (1) Lance (2003); (2) A. Parlin *et al.*, "Do
habitat characteristics influence American alligator occupancy
of barrier islands in North Carolina?" *Southeastern Naturalist* 14
(2015): 33–40.

p. 10   "Yet the weather in early March, with average lows around 10°C
(50°F), was still not quite warm enough to coax this one out of
its temporary refuge . . ." Average monthly temperature data were
not available for St. Catherines Island (Georgia), but are for Sapelo
Island, which is immediately south of St. Catherines. Data are at:
http://www.usclimatedata.com/climate/sapelo-island/georgia/
united-states/usga0501

p. 10   ". . . such as the Chinese alligator (*Alligator sinensis*), which
dig out extensive tunnels in riverbanks to make dens . . ." J.
Thorbjarnarson and X. Wang, *The Chinese Alligator: Ecology,*

Behavior, Conservation, and Culture, (Baltimore: Johns Hopkins University Press, 2010).

p. 10   "In fact, more than half of all crocodilian species (14 out of 23) dig and live in burrows during times of environmental stress, such as droughts . . ." I made this estimate after reading behavioral descriptions of each species of modern crocodilian, noting whether they were reported as digging dens (or not), and tabulating the results. Crocodilian species reported as digging dens are (in alphabetical order): *Alligator mississippiensis, A. sinensis; Caiman crododilus; Crocodylus acutus, C. intermedius, C. moreletti, C. niloticus, C. palustris; Osteolaemus tetraspis; Paleosuchus palpebrosus, P. trigonatus; Tomistoma schlegelii*. Basic information about each species is at the University of Florida site on crocodilians with the appropriately named URL of: www.crocodilian.com

p. 11   "Moreover, more than half of all penguin species make and live in burrows . . ." Surprisingly, penguin biologists do not agree on the exact number of penguin species, but it ranges from 18 to 20, with most species (at least 13) under the genera *Eudyptes* (crested penguins) and *Sphenicus* (banded penguins). Burrowing penguins will be covered in much more detail in Chapter 4, so wait for it.

p. 12   "These furry vertebrates evolved toward the end of the Triassic Period at about 220 *mya*, which was just after the start of the dinosaurs . . ." (1) T. S. Kemp, *The Origin and Evolution of Mammals*, (Oxford, England: Oxford University Press, 2005); (2) Z. Kielan-Jaworowska *et al., Mammals from the Age of Dinosaurs: Origin, Evolution, and Structure*, (New York: Columbia University Press, 2005).

p. 12   "The ancestors of mammals, synapsid reptiles, originated even farther in time to the Carboniferous Period, more than 300 *mya* . . ." (1) T. S. Kemp, "The origin of higher taxa: macroevolutionary processes, and the case of the mammals," *Acta Zoologica* 88 (2007): 3–22; (2) S. Voigt and M. Ganzelewski, "Toward the origin of amniotes: diadectomorph and synapsid footprints from the Early Late Carboniferous of Germany," *Acta Palaeontologica Polonica* 55 (2010): 57–72. Other important evolutionary divergence times to keep in mind for future chapters, such as the lungfish-tetrapod spilt, crocodile-bird split, and lizard-bird split, were proposed by: J. Müller, and R. R. Reisz, "Four well-constrained calibration points from the vertebrate fossil record for molecular clock estimates," *BioEssays* 27 (2005): 1069–1075.

p. 12   ". . . extreme global warming and other factors caused 95% of all species living on land and in the oceans to wave good-bye to their evaporated gene pools . . ." (1) Z.-Q. Chen and M. J. Benton, "The timing and pattern of biotic recovery following the end-Permian mass extinction," *Nature Geoscience* 5 (2012): 375–383; (2) J. L. Payne and M. E. Clapham, "End-Permian mass extinction in the oceans: an ancient analog for the Twenty-First Century?" *Annual Review of Earth and Planetary Sciences* 40 (2012): 89–111.

p. 13   "Skeletons of these animals have even been found in their fossil burrows . . ." (1) K. J. Carlson "The skull morphology and estivation burrows of the Permian lungfish, *Gnathorhiza serrate*," *Journal of Geology* 76 (1968): 641–663; (2) D. S. Berman, "Occurrence of *Gnathorhiza* (Osteichthyes: Dipnoi) in aestivation burrows in the Lower Permian of New Mexico with description of a new species," *Journal of Paleontology* 50 (1976): 1034–1039; (3) D. Hembree *et al.*, "Amphibian burrows and ephemeral ponds of the Lower Permian Speiser Shale, Kansas: evidence for seasonality in the midcontinent," *Palaeogeography, Palaeoclimatology, Palaeoecology* 203 (2004): 127–152.

p. 13   "Once self-buried, some modern-day lungfish, frogs, and toads can stay underground and stay torpid for months or years . . ." (1) M. E. Feder, *Environmental Physiology of the Amphibians* (Chicago: University of Chicago Press, 1992); (2) D. Randall *et al. Eckert Animal Physiology* (New York: Macmillan, 2002).

p. 14   "Their lengthy tunnels can also have nearly 400 species cohabitating in them . . ." (1) D. R. Jackson and E. G. Milstrey, "The fauna of gopher tortoise burrows." In J. E. Diemer *et al.* (eds.), *Relocation Symposium Proceedings, Technical Report No. 5* (Tallahassee, Fla.: Florida Game and Freshwater Fish Commission: 86–98, 1989); K. R. Lips, "Vertebrates associated with tortoise (*Gopherus polyphemus*) burrows in four habitats in south-central Florida," *Journal of Herpetology* 25 (1991): 477–481.

CHAPTER 2: BEYOND "CAVEMEN": A BRIEF HISTORY OF HUMANS UNDERGROUND

p. 16   ". . . a highland formed by tectonic uplift accompanied by extensive volcanism . . ." A. Ciner *et al.*, "Volcanism and evolution of the landscapes in Cappadocia." In D. Beyer *et al.* (eds.), *La Cappadoce Méridionale: de la Préhistoire à la Période Byzantine, 3èmes Rencontres d'Archéologie d'IFEA*, 2012 (Istanbul: IFEA-Institut Français d'Études Anatoliennes Georges Dumézil: 1–15, 2015).

p. 17   ". . . these rock-hewn homes were made more than 1,500 years ago by Christians trying to stay hidden from the then-reigning Romans . . ." R. V. Dam, *Becoming Christian: The Conversion of Roman Cappadocia* (Philadelphia: University of Pennsylvania Press, 2011).

p. 17   ". . . think of the Miocene Epoch, which ranged from 23 to 5 *mya*, and the Pliocene Epoch . . ." F. M. Gradstein *et al.*, *The Geologic Time Scale 2012* (Amsterdam: Elsevier, 2012).

p. 17   "... some expressed as lava flows but most of which blanketed the land as volcanic ash and mudflows ..." (1) J.-L. Le Pennec *et al.*, "Neogene ignimbrites of the Nevsehir Plateau (Central Turkey), stratigraphy, distribution and source constraints," *Journal of Volcanology and Geothermal Research* 63 (1994): 59–87; (2) E. Aydar *et al.*, "Correlation of ignimbrites in the central Anatolian volcanic province using zircon and plagioclase ages and zircon compositions," *Journal of Volcanology and Geothermal Research* 213–214 (2012): 83–97.

p. 18   "At some point after the Pliocene Epoch, weathering and erosion began working on the thick ashfall sediments ..." M. A. Sarikaya *et al.*, "Fairy chimney erosion rates on Cappadocia ignimbrites, Turkey: insights from cosmogenic nuclides," *Geomorphology* 234 (2015): 182–191.

p. 21   "The Derinkuyu underground city is the deepest known in the region of Cappadocia ..." Ö. Aydan *et al.*, "Geomechanical evaluation of Derinkuyu antique underground city and its implications in geoengineering," *Rock Mechanics and Rock Engineering*, 46 (2013): 731–754.

p. 21   "... this city might have been started well before Christians moved into the area, perhaps as long ago as when the Hittites were there ..." A. Bertini, "Underground cities, cave dwelling, cave homes: yesterday, today, tomorrow," in Alessandro Bucci and L. Mollo (eds.), *Regional Architecture in the Mediterranean Area: A Vision of Europe* (Florence, Italy: Alinea Editrice, 2010: 104–110); (2) R. Ulusay, and Ö. Aydan, *Cultural, Historical and Geo-Engineering Aspects of the Cappadocia Region (Excursion Guide)*, Eurock 2016, ISRM International Symposium, Rock Mechanics & Rock Engineering from the Past to the Future, 2016.

p. 21   "For air circulation, a 55-meter (180-foot) deep ventilation shaft connected to the surface ..." Ulusay and Aydan (2016).

p. 21   "... with the oil coming from locally grown flax (species of *Linum* and *Eruca*) ..." F. Ertuğ, "Linseed oil and oil mills in central Turkey flax/*Linum* and *Eruca*, important oil plants of Anatolia," *Anatolian Studies* 50 (2000): 171–185.

p. 22   Saying "... the centurion in charge [of a century]" is probably too simplistic, as centurions also could command two centuries (with a secondary centurion), or even an entire legion, or serve as a cohort to other centurions. As all of this is far too confusing for a mere paleontologist who did not serve in the military nor study Roman history, I suggest starting with the following reference for questions related to the structure and hierarchies of the Roman military: C. E. Brand, *Roman Military Law* (Austin: University of Texas Press, 2011).

p. 23 (1) "Özkonak had ten levels and was similarly planned for sustaining a large population of people underground . . ." (1) Ö. Aydan and R. Ulusay, "Geotechnical and geoenvironmental characteristics of man-made underground structures in Cappadocia, Turkey," *Engineering Geology* 69 (2003): 245–272; (2) Aydan *et al.* (2013). Also, a few secondary references report that Özkonak could hold as many as 60,000 people, but I could not verify this extraordinary number from primary literature. Hence I settled for the noncommittal description "large number."

p. 23 Plenty of secondary sources—especially Cappadocia tourist guidebooks—mention the "boiling oil" conduits and "communication/ventilation" pipes of Özkonak, but I could not find primary literature verifying these interpretations (most of which I also suspect is in Turkish, which I can't read). These same sources say that a Turkish farmer, Latif Acar, discovered Özkonak in 1972. So I will repeat that information and those interpretations of Özkonak for now, but hope to find documentation of when researchers made such diagnoses of Özkonak.

p. 24 ". . . this potential problem was solved by the people of Derinkuyu building an 8–9 kilometer (5 mile) long tunnel that joined with another underground city, Kaymakli . . ." Bertini (2010).

p. 24 Peer-reviewed articles about this newly discovered city were not published at the time of my writing this, but Jennifer Pinkowski (*National Geographic News*) wrote a news report about it, which includes an embedded video of laser-scanned portions of the underground complex: "Massive Underground City Found in Cappadocia Region of Turkey: Subterranean Retreat May Have Sheltered Thousands of People in Times of Trouble," *National Geographic News*, March 26, 2015: http://news.nationalgeographic. com/2015/03/150325-underground-city-cappadocia-turkey-archaeology/

p. 26 ". . . nuclear weapons were a source of dread, worth worrying about like no other menace . . ." An overview of U.S. government efforts in planning for nuclear war is in: D. F. Krugler, *This Is Only a Test: How Washington D.C. Prepared for Nuclear War* (New York: Palgrave Macmillan, 2006).

p. 26 "In these drills, schoolchildren watched training films that showed them how to get low ('duck') and below ('cover') . . ." The "Duck and Cover" campaign was started by the U.S. Federal Civil Defense Administration in 1951, which was conveyed through a short animated film (discussed more later), pamphlets, and drills. The effects of these on schoolchildren afterward (the film in particular) are discussed by: B. Jacobs, "Atomic kids: duck

FIGURE 32. Vertical burrow and trail made by an earthworm at Wormsloe Historic Site, Savannah (Georgia) following high tide in a nearby salt marsh. (Photo by Anthony J. Martin.)

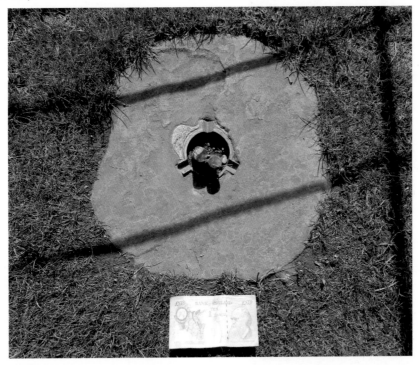

FIGURE 33. "Wormstone" used by Charles Darwin to measure rates of burrowing and soil overturn by earthworms, located in backyard of Down House near Downe Village, England; ten-pound note with Darwin on its reverse side for scale. (Photo by Anthony J. Martin.)

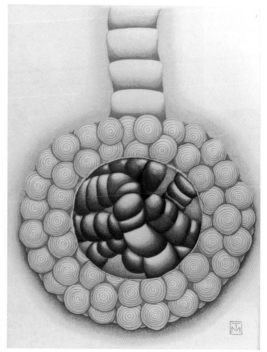

FIGURE 34. Stylized earthworm estivation chamber with "window" showing worm inside, created as a composite of modern and fossil chambers, with the latter based on research done by Verde *et al.* (2008). (Artwork by Anthony J. Martin, titled *On the Formation of Refuge from Drought* (2011).)

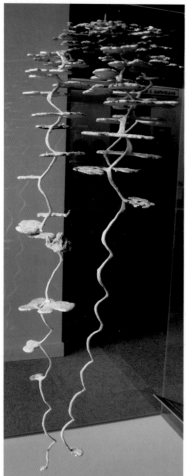

FIGURE 35. Nest of Florida harvester ant (*Pogonomyrmex badius*). Notice how the bigger chambers are concentrated toward the top of the nest, with smaller ones toward the bottom connected by spiraled shafts. The aluminum cast, which is about 2 meters (6.6 feet) tall, is displayed at the Smithsonian Institute, Washington (D.C.). (Photo by Anthony J. Martin.)

FIGURE 36. Modern crayfish at top of its burrow in a drained pond, still surviving by having its burrow intersect the local water table, and ferociously defending its burrow entrance; Atlanta (Georgia). (Photo by Anthony J. Martin.)

FIGURE 37. Modern crayfish burrow (LEFT) in Atlanta (Georgia) and Early Cretaceous (105 *mya*) crayfish burrows (RIGHT) from Victoria (Australia); scale = 1 centimeter (0.4 inches). (Photos by Anthony J. Martin.)

FIGURE 38. Early Cretaceous (130 *mya*) lobster burrow preserved as natural cast on bottom of limestone bed, Portugal. Although the lobster's body is not preserved, its leg impressions and body outline were left behind. (Photo by Anthony J. Martin.)

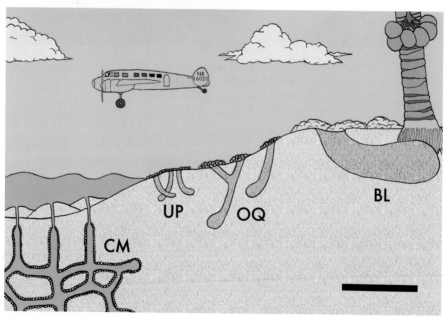

FIGURE 39. Decapod burrows from shallow marine to land (LEFT TO RIGHT): Carolina ghost shrimp (CM = *Callichirus major*), sand fiddler crab (UP = *Uca pugilator*), Atlantic ghost crab (OQ = *Ocypode quadrata*), and coconut crab (BL = *Birgus latro*); scale = 1 meter (3.3 feet). (Note: The coconut crab lives on Pacific islands, whereas the other species are in the southeastern U.S.) (Figure by Anthony J. Martin, based on various sources but mostly from Martin (2103).)

FIGURE 40. Fossil ghost crab burrow (about 4,000 years old) in longitudinal section showing full expression of "Y" shape, preserved in coastal limestone on San Salvador Island, Bahamas; scale in centimeters and held by Dottie Stearns. (Photo by Anthony J. Martin.)

FIGURE 41. Overlapping and densely populated decapod burrows, originally preserved in Late Triassic limestone of Italy. Pictured here is an epoxy resin cast made from the original burrows—the cast was part of Dolf Seilacher's *Fossil Art* display. This one he titled *Shrimp Burrow Jungle*. (Photo by Anthony J. Martin.)

Figure 42. Northern pocket gopher (*Thomomys talpoides*) emerging from its burrow. (Photo by Ty Smedes, courtesy of Washington Department of Fish and Wildlife.)

Figure 43. (ABOVE) Soil cast of northern pocket gopher burrow in central Idaho, made originally in snow but formed when snow melted and soil inside burrow was left behind; scale in centimeters. (Photo by Anthony J. Martin.). Figure 44. (BELOW) A re-vegetated landscape in 2009, nearly thirty years after devastation caused by the 1980 eruption of Mount St. Helens, an ecological recovery aided by burrowing northern pocket gophers; Mount St. Helens Volcanic National Monument, Washington. (Photo by Anthony J. Martin.)

FIGURE 45. (LEFT) Western toad (*Anaxyrus boreas*) found hopping around vicinity of John-ston Ridge Observatory, Mount St. Helens National Volcanic Monument, Washington. This toad likely descended from survivors of the 1980 eruption of Mount St. Helens, which also likely used burrows to survive. (Photo by Anthony J. Martin.) FIGURE 46. (RIGHT) Natural cast of spiraled burrow made by Oligocene beaver *Palaeocastor*, with beaver skeleton in its bottommost portion; at Smithsonian Institute, Washington (D.C.). Com-pare this burrow form with spiraled burrows made by Early Triassic cynodont *Diictodon* and modern yellow-spotted monitor lizard (*Varanus panoptes*). FIGURE 47. (BELOW) Tunnel attributed to Pleistocene giant ground sloths, 4.1 meters (13.5 feet) wide and 2.2 meters (7.2 feet) wide; in Rio Grande do Sul, Brazil. (Photo from Frank *et al.* (2013), with per-mission of the authors.)

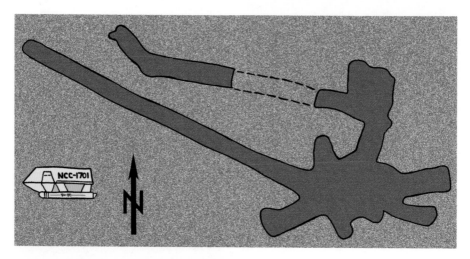

FIGURE 48. Map of tunnel system made by Pleistocene giant ground sloths, showing main gallery and branching from a main gallery. Dashed line indicates collapsed section; U.S.S. *Enterprise* shuttlecraft (7 meters/23 feet long) for scale. (Figure by Anthony J. Martin, based on figure in Frank et al. (2012).)

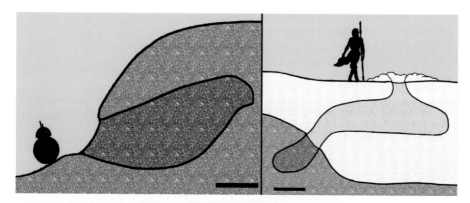

FIGURE 49. (LEFT) Grizzly bear den (*Ursus arctos*) dug in soil, longitudinal section based on figure by Servheen and Klaver (1983) in Montana; scale bar = 1 meter (3.3 feet). (RIGHT) Polar bear den (*Ursus maritimus*), first dug in soil, then the bear extended it upward after snow buried the den, longitudinal section based on figure by Jonkel *et al.* (1972) in Ontario (Canada); scale bar = 1 meter (3.3 feet).

and cover and atomic alert teach American children how to survive atomic attack," *Film & History* 40 (2010): 25–44. This and other societal responses to potential and real nuclear threats are summarized in a 2011 NPR story by Linton Weeks, "Living in the Atomic Age: Remember These Images?" (March 17, 2011): http://www.npr.org/sections/pictureshow/2011/03/17/134604352/images-of-the-atomic-age

p. 26   "In 1950, the U.S. military was authorized by the government to start excavating such a facility in Pennsylvania . . ." As one might imagine, I had to be very careful about using keywords in Internet searches about Raven Rock Mountain Complex and other U.S. government installations connected to the Department of Defense. At least one link to the Brookings Institution on U.S. government bunkers vanished when I revisited it, too. I also had to be skeptical of anything published online about it. (Hint: Many Web sites about underground government facilities revolve around themes of "secret" government conspiracies, none of which are as entertaining as *The X-Files*.) With that said, several seemingly well documented and non-crazy descriptions of Raven Rock Mountain Complex, with links to government documents and mainstream news stories about it, are at: (1) "Site-R Raven Rock Alternate Joint Communications Center (AJCC)," by the Federation of American Scientists: http://fas.org/nuke/guide/usa/c3i/raven_rock.htm; (2) "Raven Rock Mountain Complex (Site R)," by Public Intelligence: https://publicintelligence.net/raven-rock-mountain-complex-site-r/; and (3) "About Camp David: Raven Rock Mountain Complex," http://aboutcampdavid.blogspot.com/2011/08/raven-rock-mountain-complex.html (posted on August 23, 2011, updated March 2, 2016).

p. 27   ". . . the underground part of Raven Rock was hewn out of greenstone, a metamorphic rock sometimes erroneously called 'greenstone granite' . . ." The geologically contradictory term "greenstone granite" is a repeated error I've seen in nearly every source describing the geology of the site, and even shows up in Bill Gifford's article about Raven Mountain, "Bunker? What Bunker?" *New York Times Magazine*, December 3, 2000: http://partners.nytimes.com/library/magazine/home/20001203mag-gifford.html

p. 27   "To make this facility, which is located about 200 meters (650 feet) below the mountaintop, an estimated 14,000 square meters (500,000 square feet) of greenstone was removed . . ." (1) K. D. Rose, *One Nation Underground: The Fallout Shelter in American Culture* (New York: NYU Press, 2001); (2) T. Vanderbilt, *Survival City: Adventures Among the Ruins of Atomic America* (Chicago: University of Chicago Press, 2010).

p. 27   "... a dining hall (romantically named 'Granite Cove') ..." Vanderbilt (2010).

p. 27   "... an infirmary, and amenities for its employees ..." Vanderbilt (2010).

p. 27   "Raven Rock was used for joint communications between the U.S. Air Force, Navy, and Army throughout the Cold War ..." Krugler (2006).

p. 27   "... Vice President Dick Cheney's secret lair soon after the 2001 terrorist attacks ..." "Undisclosed Location Disclosed: A Visit Offers Some Insight into Cheney Hide-out," by Steve Goldstein, *Boston Globe*, July 20, 2004: http://archive.boston.com/news/nation/articles/2004/07/20/undisclosed_location_disclosed/

p. 28   Mount Weather was described in detail by Ted Gup, "Civil Defense Doomsday Hideaway," *Time*, June 24, 1991: http://content.time.com/time/magazine/article/0,9171,156041,00.html. Krugler (2006) and Vanderbilt (2010) also discuss its facilities further, and Vanderbilt wrote about it in *The Guardian*, "Is This Bush's Secret Bunker?", August 26, 2006: https://www.theguardian.com/world/2006/aug/28/usa.features11

p. 28   "... Mount Weather was kept mostly out of public view until 1974, when bad weather forced a commercial jet to crash into its side ..." On December 1, 1974, TWA Flight 514 crashed while flying from Columbus, Ohio, to Washington National Airport (now Reagan National) in D.C. It was rerouted to Dulles because of rainstorms in the area, but to no avail; all ninety-two people on board were killed. The crash resulted in reporting of the center by newspapers, and an NBC report on December 2, 1974, mentioned "a super-secret government installation" near the crash site: http://www.nbcwashington.com/news/local/309636521.html

p. 29   "The bunker ... was located underneath the Greenbrier Resort in White Sulphur Springs, West Virginia ..." A good introduction to this underground bunker is in: "A West Virginia Cold War Bunker Now a Tourist Spot," by John Strausbaugh in *New York Times*, November 12, 2006: http://www.nytimes.com/2006/11/12/travel/12heads.html?_r=0

p. 29   "... the Millboro Shale, which is in the Allegheny Front of the Appalachian Mountains ..." D. J. Soeder *et al.*, "The Devonian Marcellus Shale and Millboro Shale," *GSA Field Guides* 35 (2014): 129–160.

p. 29   "Another code name, "Project Greek Island ..." "The Secret Bunker Congress Never Used," *All Things Considered* (NPR), March 26, 2011: http://www.npr.org/2011/03/26/134379296/the-secret-bunker-congress-never-used

p. 29   ". . . its first level only about 6 meters (20 feet) below ground, and consisting of more than 10,000 square meters (120,000 square feet) of space . . ." Krugler (2006).

p. 30   ". . . Thomas 'Tip' O'Neill, said that Congress members were told they could not bring their families there . . ." "The Ultimate Congressional Hideaway," by Ted Gup, *Washington Post*, May 31, 1992: http://www.washingtonpost.com/wp-srv/local/daily/july/25/brier1.htm

p. 30   ". . . a 1992 *Washington Post* article revealed "Project Greek Island" for the needless boondoggle it was . . ." Strausburgh (*New York Times*), 2006.

p. 30   ". . . military planners suggested yet another U.S. government installation in 1962, imaginatively dubbed the Deep Underground Command Center . . ." L. Wainstein (Project Leader), *The Evolution of U.S. Strategic Command and Control and Warning: Part Three* (1961–1967), Study S-467. Institute for Defense Analyses (1975): 267–370.

p. 31   ". . . NORAD (North American Aerospace Defense Command) . . . was conceived as an early-warning/defense system against enemy missile attacks . . ." Basic information about NORAD is at their Web site, North American Aerospace Defense Command: http://www.norad.mil/

p. 31   ". . . access and side tunnels; main chambers; water reservoir; ventilation systems (with air filters); power generators; medical-treatment center . . ." "America's Fortress: Cheyenne Mountain, NORAD Live On," by Daniel Terdiman, *CNET*, July 8, 2009: https://www.cnet.com/news/americas-fortress-cheyenne-mountain-norad-live-on/

p. 31   ". . . electromagnetic pulse (EMP) emitted by a nuclear weapon, which would disable nearly all electronic-communications equipment . . ." An article with the cheery title "Failure to Protect U.S. Against Electromagnetic Pulse Threat Could Make 9/11 Look Trivial Someday," by Peter Kelly-Detwiler in *Forbes* (July 31, 2014), explains the basic principles behind EMPs generated by nuclear weapons and why these are still a threat: http://www.forbes.com/sites/peterdetwiler/2014/07/31/protecting-the-u-s-against-the-electromagnetic-pulse-threat-a-continued-failure-of-leadership-could-make-911-look-trivial-someday/#3564fd27fcdc

p. 31   "A 1991 report by the U.S. Department of Defense noted that one bunker was directly below the Kremlin . . ." United States Department of Defense. *Military Forces in Transition* (Washington: United States Department of Defense, 1991).

p. 31   ". . . which was below the Ramenki District of Moscow, was later confirmed and nicknamed the "Underground City . . ."

"Subterranean Secrets: Moscow's Metro-2 and the DIY-Subway System,' by M. Oksana, *Atlas Obscura*, February 1, 2011: http://www.atlasobscura.com/articles/subterranean-secrets-moscow-s-metro-2-and-the-diy-subway-system

p. 32   "With the end of the Cold War, at least one former bunker ("Bunker 42") . . . was converted into a tourist attraction . . ." A photo essay about this place, which I have absolutely no intention of ever visiting, is described here: "This Underground Bunker in Russia Will Let You 'Nuke' the United States for $26," by Dave Mosher, *Tech Insider*, December 7, 2015: http://www.techinsider.io/moscow-nuclear-bunker-42-tour-tagansky-2015-12/#the-entrance-looks-like-a-basic-military-checkpoint-3

p. 32   "And dig they did, as approximately 300,000 people excavated more than 20,000 bomb shelters underneath Beijing . . ." (1) S. Ye and G. R. Barmé, "Beijing Underground," *China Heritage Quarterly*, the Australian National University, 14 (June 2008): http://www.chinaheritagequarterly.org/features.php?searchterm=014_undergroundBeijing.inc&issue=014; (2) P. Mooney, *National Geographic Traveler: Beijing* (Washington: National Geographic Society, 2008).

p. 33   "Today, as many as a million people live in what has been called *Dixia Cheng* ("The Underground City") . . ." "Meet 'The Rat Tribe' Living in Beijing's Underground City," by Katie Hunt, *CNN*, February 18, 2015: http://www.cnn.com/2015/02/17/asia/china-beijing-rat-tribe/

p. 33   "Beijing's air pollution is caused by . . . coal . . . as well as millions of cars replacing bicycles as personal transportation over the past few decades . . ." E. Saikawa, *et al.* "The impact of China's vehicle emissions on regional air quality in 2000 and 2020: a scenario analysis," *Atmospheric Chemistry and Physics*, 11 (2011): 9465–9484.

p. 34   "Coober Pedy is a community of a few thousand people in the middle of the central desert of Australia . . ." This five-minute teaser of an upcoming documentary, titled *Outback Underworld: Living Coober Pedy*, provides a taste of this unique town: https://vimeo.com/138595646

p. 34   "Its unusual name comes from a corruption of *kupa-piti*, an indigenous (aboriginal) term . . ." According to *Place Names of South Australia* (http://www.slsa.sa.gov.au/manning/pn/c/c10.htm#cooberPedy), it translates as "boy's waterhole," but other sources say it was "white man's hole," "white man in a hole," and "white man's hole in the ground." Petter Naessan gives a lengthy and thorough exploration of this term, and concluded its origin was Kukata, but probably borrowed in

part from another indigenous language (Parnkalla): https://
coberpedyregionaltimes.files.wordpress.com/2008/09/the-
etymology-of-coober-pedy1.pdf

p. 34   "In 1915, when prospectors began looking for earth resources in
the Coober Pedy area . . ." A. W. Eckert, *The World of Opals* (New
York: John Wiley & Sons, 1997).

p. 34   ". . . Bulldog Shale, a geologic formation formed . . . during the
Early Cretaceous Period . . ." Geoscience Australia (Australian
Government), Australian Stratigraphic Units Database:
http://dbforms.ga.gov.au/pls/www/geodx.strat_units.
sch_full?wher=stratno=2891

p. 34   "The opal is so abundant that some of it even filled interior spaces
of Cretaceous clams, snails, and marine-reptile bones . . ." (1) B.
Pewkliang *et al.*, "Opalisation of fossil bone and wood: clues to the
formation of precious opal," *Regolith 2004, CRC LEME* (2004): 264–
268; (2) B. P. Kear *et al.*, "An archaic crested plesiosaur in opal from
the Lower Cretaceous high-latitude deposits of Australia," *Biology
Letters* 2 (2006): 615–619.

p. 35   ". . . where daytime temperatures during the summer could reach
40–45°C (104–113°F) . . ." Coober Pedy Climate: http://www.
weatherzone.com.au/climate/station.jsp?lt=site&lc=16090

p. 35   "January temperatures average around –10°C (14°F), and winter
snowfalls normally total about 2 meters (6.6 feet) thick . . ."
Statistics Canada, Weather Conditions in Capital and Major
Cities (Temperatures): http://www.statcan.gc.ca/tables-tableaux/
sum-som/l01/cst01/phys08b-eng.htm

p. 36   "Montreal was built on glacial clay, sand, pebbles, and gravel from
the Pleistocene Epoch . . ." V. K. Prest and J. Hode-Keyser, *Geology
and Engineering Characteristics of Surficial Deposits, Montreal Island and
Vicinity, Quebec.* Geological Survey Paper 75–27; 1977.

p. 36   "This complex, covering a 12 square kilometer (4.6 square mile)
area below the central business district . . ." *RESO—Montreal
Underground City (La Ville Souterraine)*: http://montrealvisitorsguide.
com/reso-underground-city-la-ville-souterraine/

p. 37   ". . . Joaquín Guzmán Loera (nicknamed *El Chapo*) said adiós
to his cell . . ." (1) "Underworld: How the Sinaloa Drug Cartel
Digs its Tunnels," by Monte Reel, *The New Yorker*, August 3, 2015:
http://www.newyorker.com/magazine/2015/08/03/underworld-
monte-reel; (2) "How Mexico's Most-Wanted Drug Lord Escaped
From Prison (Again)," by Larry Buchanan, Josh Keller, and Derek
Watkins, *New York Times*, January 8, 2016: http://www.nytimes.
com/interactive/2015/07/13/world/americas/mexico-drug-kingpin-
prison-escape.html?_r=0

p. 37 "His cartel is credited with the construction of many drug-smuggling tunnels under the border . . ." Reel (2015).

p. 38 ". . . when 76 Allied soldiers burrowed their way out in World War II Stalag Luft III . . ." T. Carroll, *The Great Escape from Stalag Luft III: The Full Story of How 76 Allied Officers Carried Out World War II's Most Remarkable Mass Escape* (New York: Simon and Schuster, 2010).

p. 38 "In June 2016, an international team of geoscientists and archaeologists . . . detected the subsurface tunnel . . ." (1) "Researchers Find Tunnel Dug by Jewish Prisoners to Escape Nazi Death Squads in Lithuania," by Melissa Etehad, *Washington Post*, June 29, 2016: https://www.washingtonpost.com/news/worldviews/wp/2016/06/29/researchers-find-tunnel-dug-by-jewish-prisoners-to-escape-nazi-death-squads-in-lithuania/; (2) "Escape Tunnel, Dug by Hand, Is Found at Holocaust Massacre Site," by Nicholas St. Fleur, *New York Times*, June 29, 2016: http://www.nytimes.com/2016/06/29/science/holocaust-ponar-tunnel-lithuania.html, which includes an embedded video of an upcoming PBS *NOVA* documentary on the discovery of the tunnel, due to be broadcast in 2017.

p. 43 "This is not surprising once you examine their [badgers'] robust claws, arms, and shoulders . . ." Badger: *Meles meles*, Arkive: http://www.arkive.org/badger/meles-meles/

p. 43 "Setts are also placed more than 5 meters (16 feet) away from an entrance, and usually are 1–2 meters (3.3–6.6 feet) deep . . ." T. J. Roper, "Badger *Meles meles* setts: architecture, internal environment and function," *Mammal Review* 22 (1992): 43–53.

p. 43 ". . . its family members will either convert a sett into a burial chamber by closing it off, or they drag the body outside . . ." Roper (1992).

p. 44 In writing this chapter, I did not read any specific instances of military or municipal planners using animal burrows as inspiration for defense or adaptability (respectively). However, military historians are aware of how underground environments still work, even with modern warfare. For an overview of this thinking, see: T. E. Eastler, "Military use of underground terrain: a brief historical perspective." In D. R. Caldwell, *et al.* (eds.), *Studies in Military Geology and Geography* (Berlin: Springer, 2004 21–37).

p. 44 "However, if one goes back to the original 1951 'Duck and Cover' propaganda film produced by the U.S. Federal Civil Defense Administration . . ." The film, which runs 9:14 minutes long and uses a combination of animation and live action, can be viewed at: https://archive.org/details/gov.ntis.ava11109vnb1

CHAPTER 3: KALEIDOSCOPES OF DUG-OUT DIVERSITY

p. 48 "This mix of pine forest and grassland savanna used to be the most common environment in this part of North America . . ." B. Finch *et al. Longleaf, as Far as the Eye Can See* (Chapel Hill: University of North Carolina Press, 2012).

p. 49 ". . . longleaf pines are fire-resistant, a superb advantage whenever a lightning-sparked conflagration sweeps through their habitat . . ." Finch *et al.* (2012).

p. 49 "Florida alone has an estimated 1.5 million electrified hits annually . . ." Although this number is repeated in many news reports—especially following a lightning-related fatality in Florida—I could not find the original National Weather Service report on this estimate. However, the Florida state government cites the same statistic in their 2010 Severe Weather Awareness Guide: http://www.floridadisaster.org/documents/2010SWAWGuide.pdf

p. 49 "Fire-wielding Native Americans also likely shaped longleaf-pine ecosystems over the past 12,000 years . . ." (1) H. R. Delcourt and P. A. Delcourt, "Pre-Columbian Native American use of fire on southern Appalachian landscapes," *Conservation Biology*, 11 (1997): 1010–1014; (2) T. Vale, *Fire Native Peoples, and the Natural Landscape* (Washington: Island Press, 2013).

p. 50 ". . . intensive deforestation and agriculture, shrank longleaf-pine habitats to less than 1% of their original realm . . ." Finch *et al.* (2012).

p. 50 ". . . more than 1,000 species of plants contribute to its ground cover . . ." This is the nice big round number I have heard cited by several field ecologists who study longleaf pine forests and read in news articles, but I have not been able to find a reference listing all (or even most) of the species. Nonetheless, it can be extrapolated from surveys that reveal the incredible biodiversity of these places, such as 40–50 plant species in just 1 square meter. A couple of references that help provide some appreciation for their species richness are: (1) R. K. Peet, and D. J. Allard, "Longleaf pine vegetation of the southern Atlantic and eastern Gulf Coast regions: a preliminary classification." In S. M. Hermann (ed.), *Proceedings of the 18th Tall Timbers Fire Ecology Conference on the Longleaf Pine Ecosystem: Ecology, Restoration, and Management*, Tallahassee, Fla., May 30–June 2, 1991: 45–82, 1993; (2) S. Jose *et al., Longleaf Pine Ecosystem: Ecology, Silviculture, and Restoration* (New York: Springer Science & Business Media, 2007); and (3) Finch *et al.* (2012).

p. 52 "Take a closer look at a gopher tortoise's front feet, and there is the first clue of what makes them such extraordinary animals . . ." K. Buhlmann *et al., Turtles of the Southeast*, (Athens: University of Georgia Press, 2008).

p. 52   "... makes the burrow just big enough to allow it room to move up and down the tunnel, but also so it can turn around while still inside ..." T. J. Doonan, and I. J. Stout, "Effects of gopher tortoise (*Gopherus polyphemus*) body size on burrow structure," *American Midland Naturalist* 131 (1994): 273–280.

p. 52   "These burrows also can have vertical depths of 6 meters (20 feet), but normally are 2–3 meters (6.6–10 feet) deep ..." (1) Doonan and Stout (1994).

p. 53   "In contrast, more clay in an otherwise sandy soil tends to retain carbon dioxide in a burrow ..." G. R. Ultsch and J. F. Anderson, "The respiratory microenvironment within the burrows of gopher tortoises (*Gopherus polyphemus*)," *Copeia* (1986): 787–795.

p. 54   "... male tortoises make as many as fifty burrows, which is two to three times as many as made by females ..." C. Guyer and S. M. Hermann, "Patterns of size and longevity for gopher tortoise burrows: implications for the longleaf pine-wiregrass ecosystem." *Bulletin of the Ecological Society of America* 78 (1997): 254.

p. 55   "... the male then commences a courtship display, unleashing a series of alluring head-bobbing moves ..." J. C. Moon *et al.*, "Multiple paternity and breeding system in the gopher tortoise, *Gopherus polyphemus*," *Journal of Heredity* 97 (2006): 150–157.

p. 55   "For her nest, she digs into the thickest part of the apron, lays a clutch of 2–10 eggs, and buries them ..." D. M. Epperson and C. D. Heise, "Nesting and hatchling ecology of gopher tortoises (*Gopherus polyphemus*) in southern Mississippi," *Journal of Herpetology* 37 (2003): 315–324.

p. 55   "... secret video footage ... of a mother tortoise showed it aggressively defending both its nest and burrow from a rapacious armadillo ..." This link is to the first of a two-part series taken at the Jones Ecological Center in southwest Georgia, titled "Gopher Tortoise Nest Defense 1 and 2": https://www.youtube.com/watch?v=Uasp5dPESuc

p. 55   "... with warmer temperatures resulting in more girl hatchlings and cooler temperatures producing more boys ..." (1) R. L. Burke *et al.*, "Temperature-dependent sex determination and hatching success in the gopher tortoise (*Gopherus polyphemus*)," *Chelonian Conservation and Biology* 2 (1996): 86–88; (2) J. P. Demuth, "The effects of constant and fluctuating incubation temperatures on sex determination, growth, and performance in the tortoise *Gopherus polyphemus*," *Canadian Journal of Zoology* 79 (2001): 1609–1620.

p. 55   "Newborn tortoises are also literally born to burrow, and ... may add their own smaller tunnels ..." (1) M. J. Aresco, "Habitat structures associated with juvenile gopher tortoise burrows on pine plantations in Alabama," *Chelonian Conservation and Biology* 3 (1999): 507–509; (2) Doonan and Stout (1994); (3) Epperson and Heise (2003).

p. 56 ". . . tortoise burrows also moderate any potential problems . . . in temperatures, maintaining even temperatures year round . . ." J. F. Douglass, and J. N. Layne, "Activity and thermoregulation of the gopher tortoise (*Gopherus polyphemus*) in southern Florida," *Herpetologica* (1978): 359–374.

p. 58 ". . . tortoises can host many roommates, and of a wide variety, consisting of nearly 400 species . . ." (1) Jackson and Milstrey (1989); Lipps (1991).

p. 58 ". . . more than 300 are invertebrates (mostly insects) and about 60 are vertebrates . . ." (1) Jackson and Milstrey (1989); (2) Lipps (1991); (3) B. W. Witz *et al.*, "Distribution of *Gopherus polyphemus* and its vertebrate symbionts in three burrow categories," *American Midland Naturalist* 126 (1991): 152–158.

p. 58 ". . . The eastern indigo snake is the longest snake native to North America . . ." In J. B. Jensen, *et al.* (eds.), *Amphibians and Reptiles of Georgia* (Athens: University of Georgia Press, 2008).

p. 59 ". . . Among the permanent occupants are gopher frogs (*Rana capito*) . . ." Jensen *et al.* (2008).

p. 60 "This burrowing mouse digs out horizontal, mouse-wide looping tunnels . . ." (1) C. A. Jones and R. Franz, "Use of gopher tortoise burrows by Florida mice (*Podomys floridanus*) in Putnam County, Florida," *Florida Field Naturalist* 18 (1990): 45–68; (2) J. N. Layne and R. J. Jackson, "Burrow use by the Florida mouse *Podomys floridanus* in south-central Florida," *American Midland Naturalist* 131 (1994): 17–23.

p. 60 ". . . tortoise burrows are the places to be, and anyone who can fit down these holes will seek sanctuary in one . . ." A. E. Kinlaw and M. Grasmueck, "Evidence for and geomorphologic consequences of a reptilian ecosystem engineer: the burrowing cascade initiated by the gopher tortoise," *Geomorphology* 157 (2012): 108–121.

p. 61 ". . . nine-banded armadillos (*Dasypus novemcinctus*), which take over abandoned gopher tortoise burrows . . ." A. J. Martin, *Life Traces of the Georgia Coast: Revealing the Unseen Lives of Plants and Animals* (Bloomington: Indiana University Press, 2013).

p. 61 ". . . burrows host a number of dung-beetle species that are well suited for using tortoise dung . . ." As one of the finest examples of proper allocations of state taxpayer funds I can imagine, the Florida Fish and Wildlife Conservation Commission (FWCC) maintains a Web site listing some of the notable insects living in gopher-tortoise burrows, titled "Invertebrate Commensals," with "bug guides" (more information) about each species: http://myfwc. com/wildlifehabitats/managed/gopher-tortoise/commensals/ invertebrates/

p. 62   "These insects are the gopher tortoise hister beetle . . ." Florida FWCC (see link in preceding note).

p. 62   "For waste control, the gopher tortoise burrow fly . . . also partakes in tortoise dung . . ." Florida FWCC.

p. 62   "One of these is the gopher tortoise robber fly . . ." Florida FWCC.

p. 63   ". . . the tiny gopher tortoise acrolophus moth . . . is well suited to these burrows . . ." Florida FWCC.

p. 63   ". . . the most specialized of all insects in a tortoise burrow, the tortoiseshell moth . . ." M. Deyrup *et al.*, "The caterpillar that eats tortoise shells," *American Entomologist* 51 (2005): 245–248.

p. 63   "The larvae that hatch from those eggs then dig a silk-lined burrow . . ." Deyrup *et al.* (2005).

p. 64   "For example, camel crickets . . . do not necessarily have to live in tortoise burrows . . ." Kinlaw and Grasmueck (2012).

p. 65   "The microwaves, acting like underground sonar, reflect and refract off substances with different densities . . ." R. Perisco, *Introduction to Ground Penetrating Radar: Inverse Scattering and Data Processing* (New York: Wiley-IEEE Press, 2014).

p. 66   ". . . and I presented our preliminary outcomes to my geologically inclined peers at the annual Geological Society of America meeting . . ." A. J. Martin *et al.*, "Ground-penetrating radar investigation of gopher-tortoise burrows: refining the characterization of modern vertebrate burrows and associated commensal traces," *Geological Society of America Abstracts with Programs* 43(5) (2011): 381.

p. 67   "The article, published by Al Kinlaw and Mark Grasmueck in 2012 in the journal *Geomorphology* . . ." Kinlaw and Grasmueck (2012).

p. 68   "It [geomorphology] is typically regarded as a subdivision of geology, and has been largely owned by geologists . . ." D. F. Ritter *et al.*, Process *Geomorphology*, 5th Edition, (Long Grove, Ill.: Waveland Press, 2011).

p. 68   ". . . *Zoogeomorphology: Animals as Geomorphic Agents*, written by David Butler and published in 1995 . . ." D. R. Butler, *Zoogeomorphology: Animals as Geomorphic Agents* (Cambridge, England: Cambridge University Press, 1995).

p. 69   ". . . a 2012 special issue of *Geomorphology*, co-edited by David Butler and Carol Sawyer . . ." In D. R. Butler and C. F. Sawyer (eds.), *Special Issue Zoogeomorphology and Ecosystem Engineering Proceedings of the 42nd Binghamton Symposium in Geomorphology; Geomorphology* 157–158 (2011): 1–192.

p. 71   ". . . but their absence in their preferred ecosystem can cause a different kind of cascade: A collapse . . ." Trophic cascades that cause ecological collapses are better documented, and from a variety of environments, such as the following examples: (1) A. H. Altieri *et al.*, "A trophic cascade triggers collapse of a salt-marsh ecosystem with intensive recreational fishing," *Ecology* 93 (2012): 1402–1410; (2) W. J. Ripple *et al.*, "Collapse of the world's largest herbivores," *Science Advances* 1 (2015): e1400103, doi: 10.1126/sciadv.1400103; and (3) D. R. Grubbs *et al.*, "Critical assessment and ramifications of a purported marine trophic cascade," *Scientific Reports* 6 (2016): 20970, doi: 10.1038/srep20970.

p. 71   ". . . it is no wonder why they [gopher tortoises] are protected under U.S. federal law . . ." Gopher Tortoise (*Gopherus polyphemus*): U.S. Fish and Wildlife Service, North Florida Ecological Services Office: https://www.fws.gov/northflorida/gophertortoise/gopher_tortoise_fact_sheet.html

p. 72   ". . . today that is where all four of the other extant species of *Gopherus* live . . ." This number is more likely five, as DNA analyses indicate that the Mojave desert and Sonoran desert tortoises are separate species: R. Murphy *et al.*, "The dazed and confused identity of Agassiz's land tortoise, *Gopherus agassizii* (Testudines: Testudinidae) with the description of a new species and its consequences for conservation," *ZooKeys* 113 (2011): 39–71.

p. 72   ". . . fossils identified as the genus *Gopherus* go back much further in time to the earliest part of the Oligocene Epoch . . ." V. H. Reynosos and M. Montellano, "A new giant turtle of the genus *Gopherus* (Chelonia: Testudinidae) from the Pleistocene of Tamaulipas, Mexico, and a review of the phylogeny and biogeography of gopher tortoises," *Journal of Vertebrate Paleontology* 24 (2004): 822–837.

p. 72   "The oldest known fossil tortoises are from nearly twice that time span . . ." H.-Y. Tong *et al.*, "A revision of *Anhuichelys* Yeh, 1979, the earliest known stem Testudinidae (Testudines: Cryptodira) from the Paleocene of China," *Vertebrata PalAsiatica* 54 (2016): 156–170.

p. 72   ". . . fossils of their immediate non-tortoise ancestors in Asia are from nearly 70 *mya*, which was during the Late Cretaceous Period . . ." M. J. Benton *et al.*, *The Age of Dinosaurs in Russia and Mongolia* (Cambridge, England: Cambridge University Press, 2004).

p. 72   "Some became gigantic, such as the Miocene-Pleistocene tortoise *Megalochelys atlas* . . ." G. L. Badam, "*Colossochelys* [*Megalochelys*] *atlas*, a giant tortoise from the Upper Siwalks of north India," *Bulletin of the Deccan College Research Institute* 40 (1981): 149–153.

p. 72   "Tortoises are also among the longest-lived of all animals . . . estimated to have lived more than 180 years . . ." "Oldest Animal

Jonathon the Tortoise is Still Going Strong at 183," by Laura Geggel, *Live Science*, January 15, 2016: http://www.livescience.com/53365-jonathan-tortoise-is-oldest-animal.html

p. 74   "Chelonia stretches back nearly 250 *mya*, whereas tortoises are a relatively recent innovation . . ." (1) T. R. Lyson *et al.*, "Transitional fossils and the origin of turtles," *Biology Letters* (2010) doi: 10.1098/rsbl.2010.0371; (2) R. R. Schoch and H.-D. Sues, "A Middle Triassic stem-turtle and the evolution of the turtle body plan," *Nature* 523 (2015): 584–587.

p. 74   ". . . they [turtles and tortoises] are now considered to have descended from diapsid reptiles . . ." R. R. Schoch and H.-D. Sues, "The diapsid origin of turtles," *Zoology* 119 (2016): 159–161.

p. 74   ". . . scientists estimate that ancestral clades of turtles, crocodilians, and dinosaurs split from one another about 260 *mya* . . ." (1) Müller and Reisz (2005); (2) Lyson *et al.* (2010).

p. 74   "*Eunotosaurus*, a 260 million-year-old reptile from Middle Permian rocks of South Africa, backs up this supposition . . ." T. R. Lyson *et al.*, "Evolutionary origin of the turtle shell," *Current Biology* 23 (2013): 1113–1119.

p. 74   "This turtle, named *Pappochelys* ('grandfather turtle'), is from is from Middle Triassic rocks of Germany . . ." Schoch and Sues (2015).

p. 74   "Both fossils are terrific examples of so-called 'transitional fossils' . . . they show many . . . traits evolutionary scientists . . . expect to find in a primitive turtle . . ." (1) Lyson *et al.* (2010); (2) G. S. Bever *et al.*, "Evolutionary origin of the turtle skull," *Nature* 525 (2015): 239–242.

p. 75   "In a 2016 study, paleontologist Tyler Lyson reexamined *Eunotosaurus* . . ." T. R. Lyson *et al.*, "Fossorial origin of the turtle shell," *Current Biology* 26 (2016): 1887–1894.

p. 75   ". . . with the greatest diversification of turtles during the Cretaceous Period at about 100–120 *mya* . . ." D. B. Nicholson *et al.*, "Climate-mediated diversification of turtles in the Cretaceous," *Nature Communications* 6 (2015): 7848; doi: 10.1038/ncomms8848.

p. 75   ". . . they [turtles] are fairly common in Cretaceous rocks of Victoria, Australia . . ." B. P. Kear and R. J. Hamilton-Bruce, *Dinosaurs in Australia: Mesozoic Life from the Southern Continent* (Clayton, Victoria, Australia: CSIRO Publishing, 2011).

p. 76   ". . . wood turtles (*Glyptemys insculpta*), which live as far north as Nova Scotia (Canada) . . ." *Glyptemys insculpta* (North American) Wood Turtle, *Animal Diversity Web*: http://animaldiversity.org/accounts/Glyptemys_insculpta/

p. 76   "Once buried, turtles go into a trance-like state of hibernation called *estivation* . . ." G. R. Ultsch, "The ecology of overwintering among turtles: where turtles overwinter and its consequences," *Biological Reviews* 81 (2006): 339–367.

p. 76   "... in 2009 I was not surprised to find turtle-sized and turtle-shaped burrows in ... Cretaceous sedimentary rocks of Victoria, Australia ..." A. J. Martin *et al.*, "The Great Cretaceous Walk: an ichnological survey of Lower Cretaceous strata in Victoria, Australia, and implications for Gondwanan paleontology," *10th North American Paleontological Convention Abstract Book, The Paleontological Society Special Publications* 13 (2014): 63.

p. 77   "These would be sea turtles, which include leatherback turtles (*Dermochelys coriacea*) ..." J. R. Spotilla, *Sea Turtles: A Complete Guide to Their Biology, Behavior, and Conservation* (Baltimore: Johns Hopkins University Press, 2014).

p. 77   "Once in a suitable place with the right sand, these turtles carve out deep, vase-like egg chambers with their rear flippers ..." Spotilla (2014).

p. 77   "... trace fossils of sea turtles' nests and trackways link this nest-digging behavior back to the middle of the Cretaceous Period ..." G. A. Bishop *et al.*, "The foundation for sea turtle geoarchaeology and zooarchaeology: morphology of recent and ancient sea turtle nests, St. Catherines Island, Georgia and Cretaceous Fox Hills Sandstone, Elber County, Colorado." In G. A. Bishop *et al.* (eds.), *Geoarchaeology of St. Catherines Island, Georgia, Anthropological Papers of the American Museum of Natural History*, No. 94 (2011): 247–269.

p. 78   "... the 2015 discovery of *Desmatochelys padillai* in Early Cretaceous (120 *mya*) rocks of Colombia ..." E. A. Cadena and J. F. Parham, "Oldest known marine turtle? A new protostegid from the Lower Cretaceous of Colombia," *PaleoBios* 32 (2015): 1–42.

p. 78   "Birds also have not been around as long as turtles in their evolutionary history, having only arisen about 160 *mya* ..." (1) Godefroit *et al.* (2013); (2) Lee *et al.* (2014).

p. 79   "... a specific episode of the PBS series *Dinosaur Train* ... they may blurt out the name "*Oryctodromeus!*" Go to the *PBS Dinosaur Train Field Guide* and click through to the O's to get to "*Oryctodromeus:*" http://pbskids.org/dinosaurtrain/fieldguide/

CHAPTER 4: HADEAN DINOSAURS AND BIRDS UNDERFOOT
p. 82   "Colonies of fairy penguins carry out these familial functions in many places over a very broad area of the Southern Hemisphere ..." (1) T. D. Roy *et al.*, *Penguins: The Ultimate Guide* (Princeton, N.J.: Princeton University Press, 2014); (2) P. G. Borboroglu and P. D. Boersma, *Penguins: Natural History and Conservation* (Seattle: University of Washington Press, 2015).

p. 83   "... a longtime fairy-penguin colony on the seashore of Phillip Island in Victoria, Australia, has become a huge tourist attraction ..." *Phillip Island Annual Report 2011–12*, Phillip Island Nature Parks Annual Report, with this and presumably other reports available online at: www.penguins.au.org

p. 84   "Burrow entrances are wider than tall and big enough to allow two penguins to pass each other on the way up or down ..." T. Williams, *The Penguins: Spheniscidae* (Oxford, England: Oxford University Press, 1995).

p. 84   "Male penguins do most of the primary burrowing or renovating of old burrows ..." A. Chung, *Eudyptula minor*: Little Penguin (Online), *Animal Diversity Web* (2011): http://animaldiversity.org/accounts/Eudyptula_minor/

p. 84   "Male-female pairs normally occupy a burrow throughout a breeding season, but burrow exchanges can happen afterward ..." (1) L. Bull, "Fidelity and breeding success of the blue penguin *Eudyptula minor* on Matiu-Somes Island, Wellington, New Zealand," *New Zealand Journal of Zoology* 27 (2000): 291–298; (2) T. Rogers and C. Knight, "Burrow and mate fidelity in the Little Penguin *Eudyptula minor* at Lion Island, New South Wales, Australia," *Ibis* 148 (2006): 801–806.

p. 84   "Unlike all other penguins, fairy-penguin mothers can lay two or three clutches of eggs in a breeding season ..." Roy *et al.* (2014).

p. 85   "Chicks stay in the burrow chamber for the next three to five weeks ..." J. Warham, "The nesting of the little penguin *Eudyptula minor*," *Ibis* 100 (1958): 605–616.

p. 85   "Ruth and I had previously read about fairy penguins in our *Lonely Planet* guidebook on Australia ..." P. Smitz *et al.*, *Australia*, 12th Edition, (Melbourne, Australia: Lonely Planet Guidebooks, 2005).

p. 87   "The Blackleaf is a geologic unit of sandstones and shales ..." T. S. Dyman and D. J. Nichols, "Stratigraphy of mid-Cretaceous Blackleaf and lower part of the Frontier Formations in parts of Beaverhead and Madison Counties, Montana," *USGS Bulletin* 1773 (1988): 1–27.

p. 88   "Herds of ... duck-billed ornithopods and big-horned ceratopsians, had spread throughout the western half of North America ..." Good overviews of ornithopod and ceratopsian dinosaur anatomy, evolution, and paleoecology are in: (1) P. Makovicky, "Marginocephalia," in M. K. Brett-Surman *et al.* (eds.), *The Complete Dinosaur*, 2nd Edition (Bloomington: Indiana University Press, 2012, 527–550); (2) R. J. Butler and P. M. Barrett, "Ornithopods," in M. K. Brett-Surman *et al.* (eds.), *The Complete Dinosaur*, 2nd Edition (Bloomington: Indiana University Press, 2012, 551–568).

p. 89   "Dave, Yoshi, and I named it *Oryctodromeus cubicularis* ("running digger of the den") . . ." D. J. Varricchio *et al.*, "First trace and body fossil evidence of a burrowing, denning dinosaur," *Proceedings of the Royal Society of London*, B 274 (2007): 1361–1368.

p. 90   ". . . zoologist Craig White . . . mathematical relationship (correlation) between body mass (weight) and area covered by an animal burrow's cross-section . . ." C. R. White, "The allometry of burrow geometry," *Journal of Zoology, London* 265 (2005): 395–493.

p. 91   "This overlapped with a weight range of 22–32 kilograms (48–70 pounds) Dave estimated for this dinosaur . . ." Varricchio *et al.* (2007).

p. 92   ". . . this is very common behavior in mammals, called denning . . ." G. A. Feldhamer *et al.*, *Wild Mammals of North America: Biology, Management, Conservation* (Baltimore: Johns Hopkins University Press, 2003).

p. 93   ". . . paleontologists have also found more *Oryctodromeus* remains associated with appropriately sized burrows . . ." Unfortunately, these tantalizing hints of additional *Oryctodromeus* specimens preserved in burrows of their own making have not been tested rigorously nor published yet, so I can't say anything more about them now, but hopefully the research will be completed sometime in 2017.

p. 94   "I claimed that not only that these were dinosaur burrows, but also that they were the oldest in the geologic record . . ." A. J. Martin, "Dinosaur burrows in the Otway Group (Albian) of Victoria, Australia, and their relation to Cretaceous polar environments," *Cretaceous Research* 30 (2009): 1223–1237.

p. 95   "One suggestion they made in a 1988 article . . . is that at least some of these dinosaurs overwintered in burrows . . ." P. V. Rich *et al.*, "Evidence for low temperatures and biologic diversity in Cretaceous high latitudes of Australia," *Science* 242 (1988): 1403–1406.

p. 95   "Among these were burrows of turtles (mentioned earlier), lungfish, and possibly mammals . . ." A. J. Martin *et al.*, "Trace fossils from the Valanginian-Albian of Victoria, Australia and what they tell us about vertebrates in Early Cretaceous polar environments," *Journal of Vertebrate Paleontology* 35 [Supplement to No. 3] (2015): 175.

p. 96   "These ornithopods, which were slightly smaller than *Oryctodromeus*, are typically found in compact masses of rock . . ." Varricchio *et al.* (2007).

p. 96   "*Psittacosaurus* is interpreted as a potential burrower based on . . . assemblage of 34 same-age juveniles all packed together . . ." Q. Meng *et al.*, "Parental care in an ornithischian dinosaur," *Nature* 431 (2004): 145–146.

p. 96   "However, other researchers suggested these juveniles were buried by a volcanic mud flow. . ." B. P. Hedrick et al., "The osteology and taphonomy of a Psittacosaurus bonebed assemblage of the Yixian Formation (Lower Cretaceous), Lioaning, China," *Cretaceous Research* 51 (2014): 321–340.

p. 98   ". . . this unduly endearing little owl . . . actually does little of its own burrowing . . ." (1) D. J. Martin, "Selected aspects of burrowing owl ecology and behavior," *The Condor* 75 (1973): 446–456; (2) R. G. Poulin *et al.*, "Factors associated with nest- and roost-burrow selection by burrowing owls (*Athene cunicularia*) on the Canadian prairies," *Canadian Journal of Zoology* 83 (2005): 1373–1380.

p. 99   ". . . burrowing owls can fly and migrate, ranging from the mid-continent and southeastern parts of North America . . ." (1) N. M. Korfanta *et al.*, "Burrowing owl population genetics: a comparison of North American forms and migratory habits," *The Auk* 122 (2005): 464–478; (2) G. L. Holroyd *et al.*, "Breeding dispersal of a burrowing owl from Arizona to Saskatchewan," *The Wilson Journal of Ornithology* 123 (2011): 378–381.

p. 99   "They seek, acquire, and eat a variety of animals . . ." (1) M. M. York *et al.*, "Diet and food-niche breadth of burrowing owls (*Athene cunicularia*) in the Imperial Valley, California," *Western North American Naturalist* 62 (2002): 280–287; (2) M. J. Nabte *et al.*, "The diet of the burrowing owl, *Athene cunicularia*, in the arid lands of northeastern Patagonia, Argentina," *Journal of Arid Environments* 72 (2008): 1526–1530.

p. 99   "Owls exploit dung beetles' weakness for fresh, sweet mammal feces . . . plopping it down as bait . . ." (1) D. J. Levey *et al.*, "Animal behaviour: use of dung as a tool by burrowing owls," *Nature* 431 (2004): 39; (2) M. D. Smith and C. J. Conway, "Use of mammal manure by nesting burrowing owls: a test of four functional hypotheses," *Animal Behavior* 73 (2007): 65–73.

p. 100   ". . . burrowing owl egg clutches can have as few as two eggs, but normally range from five to ten . . ." (1) Martin (1973); J. L. Wade and J. R. Belthoff, "Responses of female burrowing owls to alterations in clutch size: are burrowing owls determinate or indeterminate egg-layers?" *Journal of Raptor Research* 50 (2016): 84–91.

p. 100   "These birds are native to the North Atlantic . . ." M. P. Harris and S. Wanless, *The Puffin* (London: T. & A. D. Poyser, 2011 [Reprinted by A & C Black Publishers Ltd.]).

p. 100   "Burrows are normally a little less than a meter (3.3 feet) long . . ." (1) M. Hornung, "Burrows and burrowing of the puffin (*Fratercula arctica*)," *Bangor Occasional Paper 10*, Institute of Terrestrial Ecology, Bangor, Me. (1982): 1–30; (2) Harris and Wanless (2011).

THE EVOLUTION UNDERGROUND

p. 100   "... puffins are mostly quiet birds, but once in their burrows, they
          get less inhibited and make distinctive noises ..." The Web site for
          the Cornell Lab of Ornithology is a mother lode of information
          about birds and includes sound files of many species. To hear the
          sounds made by an Atlantic puffin in its burrow, go to: https://
          www.allaboutbirds.org/guide/Atlantic_Puffin/sounds

p. 100   "Once they make or choose an abandoned burrow, they stick with
          it in successive breeding seasons ..." Harris and Wanless (2011).

p. 101   "Such behavior is called *site fidelity* ..." Site fidelity, also sometimes
          called philopatry, has been studied widely in birds, but also applies
          to sea turtles and other nesting animals that return to the same
          place every year to breed. Site fidelity has even been suggested for
          dinosaurs, which is explained in detail in this book by this guy,
          and is totally worth buying for you and your friends: A. J. Martin,
          Dinosaurs Without Bones: Dinosaur Lives Revealed by Their Trace Fossils
          (New York: Pegasus Books, 2014).

p. 101   "Puffins are also well known for their elaborate walking displays,
          performed just outside of their burrows ..." Harris and Wanless (2011).

p. 101   "Manx shearwaters are not puffins ... also using burrows for
          their nesting ..." P. C. James, "How do Manx shearwaters *Puffinus
          puffinus* find their burrows?" *Ethology* 71 (1986): 287–294.

p. 101   "Shearwaters are marvelously efficient seafarers ... capable of
          migrating more than 10,000 kilometers ..." (1) B. Hoare, *Animal
          Migration: Remarkable Journeys in the Wild* (Berkeley: University
          of California Press, 2009); (2) T. Guilford *et al.*, "Migration and
          stopover in a small pelagic seabird, the Manx shearwater *Puffinus
          puffinus*: insights from machine learning," *Proceedings of the Royal
          Society of London B* 276 (2009): 1215–1223.

p. 102   "Shearwaters ... are more than able to dig their own, with some
          more than a meter (3.3 feet) long ..." (1) James (1986); (2) M.
          Brooke, *The Manx Shearwater* (London: T. & A. D. Poyser, 1990
          [Reprinted by A & C Black Publishers Ltd.]).

p. 102   "If making new burrows, they might crash into a nearby puffin
          maternity ward ..." (1) James (1986); (2) M. Brooke, (1990).

p. 102   "... these two birds [puffins and shearwaters] collectively qualify
          as ecosystem engineers in shore-side soils ..." (1) Harris and
          Wanless (2011); (2) Brooke (1990); (3) W. J. Bancroft *et al.*, "Burrow
          building in seabird colonies: a soil-forming process in island
          ecosystems," *Pedobiologia* 49 (2005): 149–165.

p. 102   "They begin their burrows by pecking with their beaks into banks
          soft enough to break apart ..." (1) M. Elbroch and E. Marks, *Bird
          Tracks and Sign of North America* (Mechanicsburg, Penn.: Stackpole
          Books, 2001); (2) A. J. Martin (2013).

p. 102   ". . . they then use a combination of beaks and feet to extend the
          burrow farther into the bank for about a meter or two . . ." (1)
          Elbroch and Martin (2001); (2) Martin (2013).

p. 103   "Burrows are placed high on bluffs for a reason, keeping . . . out
          of harm's way from predators and well above water bodies . . ."
          M. Silver, and C. R. Griffin, "Nesting habitat characteristics of
          bank swallows and belted kingfishers on the Connecticut River,"
          Northeastern Naturalist 16 (2009): 519–534; also check out the
          references cited within this article for an overview of bank nesting
          in these two species.

p. 103   "Bee-eaters consist of 27 species, with nearly all under the genus
          Merops, and they are geographically widespread . . ." C. H. Fry, The
          Bee-Eaters (London: T. & A. D. Poyser, 1984, reprinted in 2010).

p. 103   "Burrow making is a tag-team effort, with male-female pairs
          working together . . ." C. H. Fry, "The social organisation of bee-
          eaters (Meropidae) and co-operative breeding in hot-climate birds,"
          Ibis 114 (1972): 1–14.

p. 104   "For example, European bee-eaters (Merops apiaster) make 1.5–2-meter
          (5–6.6 foot) long burrows . . ." (1) Fry (1984); (2) C. H. Frey et al.,
          Kingfishers, Bee-eaters & Rollers (London: A & C Black Publishers, 1992,
          republished in 2010); (3) Some bee-eater species make incredibly long
          burrows compared to their body sizes. For example, burrows made by
          black-headed bee-eaters (Merops breweri) can be as long as 3.2 meters
          (10.5 feet): B. K. Schmidt and W. R. Branch, "Nests and eggs of the
          black-headed bee-eater (Merops breweri) in Gabon, with notes on other
          bee-eaters," Ostrich 76 (2005): 80–81.

p. 104   "Although bee-eaters are very often colonial nesters, this large
          number of "starter holes" creates . . . impression . . . many more
          birds are present . . ." Fry (1984).

p. 104   ". . . in arid environments, burrows maintain high humidity levels,
          near-constant temperatures year-round . . ." A. Casas-Crivill and F.
          Valera, "The European bee-eater (Merops apiaster) as an ecosystem
          engineer in arid environments," Journal of Arid Environments 60
          (2005): 227–238.

p. 104   "Both of these factors [soil firmness and clay content] are
          important for successful nesting . . ." (1) P. Heneberg and K.
          Šimeček, "Nesting of European bee-eaters (Merops apiaster) in
          central Europe depends on the soil characteristics of nest sites,"
          Biologia 59 (2004): 205–211; (2) P. Heneberg, "Soil penetrability
          as a key factor affecting the nesting of burrowing birds," Ecological
          Research 24 (2009): 453–459.

p. 105   "The amount of clay minerals in the sediment is also important
          . . . decreasing ventilation and the ability of a . . . burrow to drain

excess water . . ." (1) Heneberg and Šimeček (2004); Heneberg (2009).

p. 105 ". . . they [parrots] are also regarded as among the most intelligent of birds . . ." I. Pepperberg, *Alex and Me: How a Scientist and a Parrot Discovered a Hidden World of Animal Intelligence and Formed a Deep Bond in the Process* (New York: HarperCollins, 2009). I also recommend reading Virginia Morrel's chapter summarizing Irene Pepperberg's research on her parrot, Alex, and his extraordinary intelligence: V. Morrell, *Animal Wise: How We Know Animals Think and Feel* (New York: Broadway Books, 2012).

p. 105 "Known by Argentinians as *loro barranquero* ("burrowing parrot"), these birds . . . consist of four subspecies . . ." J. M. Forshaw, (2010), *Parrots of the World* (Princeton, N.J.: Princeton University Press, 2010).

p. 105 "Some seaside cliffs may host as many as 50,000 parrot burrows . . ." J. F. Masselo *et al.*, "Population size, provisioning frequency, flock size and foraging range at the largest known colony of Psittaciformes: the Burrowing Parrots of the north-eastern Patagonian coastal cliffs," *Emu* 106 (2006): 69–79.

p. 106 ". . . they all require firm substrates for digging burrows, such as poorly cemented sandstones or limestones . . ." (1) G. Leonardi and N. R. Oporto, "Biogenetic erosion structures (modern parrots' nests) on marine and fluvial cliffs in southern Argentina," *Academia Brasileira de Ciências* 55 (1983): 293–295; (2) J. F. Masello and P. Quillfeldt, "Consequences of La Niña phase of ENSO for the survival and growth of nestling Burrowing Parrots on the Atlantic coast of South America," *Emu* 104 (2006): 337–346.

p. 106 ". . . their burrows often follow bedding planes of the host sedimentary rock . . ." Marsello *et al.* (2006).

p. 106 "Each burrow has a large nesting chamber at its end, where the mother parrot lays two to five eggs . . ." J. F. Masello and P. Quillfeldt, "Chick growth and breeding success of the Burrowing Parrot," *Condor* 104 (2002): 574–586.

p. 106 "Yet another problem facing these parrots is poaching, where people seek out nest holes, capture and cage nestlings . . ." J. L. Tella *et al.*, "Anthropogenic nesting sites allow urban breeding in Burrowing Parrots *Cyanoliseus patagonus*," *Ardelo* 61 (2014): 311–321.

p. 107 ". . . they all belong to a broad clade called Ornithurae . . ." (1) L. M. Chiappe, *Mesozoic Birds: Above the Heads of the Dinosaurs* (Berkeley: University of California Press, 2002); (2) M.S.Y. Lee *et al.*, "Morphological clocks in paleontology, and a mid-Cretaceous origin of Crown Aves," *Systematic Biology* 63 (2014): 442–449.

p. 108 ". . . any birder attempting a 'Big Year,' in which she or he may try to identify as many species of birds as possible in one year . . ."

For birders, a "Big Year" is a competition to see who can identify the greatest number of bird species in a calendar year, starting January 1. Mark Obsmascik's book *The Big Year: A Tale of Man, Nature, and Fowl Obsession* (New York: Free Press, 2004) describes this quest for three birders attempting it in 1998; a movie based on the same story was released in 2011, starring Jack Black, Steve Martin, and Owen Wilson.

p. 108   "The only modern birds belonging to this clade [Palaeognathae] are the flightless ratites . . . as well as tinamous . . ." G. Mayr, *Avian Evolution: The Fossil Record of Birds* (New York: John Wiley & Sons, 2016).

p. 108   ". . . kiwis are represented by five species under the genus *Apteryx* . . ." R. Colbourne, "Kiwi (*Apteryx* spp.) on offshore New Zealand islands: populations, translocations and identification of potential release sites," *DOC Research and Development Series* 208 (2006).

p. 108   ". . . they [kiwis] descended from winged ancestors that flew to New Zealand around the start of the Miocene Epoch at about 20 *mya* . . ." T. H. Worthy *et al.*, "Miocene fossils show that kiwi (*Apteryx*, Apterygidae) are probably not phyletic dwarves." In U. B. Göhlich and A. Kroh (eds.), *Proceedings of the 8th International Meeting of the Society of Avian Paleontology and Evolution*, Verlag Naturhistorisches Museum Wien (2013): 63–80.

p. 108   ". . . Perhaps the most startling . . . adaptation . . . is in the females . . . which can hold and lay an egg one-sixth of their body weights . . ." *Why is the Kiwi's Egg So Big?* by Sam Dean, *Audubon*, February 25, 2015: http://www.audubon.org/news/why-kiwis-egg-so-big

p. 109   "For the brown kiwi (*A. australis*), their burrows are about 12–14 centimers wide and 1–1.5 meters (3.3–5 feet) long . . ." J. A. McLennan, "Breeding of the North Island brown kiwi, *Apteryx australis mantelli*, in Hawke's Bay, New Zealand," *New Zealand Journal of Ecology* 11 (1988): 89–97.

p. 109   "For the little spotted kiwi (*A. owenii*), burrows can be as long as 2.5 meters . . ." J. N. Jolly, "A field study of the breeding biology of the little spotted kiwi (*Apteryx owenii*) with emphasis on the causes of nest failures," *Journal of the Royal Society of New Zealand* 19 (1989): 433–448.

p. 109   ". . . kiwi burrows used solely as nests and dens tend to be shallower than adult-only roosting burrows . . ." (1) J. A. McLennan *et al.*, "Range size and denning behaviour of brown kiwi *Apteryx mantelli*, in Hawkes Bay, New Zealand," *New Zealand Journal of Ecology* 10 (1987): 97–107; (2) Jolly (1989).

p. 110   "Male tinamous instead make nests on ground surfaces . . ." P.L.R. Brennan, "Incubation in Great Tinamou (*Tinamus major*)," *Wilson Journal of Ornithology* 121 (2009): 506–511.

p. 110   "The oldest known fossil penguin is from the earliest part of the
Paleogene Period . . ." K. E. Slack *et al.*, "Early penguin fossils,
plus mitochondrial genomes, calibrate avian evolution," *Molecular
Biology and Evolution* 23 (2006): 1144–1155.

p. 110   ". . . penguin ancestors originated in the Late Cretaceous Period . . ."
Slack *et al.* (2006).

p. 111   "Psittaciformes . . . also probably originated in the Late Cretaceous
. . ." T. F. Wright *et al.* "A multilocus molecular phylogeny of the
parrots (Psittaciformes): support for a Gondwanan origin during
the Cretaceous," *Molecular Biology and Evolution* 25 (10) (2008):
2141–2156.

p. 111   ". . . fossil owls have been discovered in rocks from about 60 *mya*
. . ." (1) P. V. Rich and D. J. Bohaska, "The Ogygoptyngidae, a new
family of owls from the Paleocene of North America," *Alcheringa*
5 (1981): 95–102; (2) C. Mourer-Chauvire, "A large owl from the
Paleocene of France," *Palaeontology* 37 (1994): 339–348.

p. 111   ". . . kingfishers and bee-eaters . . . are geologically younger, with
their fossils dating to the middle of the Eocene Epoch . . ." (1) Mayr
(2009); (2) E. Bourdon *et al.*, "A roller-like bird (Coracii) from
the Early Eocene of Denmark," *Scientific Reports* 6 (2016): 34050;
doi:10.1038/srep34050.

p. 111   "Puffins similarly belong to a clade (Alcidae) with its oldest fossils
in Eocene rocks . . ." Mayr (2009).

p. 111   ". . . shearwaters . . . (*Puffinus*) . . . extant for the past 5 million years
or so . . ." N. Henderson and B. J. Gill, "A mid-Pliocene shearwater
skull (Aves: Procellariidae: *Puffinus*) from the Taihape Mudstone,
central North Island, New Zealand," *New Zealand Journal of Geology
and Geophysics* 53 (2009): 327–333.

p. 112   "Synapsids . . . occupied most ecological niches of land
environments by the end of the Permian Period . . ." In C. F.
Kammerer *et al.* (eds.), *Early Evolutionary History of the Synapsida*
(Dordrecht, Netherlands: Springer, 2013).

p. 113   ". . . some . . . synapsids . . . discovered in their burrows, preserved
in sediments laid down during the earliest part of the Triassic
Period . . ." (1) G. H. Groenewald *et al.*, "Vertebrate burrow
complexes from the Early Triassic *Cynognathus* Assemblage Zone
(Driekoppen Formation, Beaufort Group) of the Karoo Basin,
South Africa," *Palaios* 16 (2001): 148–160; (2) S. P. Modesto and J.
Botha-Brink, "A burrow cast with *Lystrosaurus* skeletal remains from
the Lower Triassic of South Africa," *Palaios* 25 (2010): 274–281.

p. 113   ". . . modern tinamous . . . do exactly this sort of opportunistic
behavior today by ducking (in an avian way, no less) and covering
in burrows . . ." J. M. Barnett *et al.*, "*Nothura minor* (Tinamidae) a

globally threatened Cerrado species new to Paraguay," *Ararajuba* 12 (2010):153–155.

CHAPTER 5: BOMB SHELTERS OF THE PHANEROZOIC

p. 115   "The *Lystrosaurus* woke up to a nightmare . . ." This story was at least partially inspired by an essay written by Annalee Newitz titled "*Lystrosaurus*: The Most Humble Badass of the Triassic," published May 28, 2013 in *National Geographic Phenomena*: http://phenomena. nationalgeographic.com/2013/05/28/lystrosaurus-the-most-humble-badass-of-the-triassic/. I also highly recommend reading Newitz's book (*Scatter, Adapt, and Remember: How Humans Will Survive a Mass Extinction*, New York: Anchor Books, 2013), in which she talks more about why *Lystrosaurus* deserves our admiration as a model survivor of global adversity. The rest of the story was modeled after a Joseph Campbell mythic quest, but also contains elements of the syndicated TV series *Xena: Warrior Princess* (1995–2001), as it should.

p. 115   ". . . the *Glossopteris* tree to the left . . ." *Glossopteris* was a genus of seed plant (sometimes mistakenly called "seed ferns," because they were not true ferns), with more than 70 species that lived in southern landmasses (Gondwana) during the Late Permian Period. A few species may have survived into the Early Triassic, but this has not yet been resolved. It grew to tree sizes and was extremely widespread, which allowed geologists to use it later for reconstructing Gondwanan paleogeography. For more detailed information, see: S. McLoughlin, "*Glossopteris*: insights into the architecture and relationships of an iconic Permian Gondwanan plant," *Journal of the Botanical Society of Bengal* 65 (2012): 1–14.

p. 117   "Its ramp went down at a gentle angle, but turned right, then left, making a descending semi-spiral . . ." Probable and definite *Lystrosaurus* burrows have been described with such semi-spiraled forms: (1) G. H. Groenewald, "Burrow casts from the *Lystrosaurus–Procolophon* Assemblage-zone, Karoo Sequence, South Africa," *Koedoe* 34 (1991): 13–22; (2) M. F. Miller *et al.*, "Tetrapod and large burrows of uncertain origin in Triassic high paleolatitude floodplain deposits, Antarctica," *Palaios* 16 (2001): 218–232. (3) G. J. Retallack *et al.*, "Vertebrate extinction across Permian-Triassic boundary in Karoo Basin, South Africa," *Geological Society of America Bulletin* 115 (2003): 1133–1152.

p. 117   "*Lystrosaurus* was adapted for it [low oxygen] . . ." This interpretation of *Lystrosaurus* as tolerant of low-oxygen conditions is partly speculative, but based on its probable burrowing lifestyle (similar to gopher tortoises) and its surviving a low-oxygen crisis in the earth's atmosphere, as explained by: R. B. Huey and P. D. Ward, "Hypoxia, global warming, and terrestrial Late Permian extinctions," *Science* 308 (2005): 398–401. Also, its physiology was certainly special, as indicated by a few studies, such as: M. Laaß *et al.*, "New insights into the respiration and metabolic physiology of *Lystrosaurus*," *Acta Zoologica* 92 (2010): 363–371.

p. 117   "... but a series of interconnected sinuses inside her skull received minute chemical cues ..." Again, the olfactory abilities of *Lystrosaurus* are not quite known yet, but its nasal passages are, and other parts of its skull interior have been studied: M. Laaß *et al.* (2010).

p. 118   "... as well as remains of other animals she had never seen, living or dead: *Ictidosuchoides, Moschorhinus,* and *Tetracynodon* ..." *Moschorhinus* was a contemporary of *Lystrosaurus*, and a close association between two of their skulls and a *Lystrosaurus* skeleton suggests that they were mired together, perhaps because the *Moschorhinus* wanted to feed on the *Lystrosaurus* carcass: R.M.H. Smith and J. Botha-Brinks, "Anatomy of a mass extinction: Sedimentological and taphonomic evidence for drought-induced die-offs at the Permo-Triassic boundary in the main Karoo Basin, South Africa," *Palaeogeography, Palaeoclimatology, Palaeoecology* 396 (2014): 99–118.

p. 118   "... She ate by cropping the leaves and stems with her beak, then moved her jaws forward and backward ..." To get an idea of how *Lystrosaurus* chewed, read: (1) S. C. Jasinoski *et al.*, "Comparative feeding biomechanics of *Lystrosaurus* and the generalized dicynodont *Oudenodon*," *The Anatomical Record* 292 (2009): 862–874; (2) S. C. Jasinoski *et al.*, "Functional implications of dicynodont cranial suture morphology," *Journal of Morphology* 271 (2010): 705–728.

p. 118   "Because *Lystrosaurus* was inherently a social animal ..." Because undoubted *Lystrosaurus* tracks have not yet been identified, which would show whether or not they moved about in groups, we do not know much about their gregariousness. But at least one study concluded that a bone bed composed mostly of *Lystrosaurus* remains suggests they were aggregating, which was aided by burrowing and done as a means of dealing with climatic extremes: P. A. Viglietti *et al.*, "Origin and palaeoenvironmental significance of *Lystrosaurus* bonebeds in the earliest Triassic Karoo Basin, South Africa," *Palaeogeography, Palaeoclimatology, Palaeoecology* 392 (2013): 9–21.

p. 119   "This plant [*Pleuromaia*] was taller than each of them, had a single stem topped by a cone and simple leaves ..." Plant species under this genus were evidently widespread, densely populated, and diversified in the Early Triassic Period following the end-Permian mass extinction, perhaps as a result of little competition from similar plants: C. V. Looy *et al.*, "On the ecological success of isoetalean lycopsids after the end-Permian biotic crisis," *LPP Contributions Series* 13 (2000): 63–70.

p. 122   "A slurry of mud and sand filled their newly dug shelter . . . casting them for a distant future . . ." This sort of natural casting by another sediment is normally how both invertebrate and vertebrate burrows get preserved in the geologic record. Moreover, if these burrows are cast by sediments with different grain sizes and/or colors, they often weather out of the rock or otherwise contrast with their surroundings, making it much easier for geologists and ichnologists to discern these structures as burrows.

p. 123   "These shafts and tunnels . . . effectively worked like pipes, linking nearby groundwater sources with the pond . . ." Many studies of modern mammal burrows confirm their effects on groundwater flow, soil properties, seed dispersal, plant growth, and nutrient cycling, and are discussed more in Chapter 9. However, here's a good review article on such effects of mammals in deserts: W. G. Whitford and F. R. Kay, "Biopedturbation by mammals in deserts: a review," *Journal of Arid Environments* 41 (1999): 203–230.

p. 123   "The collective action of more *Lystrosaurus* wallowing and feeding in the shallow water . . . trampled and foraged areas filled with water . . ." The common association of *Lystrosaurus* with sediments of former water bodies led paleontologists to speculate that it was partially adapted to aquatic environments, and may have even swum (depicted earlier in the story): (1) G. M. King and M. A. Cluver, "The aquatic *Lystrosaurus*: an alternative lifestyle," *Historical Biology* 4 (1990): 323–341; (2) S. Ray *et al.*, "*Lystrosaurus murrayi* (Therapsida, Dicynodontia): bone histology, growth, and lifestyle adaptations," *Palaeontology* 4 (2005): 1169–1185.

p. 124   ". . . where amphibians occasionally shared a burrow with a therapsid . . ." This specific instance is based on: V. Fernandez *et al.*, "Synchrotron reveals Early Triassic Odd Couple: injured amphibian and aestivating therapsid share burrow," *PLoS One* 8 (2013): e64978; http://dx.doi.org/10.1371/journal.pone.0064978. However, another Early Triassic example of burrow sharing, with two different vertebrate species (*Galesaurus planiceps* and *Owenetta kitchingorum*) and a millipede preserved in the same burrow, was also reported by: F. Abdala *et al.*, "Faunal aggregation in the Early Triassic Karoo Basin: earliest evidence of shelter-sharing behaviour among tetrapods," *Palaios* 21 (2006): 507–512.

p. 124   *"Proterosuchus . . . became torpid in burrows by themselves . . ."*
Whether or not *Proterosuchus* burrowed remains untested, but its
skeletal remains are in Lower Triassic strata containing *Lystrosaurus* and
burrows have been attributed to it: Modesto and Botha-Brinks (2010).

p. 124   *". . . the original cofounder of the colony . . . contributing six
offspring . . ."* Besides using burrows as a survival strategy,
Lystrosaurus also likely grew fast, reached sexual maturity at
an early age, and bred in large numbers: (1) J. Botha-Brink and
K. D. Angielczyk, "Do extraordinarily high growth rates in Permo-
Triassic dicynodonts (Therapsida, Anomodontia) explain their
success before and after the end-Permian extinction?" *Zoological
Journal of the Linnean Society* 160 (2010): 341–365. (2) J. Botha-Brink
et al., "Breeding young as a survival strategy during Earth's greatest
mass extinction," *Scientific Reports* (2016): 24053; doi: 10.1038/
srep24053.

p. 125   *"Meanwhile, their burrowing also altered ecosystems . . ."* This
ecological theme, mentioned in previous chapters, is explored in
more detail in Chapters 8 and 9, in which invertebrate burrowing
and mammal burrowing (respectively) are held up as examples of
ecosystem engineering through the mass effects of burrowing.

p. 126   *". . . Lystrosaurus mostly lived in . . . Africa, Antarctica, and India
. . . during the latest Permian Period through the earliest Triassic
Period . . ."* J. Botha and R.M.H. Smith, *"Lystrosaurus* species
composition across the Permo–Triassic boundary in the Karoo
Basin of South Africa," *Lethaia* 40 (2007): 125–137.

p. 126   "In the Permian Period . . . Africa, Antarctica, and India were
united with South America and Australia into . . . Gondwana . . ."
Every historical geology textbook mentions how the southern
continents were united as Gondwana during the Permian, but it
is also good to know some of the human historical perspective on
how this hypothesis developed, provided in a nice summary (with
pictures) by the U.S. Geological Survey, *Historical Perspective*: http://
pubs.usgs.gov/gip/dynamic/historical.html

p. 126   *". . . although Lystrosaurus* was mostly Gondwanan . . . specimens
found in China, Mongolia, and Russia . . ."* Botha and Smith (2007).

p. 127   "This bonding formed . . . Pangea, surrounded by . . . Panthalassa . . ."
Sometimes Panthalassa is called the "Paleo-Pacific" because the
Pacific plate (which holds the Pacific Ocean) originated from it. Yet
nearly all of the original oceanic crust from the time that Pangea
broke up was destroyed by subduction since: L. M. Boschmann
and D.J.J. van Hinsbergen, "On the enigmatic birth of the Pacific
Plate within the Panthalassa Ocean," *Science Advances* 2 (2016):
e1600022; doi: 10.1126/sciadv.1600022.

p. 127   ". . . fossils of *Lystrosaurus, Cynognathus* . . . a seed fern (*Glossopteris*) . . . reptile (*Mesosaurus*) . . . helped support the idea of 'continental drift' . . ." This now classic connection between fossil distributions and plate tectonics, made originally by Alfred Wegener (1880–1930), later contributed to a better understanding of modern biogeography: D. McCarthy, *Here Be Dragons: How the Study of Animal and Plant Distributions Revolutionized Our Views of Life and Earth* (Oxford, England: Oxford University Press, 2011).

p. 127   "All species of *Lystrosaurus* were synapsids . . . a pair of temporal foramens . . ." A summary of early synapsid evolutionary history and relationships, as well as anatomical traits of their skulls, is at the *Tree of Life* Web site, written by Michel Laurin and Robert R. Reisz: http://tolweb.org/Synapsida/14845

p. 127   ". . . diapsids . . . have two temporal foramens on each side of their skulls . . ." M. Benton, *Vertebrate Palaeontology*, 3rd Edition, (New York: John Wiley & Sons, 2009).

p. 127   ". . . Synapsida also has Therapsida within it . . ." Benton (2009).

p. 128   "The oldest known fossil synapsids are *Archaeothyris, Clepsydrops*, and *Echinerpeton* . . . lived . . . during the Carboniferous in . . . Nova Scotia . . ." With the caveat that I am not a vertebrate paleontologist (but could easily play one on TV), I need to state that not all vertebrate paleontologists agree these three genera are basal synapsids. For those interested, check out this discussion of early synapsid evolution: R.B.J. Benson, "Interrelationships of basal synapsids: cranial and postcranial morphological partitions suggest different topologies," *Journal of Systematic Palaeontology* 10 (2012): 601–610.

p. 128   "Not too long after that . . . fossil therapsids show up in the middle of the Permian Period . . ." A. Chinsamy-Turan, (ed.), *Forerunners of Mammals: Radiation, Histology, Biology* (Bloomington: Indiana University Press, 2011).

p. 128   ". . . pelycosaurs, a loosely defined group . . . that dominated Early Permian landscapes . . ." (1) Benton (2009); (2) Chinsamy-Turan (2011).

p. 128   ". . . *Dimetrodon*: 'two measures of teeth'), and a broad, tall sailback supported by long vertebral spines . . ." (1) T. S. Kemp, *The Origin and Evolution of Mammals* (Oxford, England: Oxford University Press, 2005); (2) Benton (2009).

p. 128   ". . . division between Anomodontia and Theriodontia also marks where . . . Dicynodontia and Cynodontia split from a common ancestor . . ." (1) Benton (2009); Kammerer *et al.* (2013).

p. 129   ". . . mammals are the only living cynodonts . . . all dicynodonts likely died by the end of the Triassic . . ." One possible exception to the generally accepted demise of the dicynodonts comes from Australia, where one lineage may have survived until the Cretaceous Period, although this was later disputed: (1) T. Thulborn, and S. Turner, "The last dicynodont: an Australian Cretaceous relict," *Proceedings of the Royal Society of London B* 270 (2003): 985–993; (2) F. Agnolin *et al.*, "A reappraisal of the Cretaceous non-avian dinosaur faunas from Australia and New Zealand: evidence for their Gondwanan affinities," *Journal of Systematic Palaeontology* 8 (2010): 257–300.

p. 129   ". . . some *Lystrosaurus* were . . . half meter (2.5 feet) to more than 2 meters (6.6 feet) long . . . average length was about a meter (3.3 feet) . . ." *Lystrosaurus* sizes varied with species, but species that survived the end-Permian extinction may have been smaller than those of their predecessors: (1) Botha and Smith (2007); (2) Smith and Botha-Brink (2014).

p. 130   "Studies by Jennifer Botha-Brink and other paleontologists have revealed that *Lystrosaurus* bones were greatly vascularized . . ." Botha-Brink *et al.* (2016).

p. 130   ". . . stout bones reinforced what must have been massive musculature and connective tissues . . ." (1) King and Cluver (1990); (2) M. V. Surkov *et al.*, "*Lystrosaurus georgi*, a dicynodont from the Lower Triassic of Russia," *Journal of Vertebrate Paleontology* 25 (2005): 402–413; (3) Botha and Smith (2007).

p. 130   ". . . these burrows had *Lystrosaurus* skeletons in them . . ." Interestingly, Modesto and Jennifer Botha-Brinks (2010) proposed that larger burrows with *Lystrosaurus* remains in them might have been made by predators of *Lystrosaurus*—such as *Moschorhinus* or *Olivierosuchus*—in which they cached *Lystrosaurus* bodies in burrows.

p. 131   ". . . *Procolophon* in South America, Antarctica, and South Africa, and *Thrinaxodon* in Antarctica and South Africa . . ." (1) J. C. Cisneros, "Taxonomic status of the reptile genus *Procolophon* from the Gondwanan Triassic," *Palaeontologia Africana* 43 (2007): 7–17; (2) Chinsamy-Turan (2011).

p. 131   ". . . *Diictodon* . . . was a burrowing dicynodont that lived in South Africa, having made spiraling burrows more than 1.5 meters (5 feet) deep . . ." R.M.H. Smith, "Helical burrow casts of therapsid origin from the Beaufort Group (Permian) of South Africa," *Palaeogeography, Palaeoclimatology, Palaeoecology* 60 (1987): 155–170.

p. 131   ". . . *Thrinaxodon* was . . . first discovered example of . . . burrowing synapsid from . . . earliest part of the Triassic . . ." R. Damiani *et al.*, "Earliest evidence of cynodont burrowing," *Proceedings of the Royal Society of London, B* 270 (2003: 1747–1751.

p. 131   "... *Thrinaxodon* burrow ... contained ... salamander-like amphibian called *Broomistega* ..." Fernandez *et al.* (2013).

p. 131   "*Procolophon* ... is also interpreted as a burrower ..." Cisneros (2007).

p. 131   "... *Lystrosaurus* may have composed 95% of all land vertebrates for a short time ..." H. Dieter-Sues and N. C. Fraser, *Triassic Life on Land: The Great Transition* (New York: Columbia University Press, 2013).

p. 132   "In what is now Siberia, volcanism dominated the last million years or so of the Permian ..." (1) M. Sharma, "Siberian traps." In J. J. Mahoney, and M. F. Coffin (eds.), *Large Igneous Provinces: Continental, Oceanic, and Planetary Flood Volcanism,* American Geophysical Union, Geophysical Monograph Series, 100 (1997): 273–295; (2) B. A. Black *et al.,* "Environmental effects of large igneous province magmatism: a Siberian perspective," In A. Schmidt *et al.* (eds.), *Volcanism and Global Environmental Change* (Cambridge, England: Cambridge University Press, 2015, 307–320).

p. 132   "These sheets of lava flowed across Siberia ... deposited ... several million cubic kilometers of basalt ..." (1) M. I. Kuzmin *et al.,* "Phanerozoic hot spot traces and paleogeographic reconstructions of the Siberian continent based on interaction with the African large low shear velocity province," *Earth-Science Reviews* 102 (2010): 29–59.

p. 133   "These eruptions were accompanied by immense amounts of carbon dioxide ..." (1) V. E. Courtillot and P. R. Renne, "On the ages of flood basalt events," *Comptes Rendus Geoscience* 335 (2003): 113–140; (2) Black *et al.* (2015).

p. 133   "Greenhouse gases are like blankets, retaining more heat energy ... than is reflected ..." Just in case anyone still needs educating about "the greenhouse effect" and is willing to acknowledge the reality of climate change, the U.S. Environmental Protection Agency has a little primer on it, *Overview of Greenhouse Gases*: https://www.epa.gov/ghgemissions/overview-greenhouse-gases

p. 133   "... flood basalts made matters worse by burning Carboniferous-age coal deposits lying near the surface ..." D. E. Ogden and N. H. Sleep, "Explosive eruption of coal and basalt and the end-Permian mass extinction," *Proceedings of the National Academy of Sciences* 109 (2011): 50–62.

p. 133   "This compound [sulfur dioxide] in turn hooks up with atmospheric water to form sulfuric acid ..." Like the "greenhouse effect," acidic rain caused by combination with sulfur dioxide— which we cause by burning sulfur-rich fossil fuels (especially coal) is long studied and well established as a phenomenon. To learn more about the basic principles of it, go to the U.S. Environmental Protection Agency site, *What is Acid Rain?*: https://www.epa.gov/acidrain/what-acid-rain

p. 133   "... geoscientists ... proposed that freshwater lakes, streams, and soils became more acidic toward the end of the Permian ..." B. A. Black *et al.*, "Acid rain and ozone depletion from pulsed Siberian Traps magmatism," *Geology* 42 (2014): 67–70.

p. 134   "Like carbon dioxide, methane is a greenhouse gas, but much more effective in holding heat ..." U.S. Environmental Protection Agency, *Understanding Global Warming Potentials*: https://www.epa.gov/ghgemissions/understanding-global-warming-potentials

p. 134   "... as this formerly solid methane thawed and released big bubbles from the seafloor ..." (1) G. Ryskin, "Methane-driven oceanic eruptions and mass extinctions," *Geology* 31 (2003): 741–744; (2) G. J. Retallack and E. S. Krull, "Carbon isotopic evidence for terminal-Permian methane outbursts and their role in extinctions of animals, plants, coral reefs, and peat swamps," *Geological Society of America Special Papers* 399 (2006): 249–268.

p. 134   "Oceanic circulation stagnated toward the end of the Permian, leading to ... expansion of oxygen-poor zones ..." (1) P. B. Wignall and A. Hallam, "Anoxia as a cause of the Permian/Triassic mass extinction: facies evidence from northern Italy and the western United States," *Palaeogeography, Palaeoclimatology, Palaeoecology* 93 (1992): 21–46; (2) P. B. Wignall and R. J. Twitchett, "Oceanic anoxia and the end Permian mass extinction," *Science* 272 (1996): 1155–1158; (3) G. A. Brennecka *et al.*, "Rapid expansion of oceanic anoxia immediately before the end-Permian mass extinction." *Proceedings of the National Academy of Sciences* 108 (2011): 17631–17634.

p. 135   "Ocean acidification would have hindered marine algae and animals from making shells or skeletons ..." M. E. Clapham and J. L. Payne, "Acidification, anoxia, and extinction: A multiple logistic regression analysis of extinction selectivity during the Middle and Late Permian," *Geology* 39 (2011): 1059–1062.

p. 135   "The weird mix of gases in the Late Permian atmosphere likely caused such a depletion of [ozone] ..." Black *et al.* (2014).

p. 135   "This layer blocks or otherwise inhibits UV-B radiation from reaching the earth's surface ..." NOAA (National Oceanic and Atmospheric Administration) provides a basic explainer of the ozone layer in the stratosphere and its importance to life, *Stratospheric Ozone: Monitoring and Research in NOAA*: http://www.ozonelayer.noaa.gov/science/basics.htm

p. 135   "Modern experiments with UV-B radiation demonstrate ... it tends to stunt plant growth ..." T. A. Day and P. J. Neale, "Effects of UV-B radiation on terrestrial and aquatic primary producers," *Annual Review of Ecology and Systematics* 33 (2002): 371–396.

p. 136   "This continental coziness slowed the movement of moisture to the interior of Pangea . . ." (1) F. Fluteau *et al.*, "The Late Permian climate. What can be inferred from climate modelling concerning Pangea scenarios and Hercynian range altitude?" *Palaeogeography, Palaeoclimatology, Palaeoecology* 167 (2001): 39–71; (2) N. J. Tabor, "Wastelands of tropical Pangea: high heat in the Permian," *Geology* 41 (2013): 623–624.

p. 136   "Having all of the continents together also decreased the number and areas of coastlines . . ." D. H. Erwin, "The Permo-Triassic extinction." In H. Gee (ed.), *Shaking the Tree: Readings from Nature in the History of Life* (Chicago: University of Chicago Press, 2000).

p. 136   "Considering these . . . conditions lasted for a minimum of 50,000 years . . ." The latest age dates of igneous rocks before and after the end-Permian extinction are now indicating that this massive volcanism may have happened for 300,000 years before the end, then persisted from nearly 500,000 years afterward. This certainly would have delayed any recovery by plant or animal life, meaning that the Early Triassic was still not a very nice time to be alive: S. D. Burgess and S. A. Bowring, "High-precision geochronology confirms voluminous magmatism before, during, and after Earth's most severe extinction," *Science Advances* 28 (2015): e1500470; 10.1126/sciadv.1500470.

p. 137   ". . . this evolutionary party came crashing down with another mass extinction at the end of the Triassic . . ." A. Hallam, "How catastrophic was the end-Triassic mass extinction?" *Lethaia* 35 (2002): 147–157; (2) L. H. Tanner *et al.*, "Assessing the record and causes of the Late Triassic extinctions," *Earth-Science Reviews* 65 (2004): 103–139.

p. 137   "Enough organisms survived the . . . end of the Permian for life to start over in the Triassic . . ." (1) S. Sahney and M. J. Benton, "Recovery from the most profound mass extinction of all time," *Proceedings of the Royal Society of London, B* 275 (2008): 759–765; (2) T. J. Algeo, *et al.*, "Terrestrial–marine teleconnections in the collapse and rebuilding of Early Triassic marine ecosystems," *Palaeogeography, Palaeoclimatology, Palaeoecology* 308 (2011): 1–11.

p. 137   ". . . placodonts, which had shell-crushing teeth just right for masticating mollusks . . ." Benton (2009).

p. 137   "These mats persisted until burrowers and grazers evolved . . . to disrupt them later in the Triassic . . ." (1) S. Pruss *et al.*, "Proliferation of Early Triassic wrinkle structures: Implications for environmental stress following the end-Permian mass extinction," *Geology* 32 (2004): 461–464; (2) D. Chu *et al.*, "Early Triassic wrinkle

structures on land: stressed environments and oases for life,"
Scientific Reports 5 (2015): 10109; doi:10.1038/srep10109.

p. 138   "The first turtles either evolved on land or in shallow marine
environments early on in the Triassic Period . . ." The immediate
ancestors of turtles probably evolved during the Permian Period,
but their descendants became turtles by the Early Triassic: see the
previous discussion of this and relevant references associated with
Chapter 3.

p. 138   "Among these were . . . rauisuchians . . . crocodile-like phytosaurs,
and aetosaurs . . ." N. Fraser and D. Henderson, *Dawn of the
Dinosaurs: Life in the Triassic* (Bloomington: Indiana University
Press, 2006).

p. 138   "Pterosaurs . . . represented a completely new strategy for
vertebrates, becoming the first to occupy the skies . . ." M. P.
Witton, *Pterosaurs: Natural History, Evolution, Anatomy* (Princeton,
N.J.: Princeton University Press, 2013).

p. 139   "Much of this split was prompted by . . . volcanic hot spots that
pushed apart North America, South America, and Africa . . ." M.
Manspeizer (ed.), *Triassic-Jurassic Rifting: Continental Breakup and
the Origin of the Atlantic Ocean and Passive Margins* (Amsterdam:
Elsevier, 2015).

p. 139   "Some ocean-adapted clades made it into the Jurassic Period,
such as ammonites, sharks, bony fish, marine reptiles . . ."
(1) A. Brayard *et al.*, "The Early Triassic ammonoid recovery:
paleoclimatic significance of diversity gradients," *Palaeogeography,
Palaeoclimatology, Palaeoecology* 239 (2006): 374–395; (2) P. M.
Thorne *et al.*, "Resetting the evolution of marine reptiles at the
Triassic-Jurassic boundary," *Proceedings of the National Academy of
Sciences* 108 (2011): 8339–8344.

p. 139   ". . . another trophic cascade began, toppling nearly every big-
bodied animal . . . toward the end of the Triassic . . ." P. D. Ward
et al., "Sudden productivity collapse associated with the Triassic-
Jurassic boundary mass extinction," *Science* 292 (2001): 1148–1151;
(2) Tanner *et al.* (2004).

p. 140   "The evidence for a meteorite impact at the end of the Triassic
includes . . . chemical signatures and . . . impact crater in France
. . ." M. Schmieder *et al.*, "A Rhaetian 40Ar/39Ar age for the
Rochechouart impact structure (France) and implications for the
latest Triassic sedimentary record," *Meteoritics and Planetary Science*
45 (2010): 1225–1242.

p. 141   ". . . change in climate caused shallow marine habitats . . . to
diminish . . ." (1) C. Linnert *et al.*, "Evidence for global cooling in
the Late Cretaceous," *Nature Communications* 5 (2016): 4194; doi:

10.1038/ncomms5194; (2) N. Thibault *et al.*, "Late Cretaceous (late Campanian-Maastrichtian) sea-surface temperature record of the Boreal Chalk Sea," *Climate of the Past* 12 (2016): 429–438.

p. 141   "These basalts now compose the Deccan Plateau . . . of western India . . ." B. Schoene *et al.*, "U-Pb geochronology of the Deccan Traps and relation to the end-Cretaceous mass extinction," *Science* 347 (2015): 182–184; doi: 10.1126/science.aaa0118.

p. 141   ". . . volcanism responsible for the Deccan Traps very likely . . . released huge volumes of carbon dioxide, initiating rapid global warming . . ." T. S. Tobin *et al.*, "Extinction patterns, δ^{18} O trends, and magnetostratigraphy from a southern high-latitude Cretaceous–Paleogene section: links with Deccan volcanism," *Palaeogeography, Palaeoclimatology, Palaeoecology* 350–352 (2012): 180–188.

p. 142   ". . . impact from this estimated 10-kilometer (6-mile) wide bolide . . . converted its potential energy into kinetic energy . . . heat equivalent to about five billion atomic bombs . . ." Alvarez (1997).

p. 142   ". . . it [impact] caused a gigantic tsunami, which would have inundated all low-lying coastal areas . . ." T. Matsui *et al.*, "Generation and propagation of a tsunami from the Cretaceous-Tertiary impact event," In C. Koeberl and K. G. MacLeod (eds.), *Catastrophic Events and Mass Extinctions: Impacts and Beyond* (Boulder, Colo.: Geological Society of America Special Papers 356 (2002): 69–77).

p. 142   "That's right, acid rain, which adversely affects soils and freshwater ecosystems . . ." (1) P. Schulte *et al.*, "The Chicxulub asteroid impact and mass extinction at the Cretaceous-Paleogene boundary," *Science* 327 (2010): 1214–1218; (2) S. Ohno *et al.*, "Production of sulphate-rich vapour during the Chicxulub impact and implications for ocean acidification," *Nature Geoscience* 7 (2014): 279–282.

p. 142   "These conditions were analogous to a 'nuclear winter,' first envisaged by scientists who opposed nuclear war in the 1980s . . ." (1) R. P. Turco *et al.*, "Nuclear winter: global consequences of multiple nuclear explosions," *Science* 222 (1983): 1283–1292; (2) K. O. Pope *et al.*, "Impact winter and the Cretaceous/Tertiary extinctions: results of a Chicxulub asteroid impact model," *Earth and Planetary Science Letters* 128 (1994): 719–725; (3) J. Vellekoop *et al.*, "Rapid short-term cooling following the Chicxulub impact at the Cretaceous-Paleogene boundary," *Proceedings of the National Academy of Sciences* 111 (2014): 7537–7541.

p. 142   ". . . an estimated 80% of all species died, and quickly . . ." Robertson *et al.* (2004).

p. 143   "Raup liked this question so much . . . he . . . used it as the title of a short book he wrote about mass extinctions . . ." D. Raup, *Extinction: Bad Genes or Bad Luck?* (New York: W.W. Norton, 1992).

p. 144   "This ability to make it past an extinction event is often called *extinction selectivity* . . ." S. E. Peters, "Environmental determinants of extinction selectivity in the fossil record," *Nature* 454 (2008): 626–629.

CHAPTER 6: TERRAFORMING A PLANET, ONE HOLE AT A TIME

p. 148   ". . . these animals were also marvelously varied in form and function during their 250 million years of living on, above, and below seafloors . . ." (1) R. Levi-Setti, *Trilobites* (Chicago: University of Chicago Press, 1995). (2) R. Fortey, *Trilobites: Eyewitness to Evolution* (New York: Knopf Doubleday, 2010).

p. 148   "Some were tiny, measuring less than 2 millimeters (0.08 inches) long, whereas others were longer than most domestic cats . . ." Fortey (2010).

p. 148   "Some trilobite eyes were incredibly complex, comprised of a myriad of calcite lenses . . ." Levi-Setti (1995).

p. 149   "The closest living relatives of trilobites today are modern horseshoe crabs . . ." Horseshoe crabs, also known as limulids, belong to the clade Chelicerata, which includes sea spiders and arachnids, such as scorpions and spiders. Trilobites were originally classified as crustaceans, but are not, although crustaceans, trilobites, and chelicerates probably all share a common ancestor: G. Scholtz and G. D. Edgecombe, "Heads, Hox and the phylogenetic position of trilobites," In S. Koenemann and R. A. Jenner (eds.), *Crustacea and Arthropod Relationships* (Boca Raton, Fla.: CRC Press, 2005, 139–165).

p. 150   ". . . animal settlement of continental environments began in about the middle of the Silurian, perhaps 430 *mya* . . ." (1) W. A. Shear *et al.* (1984), "Early land animals in North America," *Science* 224 (2005): 492–494; (2) A. J. Jeram, "Land animals in the Silurian: arachnids and myriapods from Shropshire, England," *Science* 250 (1990): 658–661.

p. 151   "On the other hand, Ordovician seas teemed with animal life . . ." D. L. Meyer and R. A. Davis, *A Sea Without Fish: Life in the Ordovician Sea of the Cincinnati Region* (Bloomington: Indiana University Press, 2009).

p. 151   "In between all of this labor, my PhD adviser, renowned ichnologist Robert (Bob) Frey . . . " (1) R. W. Frey (ed.), *The Study of Trace Fossils: A Synthesis of Principles, Problems, and Procedures in Ichnology* (Berlin: Springer-Verlag, 1975); S. G. Pemberton,

"Memorial to Robert Wayne Frey (1938–1992)," *Ichnos* 2 (1992): 1–6.

p. 151 ". . . all of these papers were about his seminal and voluminous work on the ichnology of the modern Georgia coast . . ." Martin (2013).

p. 152 ". . . greatly outnumbered Confederate soldiers pushed boulders of Ordovician sandstone down the steep slopes . . ." M. H. Dunkelman, *War's Relentless Hand; Twelve Tales of Civil War Soldiers* (Baton Rouge: Louisiana University Press, 2006).

p. 153 "Estuaries are more like open bays, or have marshes or lagoons that are sometimes behind a string of barrier islands . . ." E. Wolanski and M. Elliott, *Estuarine Ecohydrology: An Introduction* (Amsterdam: Elsevier, 2015).

p. 153 "Here these environments are more dominated by marine processes and organisms . . ." Wolanski and Elliott (2015).

p. 154 ". . . salt-adapted animals may stay near the estuary bottom, where they also might burrow . . ." Martin (2013).

p. 154 ". . . droughts have allowed barnacles and other marine animals to live more than 30 kilometers (19 miles) inland . . ." R. W. Frey and J. D. Howard, "Mesotidal estuarine sequences: a perspective from the Georgia Bight," *Journal of Sedimentary Petrology* 56 (1986): 911–924.

p. 154 ". . . Another is cross-bedding angled one way, then the opposite way, a pattern called 'herringbone' cross-bedding . . ." Sometimes this pattern is also called herringbone cross-stratification: G. Nicols, *Sedimentology and Stratigraphy* (New York: John Wiley & Sons, 2009).

p. 155 ". . . these trace fossils and their enclosing rocks were helping me to document the oldest known estuary recorded in the geologic record . . ." (1) A. J. Martin, "Semiquantitative and statistical analysis of bioturbate textures in the Sequatchie Formation (Upper Ordovician), Georgia and Tennessee, U.S.A." *Ichnos* 2 (1992): 117–136; (2) A. J. Martin, "Peritidal and estuarine facies in the Shellmound and Mannie Shale Members, Sequatchie Formation," In T. M. Chowns and B. J. O'Connor (eds.), *Cambro-Ordovician Strata in Northwest Georgia and Southeast Tennessee; The Knox Group and the Sequatchie Formation*, Georgia Geological Society Guidebook, 12 (1992): 107–124. My original studies were also included in a more comprehensive look at burrowing in estuaries through geologic time: L. A. Buatois *et al.*, "Colonization of brackish-water systems through time: evidence from the trace-fossil record," *Palaios* 20 (2005): 321–347.

p. 158 ". . . some arthropod trackways are preserved in likely beach deposits from the Early Cambrian Period . . ." (1) M. G. Mángano *et al.*, "Trilobites in early Cambrian tidal flats and the landward expansion of the Cambrian explosion," *Geology* 42 (2014): 143–146; (2) R. B. McNaughton *et al.*, "First steps on land: Arthropod trackways in Cambrian-Ordovician eolian sandstone, southeastern Ontario, Canada," *Geology* 30 (2002): 391–394.

p. 158   ". . . and boldly going where no invertebrates had gone before then . . ." Yes, that's a *Star Trek* reference, but I also made several *Star Wars* references earlier. I'm all about inclusion when it comes to geek culture.

p. 158   ". . . trilobite burrows from the latest Cambrian and earliest Ordovician Periods . . . in Spain and Portugal . . . long, deep, and looping, with beautifully expressed scratches from those legs . . ." (1) R. N. Pickerell *et al.*, "Arenig trace fossils from the Salamanca area, western Spain," *Geological Journal* 19 (1984): 249–269. (2) C. N. de Carvalho, "Roller coaster behavior in the *Cruziana Rugosa* Group from Penha Garcia (Portugal): implications for the feeding program of trilobites," *Ichnos* 13 (2007): 255–265.

p. 159   ". . . helping to reserve water and hence delaying desiccation in their owners whenever they were out of the water . . ." E. B. Edney, *Water Balance in Land Arthropods* (New York: Springer, 2012).

p. 159   "Classified as merostomes—the same group containing spiders, scorpions, and other arachnids . . ." In R. A. Fortey and R. H. Thomas (eds.), *Arthropod Relationships* (New York: Springer:, 2012).

p. 159   "Four species of horseshoe crabs are known, with three in Southeast Asia . . ." J. T. Tanacredi *et al.*, *Biology and Conservation of Horseshoe Crabs* (New York: Springer, 2009).

p. 159   "The largest horseshoe crabs are on the Georgia coast, where they get huge . . ." Martin (2013).

p. 159   "This mass movement of female and male horseshoe crabs happens seasonally . . ." (1) H. J. Brockmann, "Mating behavior of horseshoe crabs, *Limulus polyphemus*," *Behaviour* 114 (1990): 206–220; (2) In C. N. Shuster, Jr., *et al.* (eds.), *The American Horseshoe Crab* (Cambridge, Mass.: Harvard University Press, 2003).

p. 160   "This little notch in the tail section of her body evolved for just the purpose of a male horseshoe crab fitting . . . onto it . . ." Brockmann (1990).

p. 160   "I was further struck by how closely these traces resembled Paleozoic trace fossils that were often attributed to trilobites or marine worms . . ." A. J. Martin and A. K. Rindsberg, "Arthropod tracemakers of *Nereites*? Neoichnological observations of juvenile limulids and their paleoichnological applications," In W. M. Miller, III (ed.), *Trace Fossils: Concepts, Problems, Prospects* (Amsterdam: Elsevier, 2007, 478–491).

p. 161   "... the massive ones on the Georgia coast are living in water with some salt in it, but not quite at full marine salinity ..." Martin (2013).

p. 161   "Interestingly, the oldest known body fossils of horseshoe crabs come from Ordovician rocks (about 450 *mya*) ..." D. M. Rudkin *et al.*, "The oldest horseshoe crab: a new xiphosurid from Late Ordovician konservat-lagerstätten deposits, Manitoba, Ontario," *Palaeontology* 51 (2008): 1–9.

p. 161   "... trackways are in estuarine and even freshwater deposits from ... Paleozoic Era and ... Mesozoic Era ..." (1) A. J. Boucot, *Evolutionary Paleobiology of Behavior and Coevolution* (Amsterdam: Elsevier, 2013), and references therein; (2) C. G. Diedrich, "Middle Triassic horseshoe crab reproduction areas on intertidal flats of Europe with evidence of predation by archosaurs," *Biological Journal of the Linnean Society* 103 (2011): 76–105; (3) M. Romano and M. A. Whyte, "The first record of xiphosurid (arthropod) trackways from the Saltwick Formation, Middle Jurassic of the Cleveland Basin, Yorkshire," *Palaeontology* 46 (2003): 257–269.

p. 161   "... migrating shorebirds, such as red knots (*Calidris canutus*), gobble up eggs and newly hatched 'trilobites' alike ..." (1) G. Castro and J. P. Meyers, "Shorebird predation on eggs of horseshoe crabs during spring stopover on Delaware Bay," *Auk* 110 (1993): 927–930; Martin (2013).

p. 161   "These limulids cause the depressions by rapidly moving their legs, which allows more water in between the sand grains ..." (1) Martin and Rindsberg (2007); (2) Martin (2013).

p. 162   "These small burrows are in fossil soils (called *paleosols*), preserved in Late Ordovician rocks of Pennsylvania ..." (1) G. J. Retallack and C. R. Feakes, "Trace fossil evidence for Late Ordovician animals on land," *Science* 235 (1987): 61–63; (2) G. J. Retallack, "*Scoyenia* burrows from Ordovician palaeosols of the Juniata Formation in Pennsylvania," *Palaeontology* 44 (2001): 209–235.

p. 162   "For instance, by the Silurian, large animals called myriapods ... occupied freshwater streams and land environments ..." A. J. Jeram *et al.*, "Land animals in the Silurian: arachnids and myriapods from Shropshire, England," *Science* 250 (1990): 658–661; (2) J. Gray and A. J. Boucot, "Early Silurian nonmarine animal remains and the nature of the early continental ecosystem," *Acta Palaeontologica Polonica* 38 (1993): 303–328.

p. 162   "Similar fossil burrows are in rocks of the same age in Scotland, Antarctica, and other parts of the world ..." (1) T. W. Gevers *et al.*, "Trace Fossils in the Lower Beacon Sediments (Devonian), Darwin Mountains, Southern Victoria Land, Antarctica," *Journal of Paleontology* 45 (1971): 81–94; (2) J.R.L. Allen and B.P.J. Williams,

"*Beaconites antarcticus*: a giant channel-associated trace fossil from the Lower Old Red Sandstone of South Wales and the Welsh borders," *Geological Journal* 16 (1981): 255–269.

p. 163 ". . . adults [myriapods] . . . left trackways about 40–50 centimeters (16–20 inches) wide . . ." (1) P. N. Pearson, "Walking traces of the giant myriapod *Arthropleura* from the Strathclyde Group (Lower Carboniferous) of Fife," *Scottish Journal of Geology* 28 (1992): 127–133; (2) R. L. Martino and S. F. Greb, "Walking trails of the giant terrestrial arthropod *Arthropleura* from the Upper Carboniferous of Kentucky," *Journal of Paleontology* 83 (2009): 140–146.

p. 164 ". . . all four-limbed vertebrates (called *tetrapods*) . . . arose from lobe-finned fishes, called *sarcopterygians* . . ." (1) Benton (2009); (2) N. Shubin, *Your Inner Fish: A Journey into the 3.5-Billion-Year History of the Human Body* (New York: Vintage, 2009).

p. 164 "The best representative found thus far of a fossil marking this transition from a swimming-to-walking fish is *Tiktaalik* . . ." (1) E. B. Daeschler *et al.*, "A Devonian tetrapod-like fish and the evolution of the tetrapod body plan," *Nature* 440 (2006): 757–763; (2) N. H. Shubin *et al.*, "The pectoral fin of Tiktaalik roseae and the origin of the tetrapod limb," *Nature* 440 (2006): 764–771.

p. 164 "The body fossil, anatomical, and genetic evidence for the evolutionary transition of lobe-finned fish to tetrapods has been the topic of much research . . ." Following the popularity of *Your Inner Fish* as a book, in 2014 PBS broadcast a very good three-part series based on the book and by the same title, with Neil Shubin as its "star" (although he is understandably overshadowed by *Tiktaalik* in places): http://www.pbs.org/your-inner-fish/watch/

p. 165 ". . . in 2009, paleontologists in Poland found a series of depressions in strata . . . interpreted as trackways made by primitive tetrapods . . ." G. Niedźwiedzki *et al.*, "Tetrapod trackways from the early Middle Devonian period of Poland," *Nature* 463 (2010): 43–48.

p. 165 ". . . Other paleontologists have disputed these tracks since, claiming they actually are fish nests . . ." S. G. Lucas, "*Thinopus* and a critical review of Devonian tetrapod footprints," *Ichnos* 22 (2015): 136–154.

p. 165 "Lungfishes are classified as sarcopterygians under the clade Dipnoi, and originated from other bony fish toward the start of the Devonian Period . . ." Benton (2009).

p. 166 "Their air breathing comes from a swim bladder functioning as a sort of 'lung' . . ." J. N. Maina, "The morphology of the lung of the African lungfish, *Protopterus aethiopicus*," *Cell and Tissue Research* 250 (1987): 191–196; (2) J. M. Jørgensen *et al.*, *The Biology of Lungfishes* (Boca Raton, Fla.: CRC Press, 2011).

p. 166   "Lungfishes are also tolerant of low-oxygen conditions in their environments . . ." J. M. Jørgensen, *et al.* (2011).

p. 166   "For example, a single, fist-sized Cretaceous lungfish tooth from the western U.S. . . ." K. Shimada and J. K. Kirkland, "A mysterious king-sized Mesozoic lungfish from North America," *Transactions of the Kansas Academy of Science* 114 (2011): 135–141.

p. 166   "Lungfishes today are all Gondwanan animals, with four species in Africa . . . one species each in South America . . . and Australia . . ." (1) J. B. Graham, *Air-Breathing Fishes: Evolution, Diversity, and Adaptation* (Cambridge, Mass.: Academic Press, 1997); J. M. Jørgensen *et al.* (2011).

p. 167   "A lungfish accomplishes this feat by first poking its head into still-soft mud of a former pond or river . . ." C. A. Navas and J. E. Carvalho, *Aestivation: Molecular and Physiological Aspects* (New York: Springer, 2010).

p. 167   "Once properly ensconced in this burrow, a lungfish secretes mucus from its skin . . ." (1) A. P. Fishman *et al.*, "Estivation in *Protopterus*," *Journal of Morpholology* 190 Supplement 1, (1986): 237–248; (2) M. P. Wilkie *et al.*, "The African Lungfish (*Protopterus dolloi*): ionoregulation and osmoregulation in a fish out of water," *Physiological and Biochemical Zoology* 80 (2006): 99–112.

p. 167   ". . . it also absorbs some of its muscular tail as a form of self-cannibalism . . ." Fishman *et al.* (1986).

p. 168   "One BBC documentary film featuring African lungfishes showed this rejuvenation in a dramatic way . . ." *Fish and Frogs Living Out of Water—BBC*, excerpt from BBC Earth's *Supernatural: The Unseen Powers of Animals* (2008): https://www.youtube.com/watch?v=ZUsARF-CBcI

p. 168   ". . . lungfishes have burrowed since near their origin because of lungfish-like burrows from Devonian rocks in Pennsylvania . . ." M. Friedman and E. B. Daeschler, "Late Devonian (Famennian) lungfishes from the Catskill Formation of Pennsylvania, USA," *Palaeontology* 49 (2006): 1167–1183.

p. 168   "Lungfish burrows containing their skeletons are in Late Permian (about 255 *mya*) rocks of Kansas . . ." T. J. McCahon and K. B. Miller, "Environmental significance of lungfish burrows (*Gnathorhiza*) within Lower Permian (Wolfcampian) paleosols of the US midcontinent," *Palaeogeography, Palaeoclimatology, Palaeoecology* 435 (2015): 1–12.

p. 168   "I have also interpreted lungfish burrows in Early Cretaceous strata (about 105 *mya*) of Australia . . ." Martin *et al.* (2015).

p. 169   "This variation in patterns was well documented in a study of Cretaceous lungfish burrows in Madagascar . . ." M. S. Marshall and R. R. Rogers, "Lungfish burrows from the Upper Cretaceous Maevarano Formation, Mahajanga Basin, northwestern Madagascar," *Palaios* 27 (2012): 857–866.

p. 170 "... *Koolasuchus* (named after Australian paleontologist Lesley Kool) lived in polar Australia during the Early Cretaceous ..." A. Warren *et al.*, "The last last labyrinthodonts?" *Palaeontographica A* 247 (1997): 1–24.

p. 171 "Hibernation is dormancy ... so an animal stays warm, whereas estivation is a summer plan for staying cool ..." D. I. Hembree, "Aestivation in the fossil record: evidence from ichnology," *Progress in Molecular and Subcellular Biology* 49 (2010): 245–262.

p. 171 "The highest latitude amphibian is the Siberian wood frog (*Rana amurensis*) ..." K. D. Wells, *The Ecology and Behavior of Amphibians* (Chicago: University of Chicago Press, 2010).

p. 171 "... the Alaskan wood frog (*Lithobates sylvaticus*) has natural antifreeze in its tissues ..." D. L. Larson *et al.*, "Wood frog adaptations to overwintering in Alaska: new limits to freezing tolerance," *Journal of Experimental Biology* 217 (2014): 2193–2200.

p. 171 "... The northwestern salamander ... and long-toed ... salamander live in burrows year-round ..." R. L. Hoffman *et al.*, "Habitat segregation of *Ambystoma gracile* and *Ambystoma macrodactylum* in mountain ponds and lakes, Mount Rainier National Park, Washington, USA," *Journal of Herpetology* 37 (2003): 24–34.

p. 171 "... they [mole salamanders] are relatively large, and are geographically widespread, occupying a wide range of ecosystems ..." (1) J. P. Gibbs *et al.*, *Amphibians and Reptiles of New York State: Identification, Natural History, and Conservation* (Oxford, England: Oxford University Press, 2007); (2) Jensen *et al.* (2008).

p. 171 "Mole salamander burrows can be impressive, reaching 50–60 centimeters (20–24 inches) below the surface ..." (1) R. D. Semlitsch, "Burrowing ability and behavior of salamanders of the genus *Ambystoma*," *Canadian Journal of Zoology* 61(3) (1983): 616–620; (2) D. I. Hembree, "Using experimental neoichnology and quantitative analyses to improve the interpretation of continental trace fossils," *Ichnos* 23 (2016): 262–297.

p. 172 "One of these is the desert rain frog (*Breviceps macrops*), which lives in a thin area along the southwestern coast of Africa ..." A. Channing and K. Wahlberg, "Distribution and conservation status of the desert rain frog *Breviceps macrops*," *African Journal of Herpetology* 60 (2011): 101–112.

p. 172 "Incidentally, this species of burrowing frog became a YouTube star in 2013 ..." *Worlds* [sic] *Cutest Frog—Desert Rain Frog*: https://www.youtube.com/watch?v=cBkWhkAZ9ds

p. 172   ". . . females lay fertilized eggs in a burrow, where the young skip become aquatic tadpoles by developing in the eggs . . ." Channing and Wahlberg (2011).

p. 172   ". . . Similarly dry-adapted species of frogs are in Australia . . . the trilling frog . . . painted burrowing frog . . . Spencer's burrowing frog . . . and ornate burrowing frog . . ." M. Clayton, *et al.*, *CSIRO List of Australian Vertebrates: A Reference with Conservation Status* (Clayton, Victoria, Australia: CSIRO, 2006).

p. 173   "These fabulous toads estivate in burrows of their own making during the winter and part of the spring . . ." Martin (2013).

p. 173   "The resulting burrows are either shallow oval chambers, or inclined to vertical shafts with oval chambers at their ends . . ." (1) Martin (2013); (2) L. M. Johnson and D. I. Hembree, "Neoichnology of the eastern spadefoot toad, *Scaphiopus holbrookii* (Anura: Scaphiopodidae): criteria for recognizing anuran burrows in the fossil record," *Palaeontologia Electronica* 18.2.43A (2015): 1–29.

p. 173   "These impressive animals . . . reach more than a meter (3.3 feet) long, live in wetlands of the southeastern U.S. . . ." Jensen *et al.* (2008).

p. 174   "However, females are also known to make nesting burrows, where they lay their fertilized eggs . . ." (1) J. C. Knepton, Jr., "A note on the burrowing habits of the salamander *Amphiuma means means*," *Copeia* (1954): 68; (2) Martin (2013).

p. 174   "Although tied to aquatic environments because of their reproductive needs, adults of these amphibians [caecilians] live underground . . ." Wells (2010).

p. 174   "This body movement is similar to that used by earthworms, in which they use hydrostatic expansion and contraction along their lengths . . ." P. K. Ducey *et al.*, "Experimental examination of burrowing behavior in caecilians (Amphibia: Gymnophiona): effects of soil compaction on burrowing ability of four species," *Herpetologica* 49 (1993): 450–457; Wells (2010).

p. 175   "The oldest interpreted tetrapod burrow, credited to an amphibian, is in a Carboniferous Period (330 *mya*) sandstone bed . . ." L. Storm *et al.*, "Large vertebrate burrow from the Upper Mississippian Mauch Chunk Formation, eastern Pennsylvania, USA," *Palaeogeography, Palaeoclimatology, Palaeoecology* 298 (2014): 341–347.

p. 175   "The oldest undoubted amphibian burrows are in rocks of Kansas formed during the early part of the Permian Period . . ." D. I. Hembree *et al.*, "*Torridorefugium eskridgensis* (new ichnogenus and ichnospecies): amphibian aestivation burrows from the Lower Permian Speiser Shale of Kansas," *Journal of Paleontology* 79 (2005): 583–593.

p. 176 ". . . population of the rare San Salvador rock iguana (*Cyclura rileyi rileyi*) . . ."	W. K. Hayes *et al.*, "Conservation of an endangered Bahamian rock iguana," in A. C. Alberts, *et al.* (eds.), *Iguanas: Biology and Conservation* (Berkeley: California University Press, 2004, 232–257).

p. 176   "On October 2, 2015, San Salvador and all of the low-lying cays around it suffered a direct hit from Hurricane Joaquin . . ." R. Berg, "Hurricane Jaoquin (AL112015) 28 September–7 October 2015," National Hurricane Center, January 12, 2016: http://www.nhc. noaa.gov/data/tcr/AL112015_Joaquin.pdf

p. 178   "Based on fossils and genetics, herpetologists estimate lizards have been around at least since the Triassic Period, but possibly since the latest part of the Permian . . ." M.E.H. Jones *et al.*, "Integration of molecules and new fossils supports a Triassic origin for Lepidosauria (lizards, snakes, and tuatara)," *BMC Evolutionary Biology* 13 (2013): 208; doi: 10.1186/1471-2148-13-208; (2) E. R. Pianka and L. J. Vitt, *Lizards: Windows to the Evolution of Diversity* (Berkeley: University of California Press, 2006).

p. 178   "Among the most noteworthy of modern burrowing lizards are those of the genus *Uromastyx* . . ." T. M. Wilms *et al.*, "Aspects of the ecology of the Arabian spiny-tailed lizard (*Uromastyx aegyptia microlepis* Blanford, 1875) at Mahazat as-Sayd protected area, Saudi Arabia," *Salamandra* 46 (2010): 131–140.

p. 179   ". . . *U. aegyptia* . . . excavates burrows as long as 5.3 meters (17.4 feet), and with vertical depths of 1.2 meters . . ." Wilms *et al.* (2010).

p. 179   "These lizards . . . construct spiraling burrows that are the deepest nests dug by any vertebrate . . ." J. S. Doody *et al.*, "Deep nesting in a lizard, *déjà vu* devil's corkscrews: first helical reptile burrow and deepest vertebrate nest," *Biological Journal of the Linnean Society* 116 (2015): 13–26.

p. 180   ". . . snakes represent enduring and powerful myths and symbols in cultures worldwide, many of which connect snakes with the earth or an 'underworld' . . ." Such myths relating snakes to an underground realm are so numerous and cross-cultural they could constitute another book. You might start looking at the following, and go on from there: K.D.S. Lapatin, *Mysteries of the Snake Goddess: Art, Desire, and the Forging of History* (Cambridge, Mass.: Da Capo Press, 2003).

p. 180   "They evolved during the same time as dinosaurs, descending from short-limbed lizards during the Mesozoic Era . . . probably in the Middle Jurassic . . ." (1) N. Vidal and S. B. Hedges, "The molecular evolutionary tree of lizards, snakes, and amphisbaenians,"

Comptes Rendus Biologies 332 (2009): 129–139; (2) M. W. Caldwell *et al.*, "The oldest known snakes from the Middle Jurassic-Lower Cretaceous provide insights on snake evolution," *Nature Communications* 6 (2015): 5996: doi: 10.1038/ncomms6996.

p. 180   ". . . snakes either gradually lost their legs by adapting to marine environments as swimmers, or they went underground . . ." R. Bauchot, *Snakes: A Natural History* (New York: Sterling Publishing, 2006).

p. 181   "Indeed, mosasaurs . . . were Cretaceous lizards fully adapted to marine conditions . . ." (1) M. J. Everhart, *Oceans of Kansas: A Natural History of the Western Interior Sea* (Bloomington: Indiana University Press, 2005); (2) Benton (2009).

p. 181   ". . . a 2015 article by Hongyu Yi and Mark Norell bearing the unambiguous title of 'The Burrowing Origin of Modern Snakes' . . ." Yi and Norell (2015).

p. 181   "This evolutionary direction taken by lizards also implies that snake ancestors were descended from predatory burrowing lizards . . ." Yi and Norell (2015).

CHAPTER 7: PLAYING HIDE AND SEEK FOR KEEPS

p. 184   ". . . the Ediacaran Period, a time interval from 630 to 542 *mya* . . ." F. M. Gradstein *et al.* (2012).

p. 185   ". . . 'Ediacara' is a transliteration of the word *idiyakra*, meaning 'spring of water' . . ." A brief discussion of this name is in: M. A. Fedonkin *et al.*, *The Rise of Animals: Evolution and Diversification of the Kingdom Animalia* (Baltimore: Johns Hopkins University Press, 2007).

p. 185   "In 1946 . . . geologist Reginald C. Sprigg first noticed some of these fossils . . ." Fortunately, Spriggs reported these fossils in a couple of articles in a regional journal soon afterward, which helped with their documentation early on: (1) R. C. Sprigg, "Early Cambrian (?) jellyfishes from the Flinders Ranges, South Australia," *Transactions of the Royal Society of South Australia* 71 (1947): 212–224; (2) R. C. Sprigg, "Early Cambrian 'Jellyfishes' of Ediacara, South Australia, and Mt. John, Kimberley District, Western Australia," *Transactions of the Royal Society of South Australia* 73 (1949): 72–99.

p. 185   ". . . at least one paleontologist proposed the far-flung idea that they were lichens and lived on land . . ." G. J. Retallack, "Were the Ediacaran fossils lichens?" *Paleobiology* 20 (1994): 523–544. (We shall never speak of this again.)

p. 186   "In their minds, either their geologic age was wrong . . . or they were not fossils, and certainly not of animals . . ." Fedonkin *et al.* (2007).

p. 186   ". . . what Australian historian Geoffrey Blainey termed 'the tyranny of distance' . . ." G. Blainey, *The Tyranny of Distance: How Distance*

Shaped Australia's History (New York: Macmillan, 1968). Originally published by Sun Books in Melbourne, Australia in 1966.

p. 186   ". . . Martin Glaessner and Mary Wade studied them later in the 1960s . . ." (1) M. F. Glaessner and M. Wade, "The late Precambrian fossils from Ediacara, South Australia," *Palaeontology* 9 (1966): 599–628; (2) M. F. Glaessner, "Geographic distribution and time range of the Ediacara Precambian fauna," *Geological Society of America Bulletin* 82 (1971): 509–514.

p. 186   ". . . paleontologists and geologists recognized Ediacaran fossils in England, Russia, Namibia, western Canada, and even in . . . North Carolina . . ." Fedonkin *et al.* (2007).

p. 186   ". . . a stratigraphic commission decided that a geologic time unit . . . would be named after the Ediacara Hills . . ." The International Union of Geological Sciences decreed the Ediacaran as an official geological period in 2004, defined by the end of the last major glaciation of the Proterozoic (635 *mya*) and the first occurrence of the trace fossil *Treptichnus pedum* (542 *mya*). The full proposal and discussion of this proposed period (including votes and objections), submitted by the International Commission on Stratigraphy, is: A. H. Knoll, *et al.*, *The Ediacaran Period: A New Addition to the Geologic Time Scale* (2004): http://www.stratigraphy.org/bak/ediacaran/Knoll_et_al_2004a.pdf

p. 187   "The Proterozoic and . . . Archean Eon . . . represent about 87% of the earth's 4.5 billion-year history . . ." F. M. Gradstein *et al.* (2012).

p. 187   "Paleontologist Elkanah Billings . . . interpreted these peculiar forms as fossils in 1872 . . ." E. Billings, "Fossils in Huronian rocks," *Canadian Naturalist and Quarterly Journal of Science* 6 (1872): 478.

p. 187   "Billings . . . also gave them a fossil species name, *Aspidella terranovica* . . ." (1) Billings (1872); J. G. Gehling *et al.*, "The first named Ediacaran body fossil, *Aspidella terranovica*," *Palaeontology* 43 (2000): 427–456.

p. 187   ". . . Billings's peers at the time scorned his interpretation, countering that these fossils must be inorganic structures . . ." Fedonkin *et al.* (2007).

p. 187   "In 1967 . . . Shiva Balak Misra . . . found the . . . Mistaken Point fossils . . ." Misra provides a firsthand account of his discovery of the Mistaken Point fossils on his Web site (*Mistaken Point Fauna: The Discovery*), which also tells about his post-geological career as a community leader and educator in his home town of Lucknow, India: http://mistakenpointfauna.com/index.html. His inspiring biography is: S. B. Misra, *Dream Chasing: One Man's Remarkable, True Life Story* (New Delhi, India: Roli Books, 2011).

p. 188   "Fortunately, he quickly documented his find in several peer-reviewed articles in 1968 . . ." Actually, one of these articles was published in 1968 and the next in 1969: (1) M. M. Anderson and S. B. Misra, "Fossils found in pre-Cambrian Conception Group of Southeastern Newfoundland," *Nature* 220 (1968): 680–681; (2) S. B. Misra "Late Precambrian (?) fossils from southeastern Newfoundland," *Geological Society of America Bulletin* 80 (1969): 2133–2140.

p. 188   "Mistaken Point fossils have been described as leaf-like, spindle-like, lobate . . ." Fedonkin *et al.* (2007).

p. 188   ". . . sea bottom hosting the Ediacaran organisms may have been 1,000 meters (3,300 feet) below the ocean surface . . ." When I asked "How deep?" this was the number I was given by the field-trip leaders during our visit to Mistaken Point, but I could not find a reference that stated it. Nonetheless, every geological study of the stratigraphic sequence has led to the same conclusion that this seafloor was far below the zone where visible light would penetrate, which today is about 200 meters (about 660 feet). Here's one such study: D. A. Wood *et al.*, "Paleoenvironmental analysis of the late Neoproterozoic Mistaken Point and Trepassey formations, southeastern Newfoundland," *Canadian Journal of Earth Sciences* 40 (2003): 1375–1391.

p. 189   ". . . as early animal life was always assumed to have first evolved in warm, well-lit, shallow waters . . ." The "shallow-water" origin of animal life was the reigning hypothesis when I was in graduate school during the 1980s, which was dutifully repeated in paleontology and geology textbooks then. The deep-water paleoenvironments of the Mistaken Point fossils and a few other Ediacaran assemblages elsewhere contradict this, so now scientists are looking more closely at when deep oceans became oxygenated enough to support animals: D. B. Mills *et al.*, "Oxygen requirements of the earliest animals," *Proceedings of the National Academy of Sciences* 111 (2014): 4168–4172.

p. 189   "The ash erupted from a nearby volcanic island arc . . ." (1) P. M. Myrow, "Neoproterozoic rocks of the Newfoundland Avalon Zone," *Precambrian Research* 73 (1995): 123–136; (2) S. J. Mason *et al.*, "Paleoenvironmental analysis of Ediacaran strata in the Catalina Dome, Bonavista Peninsula, Newfoundland," *Canadian Journal of Earth Sciences* 50 (2013): 197–212.

p. 189   ". . . these sediments settled and buried the fossils in place, acting like an underwater . . . Pompeii . . ." This evocative description of fossil preservation was evidently coined first by Dolf Seilacher in 1992: A. Seilacher, "Vendobionta and Psammocorallia: lost constructions of Precambrian evolution," *Journal of Geological*

Society of London 149 (1992): 607–613. Current estimates are that ash flows may have killed 15–40% of the fossils at Mistaken Point biota: J. B. Antcliffe *et al.*, "A new ecological model for the ~565 Ma Ediacaran biota of Mistaken Point, Newfoundland," *Precambrian Research* 268 (2015): 227–242.

p. 189   "Dates revealed ages just on either side of 565 *mya* for most of the layers there . . ." A. G. Liu *et al.*, "First evidence for locomotion in the Ediacara biota from the 565 Ma Mistaken Point Formation, Newfoundland," *Geology* 38 (2010): 123–126.

p. 190   "All had discs on their bottoms, which acted like plant roots, anchoring the organisms to the seafloor . . ." M. E. Clapham *et al.*, "Paleoecology of the oldest known animal communities: Ediacaran assemblages at Mistaken Point, Newfoundland," *Paleobiology* 29 (2003): 527–544.

p. 190   ". . . further scrutiny revealed these were either body fossils or strange sedimentary structures that fooled people into thinking they were trace fossils . . ." A. Seilacher *et al.* "Trace fossils in the Ediacaran-Cambrian transition: behavioral diversification, ecological turnover and environmental shift," *Palaeogeography, Palaeoclimatology, Palaeoecology* 227 (2005): 323–356. Yes, that's right, my coauthors and I were bold enough to disagree with Dolf Seilacher. Sadly, I never got to discuss this with him in person. It would have been a hellacious argument, followed by our buying a round of beers afterward to celebrate.

p. 191   "The researchers studying this trace fossil suspected this and similar fossil trails might be from anemone-like animals . . ." (1) Liu *et al.* (2010); (2) L. R. Menon *et al.*, "Evidence for Cnidaria-like behavior in ca. 560 Ma Ediacaran *Aspidella*," *Geology* 41 (2013): 895–898.

p. 193   "Our group of twenty-one ichnologists represented nine countries . . ." Ichnologist-nation representatives on this field trip were (in alphabetical order): Brazil, Canada, Germany, Japan, Norway, Poland, Spain, the then-United Kingdom, and the United States.

p. 193   ". . . Fortune Head is regarded as the international stratigraphic standard for the Precambrian-Cambrian boundary . . ." G. M Narbonne *et al.*, "A candidate stratotype for the Precambrian–Cambrian boundary, Fortune Head, Burin Peninsula, southeastern Newfoundland," *Canadian Journal of Earth Sciences* 24 (1987): 1277–1293.

p. 194   "*Treptichnus* is also considered as an index fossil, telling us . . . rocks . . . were unequivocally Early Cambrian in age . . ." Perhaps "unequivocally" was too strict a word, as a few examples of this fossil burrow occur in the last of Ediacaran strata, including those at Fortune Head: J. G. Gehling *et al.*, "Burrowing below the basal Cambrian GSSP, Fortune Head, Newfoundland," *Geological Magazine* 138 (2001): 213–218.

p. 195   "Most of the gooey surfaces were made by colonies of microbial life . . ." (1) J. G. Gehling, "Microbial mats in terminal Proterozoic siliciclastics: Ediacaran death masks," *Palaios* 14 (1999): 40–57; (2) M. Steiner and J. Reitner, "Evidence of organic structures in Ediacara-type fossils and associated microbial mats," *Geology* 29 (2001): 1119–1122.

p. 195   ". . . *microbially induced sedimentary structures*, which is often shortened to the memorable . . . acronym MISS . . ." N. Noffke *et al.*, "Microbially induced sedimentary structures: a new category within the classification of primary sedimentary structures," *Journal of Sedimentary Research* 71 (2001): 649–656.

p. 195   ". . . stromatolites dominated the shallow seas . . . more than three billion years before the first corals . . ." A. C. Allwood *et al.*, "Stromatolite reef from the Early Archaean era of Australia," *Nature* 441 (2006): 714–718.

p. 195   "Stromatolites . . . still exist in the Bahamas, Abu Dhabi, western Australia . . ." J. Seckbach and A. Oren, *Microbial Mats: Modern and Ancient Microorganisms in Stratified Systems* (New York: Springer, 2010).

p. 196   "Geoscientists estimate the equator had an average temperature of about –20°C (–4°F) . . ." This already low temperature may be a conservative estimate, as some models indicate equatorial lows of –45°C (–49°F): A. Micheels, and M. Montenari, "A snowball Earth versus a slushball Earth: results from Neoproterozoic climate modeling sensitivity experiments," *Geosphere* 4 (2008): 401–410.

p. 196   "Geologists have affectionately nicknamed this interval 'Snowball Earth' . . ." (1) J. L. Kirschvink, "Late Proterozoic low-latitude global glaciation: the Snowball Earth," In *The Proterozoic Biosphere: A Multidisciplinary Study* (New York: Cambridge University Press, 1992, 51–52); (2) P. F. Hoffman *et al.*, "A Neoproterozoic Snowball Earth," *Science* 281 (1998): 1342–1346.

p. 196   "These rocks, which geologists call *dropstones* . . ." M. R. Bennett *et al.*, "Dropstones: their origin and significance," *Palaeogeography, Palaeoclimatology, Palaeoecology* 121 (1996): 331–339.

p. 197   ". . . molecular clocks predict that the first tetrapods probably evolved about 400 million years ago . . ." P. R. Alexander, "Divergence time estimation using fossils as terminal taxa and the origins of Lissamphibia," *Systematic Biology* 60 (2011): 466–481.

p. 197   "When applied to the origin of animals, molecular clocks suggest a start date as long ago as 800–700 *mya* . . ." Some of the initial attempts at using molecular clocks to figure out the start date for animals were so high, and so far removed from the realities of the fossil record, that paleontologists began to wonder if the molecular biologists were high. Anyway, 800 *mya* is probably too far back for the first animals, and more recent calculations put their origins closer to 700 *mya*: K. J. Peterson *et al.*, "The Ediacaran emergence of bilaterians: congruence between the genetic and the geological fossil records," *Philosophical Transactions of the Royal Society of London B* 363 (2008): 1435–1443.

p. 197 ". . . organic compounds called steranes, indicate a presence of animals at a younger date, at about 650 *mya* . . ." (1) G. D. Love *et al.*, "Fossil steroids record the appearance of Demospongiae during the Cryogenian period," *Nature* 457 (2008): 718–721; (2) D. A. Gold *et al.*, "Sterol and genomic analyses validate the sponge biomarker hypothesis," *Proceedings of the National Academy of Sciences* 113 (2016): 2684–2689.

p. 197 "These microscopic . . . fossils recovered from Proterozoic rocks of China are evidently from dividing cells . . ." S. Xioa *et al.*, "Three-dimensional preservation of algae and animal embryos in a Neoproterozoic phosphorite," *Nature* 391 (1998): 553–558. However, the identity of these visually striking microfossils has been disputed recently, in which other scientists have proposed these structures are not animal fossils, but those of a clade more closely related to fungi: T. Huldtgren *et al.*, "Fossilized nuclei and germination structures identify Ediacaran 'animal embryos' as encysting protists," *Science* 334 (2011): 1696–1699. As is typical in science (and the journal *Science*), the authors of the original study later delivered a smack-down response: S. Xiao *et al.*, "Comment on 'Fossilized nuclei and germination structures identify Ediacaran "animal embryos" as encysting protists.'" *Science* 335 (2012): 1169.

p. 198 "The rocks were dated from the latest part of the Ediacaran Period, at about 545 *mya* . . ." C. Tacker *et al.*, "Trace fossils versus body fossils: *Oldhamia recta* revisited," *Precambrian Research* 178 (2010): 43–50.

p. 199 ". . . when we published our research article in 2010, we stated that these Ediacaran entities were not trace fossils . . ." Tacker *et al.* (2010).

p. 199 "So three years later, Trish Weaver and I published an article describing the trails . . ." A. J. Martin and P. W. Weaver, "Ediacaran trace fossils from the Albemarle Group of the Carolina Terrane, North Carolina (USA): marks of a mobile lifestyle on a Precambrian sea bottom," In Hibbard, J. (ed.), *One Arc, Two Arcs, Old Arc, New Arc: A 21st Century Perspective on the Geology of the Carolina Terrane in Central North Carolina* (Carolina Geological Society, Field Trip Guidebook, 2013, 185–192).

p. 199 ". . . paleontologists scrutinizing Ediacaran fossils have likewise identified body fossils that were formerly regarded as trace fossils . . ." S. Jensen *et al.*, "A critical look at the Ediacaran trace fossil record," In S. Xiao and A. J. Kaufman (eds.), *Neoproterozoic Geobiologya and Paleobiology* (New York: Springer, 2006, 115–157); (2) A. Sappenfield *et al.*, "Problematica, trace fossils, and tubes within the Ediacara Member (South Australia): redefining the Ediacaran trace fossil record one tube at a time," *Journal of Paleontology* 85 (2011): 256–265.

p. 200 "These trails consist of many overlapping scratches on biomat surfaces, showing how *Kimberella* lived a life of mat scratching . . ." (1) M. A. Fedonkin and B. M. Waggoner, "The Late Precambrian fossil *Kimberella* is a mollusc-like bilaterian organism," *Nature* 388 (1997): 868–871; (2) A. Y. Ivantsov, "Trace fossils of Precambrian metazoans 'Vendobionta' and 'Mollusks,'" *Stratigraphy and Geological Correlation* 21 (2013): 252–264.

p. 200 ". . . *Dickinsonia* and *Yorgia* are connected to a series of overlapping body impressions on biomat surfaces . . ." (1) A. Y. Ivantsov, "Movement traces of large Upper Vendian Metazoa on the sediment surface." In *Ekosistemnye Perestroiki i Evolyutsiya Biosfery*, 4 (Ecosystem Rearrangements and Evolution of the Biosphere), Moscow: Palaeontological Institute, 2001, 119–120. (2) Ivantsov (2013).

p. 200 "They are also quite shallow, not plumbing depths beyond a centimeter . . ." M. Laflamme *et al.*, "The end of the Ediacara biota: Extinction, biotic replacement, or Cheshire Cat?" *Gondwana Research* 23 (2013): 558–573.

p. 201 "Burrowing animals that punched through biomats . . . abruptly . . . introduced oxygenated water from above . . ." (1) D. H. Erwin, "Macroevolution of ecosystem engineering, niche construction and diversity," *Trends in Ecology & Evolution* 23 (2008): 304–310; (2) D. H. Erwin and S. M. Tweedt, "Ecological drivers of the Ediacaran–Cambrian diversification of Metazoa," *Evolutionary Ecology* 26 (2012): 417–433.

p. 202 ". . . all animals poop . . . burrowing introduced their bountiful, nutrient-rich feces to the subsurface . . ." (1) T. Gomi, *Everyone Poops* (St. Louis, Mo.: Turtleback Books, 2001); (2) Laflamme *et al.* (2013).

p. 203 "This allusion to an Eden-like paradise was applied . . . because of the apparent paucity of predators in the Ediacaran . . ." M. McMenamin, "The Garden of Ediacara," *Palaios* 1 (1986): 178–182.

p. 203 "Soon Tennyson's 'nature red in tooth and claw' . . ." This quotation is from Canto 56 of Alfred Lord Tennyson's poem "In Memoriam A.H.H.," finished in 1849. Stephen Jay Gould wrote about this phrase and the poem in an essay, "The Tooth and Claw

Centennial," published in his book *Dinosaur in a Haystack* (New York: Harmony Books, 1995, 63–75).

p. 203   "The body fossils, called *Cloudina*, look like a stack of tiny flat-bottomed ice cream cones . . ." S.F.W. Grant, "Shell structure and distribution of *Cloudina*, a potential index fossil for the terminal Proterozoic," *American Journal of Science* 290 (1989): 261–294.

p. 203   "Yet predation is the more likely explanation, as hole diameters increase in proportion with the sizes of drilled *Cloudina* . . ." H. Hua *et al.*, "Borings in *Cloudina* Shells: complex predator-prey dynamics in the terminal Neoproterozoic," *Palaios* 18 (2003): 454–459.

p. 204   ". . . once the oceans warmed after 'Snowball Earth,' their waters contained greater amounts of calcium and bicarbonate . . ." (1) P. F. Hoffman *et al.*, "A Neoproterozoic Snowball Earth," *Science* 281 (1998): 1342–1346. (2) Y. Sawaki *et al.*, "The anomalous Ca cycle in the Ediacaran ocean: Evidence from Ca isotopes preserved in carbonates in the Three Gorges area, South China," *Gondwana Research* 25 (2014): 1070–1089.

p.204   "These earlier shell makers presumably did this . . . by combining calcium with bicarbonate in their tissues and forming calcium carbonate . . ." R. Wood and A. Y. Zhuravlev, "Escalation and ecological selectively of mineralogy in the Cambrian Radiation of skeletons," *Earth-Science Reviews* 115 (2012): 249–261.

p. 204   "Soon after this type of biomineralization evolved, vertebrates then began making calcium phosphate (apatite) . . ." Biomineralization of apatite probably happened in invertebrates before evolution of the first vertebrates, but it was certainly composing hard parts in chordates (such as hagfish-like animals called conodonts) by the end of the Cambrian Period: D.J.E. Murdock and P.C.J. Donoghue, "Evolutionary origins of animal skeletal biomineralization," *Cell Tissues Organs* 194 (2011): 98–102.

p. 204   "First proposed and named [the Red Queen Hypothesis] in 1973 by evolutionary biologist Leigh Van Valen . . ." L. Van Valen, "A new evolutionary law," *Evolutionary Theory* 1 1973: 1–30.

p. 204   "A close reading of the novel shows that the Red Queen is methodical and persistent, unlike the White Queen . . ." L. Carroll, *Through the Looking Glass and What Alice Found There* (Philadelphia: Henry Altemus Company, 1897).

p. 205   "This inequity means that prey animals have greater selection pressures applied to them than predators . . ." (1) E. D. Brodie, III, and E. D. Brodie, Jr., "Predator-prey arms races," *Bioscience* 49 (1999): 557–568; (2) A. Sih *et al.*, "Predator-prey naïveté, antipredator behavior, and the ecology of predator invasions," *Oikos* 119 (2010): 610–621.

p. 206   "Hard parts broke the rules, as animals then carried their own spades on their heads, sides, and backs . . ." (1) S. M. Stanley, "Why clams have the shape they have: an experimental analysis of burrowing," *Paleobiology* 1 (1975): 48–58; (2) G. J. Vermeiji, *A Natural History of Shells* (Princeton, N.J.: Princeton University Press, 1995).

p. 206   ". . . some modern snails have a radula that scrapes algae off hard surfaces, whereas others use a radula to drill into clams and other snails . . ." (1) Vermeiji (1995); P. H. Kelley, "The fossil record of drilling predation on bivalves and gastropods." In P. H. Kelley *et al.* (eds.), *Predator-Prey Interactions in the Fossil Record: Topics in Geobiology*, 20, (New York: Springer, 2003, 113–139).

p. 206   "This hypothesis [the Verdun Syndrome], proposed in 2007 by Polish paleontologist Jerzy Dzik . . ." J. Dzik, "The Verdun Syndrome: simultaneous origin of protective armour and infaunal shelters at the Precambrian–Cambrian transition." In P. Vickers-Rich and P. Komarower (eds.), *The Rise and Fall of the Ediacaran Biota* (London: Geological Society, London, Special Publications, 286, 2007, 405–414). However, Dzik coined this term in an earlier article: J. Dzik, "Behavioral and anatomical unity of the earliest burrowing animals and the cause of the 'Cambrian explosion.'" *Paleobiology* 31 (2005): 507–525.

p. 206   "The main strategies of French soldiers included the use of trenches, tunnels, underground shelters, and fortifications . . . used as . . . main defenses against German artillery . . ." P. Jankowski, *Verdun: The Longest Battle of the Great War* (Oxford, England: Oxford University Press, 2014).

p. 207   ". . . what paleontologists call the 'Cambrian Explosion,' a twenty-million-year burst of animal diversification . . ." (1) D. H. Erwin and J. W. Valentine, *The Cambrian Explosion and the Construction of Animal Biodiversity* (New York: W.H. Freeman, 2012); (2) D. Fox, "What sparked the Cambrian explosion?" *Nature* 530 (2016): 268–270.

p. 207   ". . . future fossils with hard parts also buried themselves deeper, greatly increasing their odds of preservation . . ." Dzik (2007).

p. 207   ". . . convergent evolution, in which certain traits . . . emerge in many lineages of . . . animals adapted for the same purpose . . ." G. R. McGhee, *Convergent Evolution: Limited Forms Most Beautiful* (Cambridge, Mass.: MIT Press, 2011).

p. 207   "Trilobite eyes were aided by the development of calcite-crystal lenses . . ." (1) Levi-Setti (1995); (2) Fortey (2010).

p. 208   "In a 2016 study of Ediacaran body fossils and trace fossils from Namibia done by Simon Darroch and his colleagues . . ." S.A.F. Darroch *et al.*, "A mixed Ediacaran-metazoan assemblage from the Zaris Sub-basin, Namibia," *Palaeogeography, Palaeoclimatology, Palaeoecology* 459 (2016): 198–208.

p. 208   ". . . animals employed a combination of mineralized bodies and acid to bore rocks, or into one another . . ." (1) N. P. James *et al.,* "The oldest macroborers: Lower Cambrian of Labrador," *Science* 197 (1977): 980–983; (2) M. A. Wilson and T. J. Palmer, "Patterns and processes in the Ordovician bioerosion revolution," *Ichnos* 13 (2006): 109–112.

p. 210   ". . . sulfur went from mostly reduced states . . . to oxidized states . . ." D. E. Canfield and J. Farquhar (2009), "Animal evolution, bioturbation, and the sulfate concentration of the oceans," *Proceedings of the National Academy of Sciences,* 106: 8123–8127.

p. 210   ". . . burrowers industriously mixed these neatly packaged and chemically altered wastes back into the sediments . . ." Laflamme *et al.* (2013).

p. 210   ". . . Dolf Seilacher and Friedrich Pflüger first proposed this hypothesis and termed it the *agronomic revolution* . . ." A. Seilacher, F. Pflüger, "From biomats to benthic agriculture: A biohistoric revolution." In W. E. Krumbein, *et al.* (eds.), *Biostabilization of Sediments* (Oldenburg, Germany: Universität Oldenburg, 1994, 97–105).

p. 210   "Other paleontologists since have also used the less catchy phrase 'Cambrian substrate revolution' . . ." D. J. Bottjer *et al.,* "The Cambrian substrate revolution," *GSA Today* 10 (2000): 1–7.

p. 211   ". . . but he was also widely acknowledged as the world's most brilliant ichnologist . . ." S. G. Pemberton *et al.,* "Curious mind: a celebration of the extraordinary life and ichnological contributions of Adolf 'Dolf' Seilacher (1925–2014)," *Ichnos* 23 (2016): 1–24.

p. 211   "Soon after the beginning of the Cambrian Period, nearly all Ediacaran-style biota vanished quickly . . ." (1) Fedonkin *et al.* (2007); (2) S.A.F. Darroch *et al.,* "Biotic replacement and mass extinction of the Ediacara biota," *Proceedings of the Royal Society of London B* 282 (2015): 20151003; http://dx.doi.org/10.1098/rspb.2015.1003

p. 211   "The presence of vertical burrows in Ediacaran rocks of Namibia signal the start of this disappearance . . ." Darroch *et al.* (2016).

p. 212   "Cambrian seas had only a few deep diggers . . . but the Middle to Late Ordovician (about 470–440 *mya*) had more . . ." Probably the most influential study for those paleontologists interested in the evolution of burrowing marine bottom-dwelling animals was: C. W. Thayer, "Biological bulldozers and the evolution of marine benthic communities," *Science* 203 (1979): 458–461. Thayer's article later inspired many paleontologists to disprove his assertion that deep and significant amounts of burrowing did not happen until the Carboniferous-Permian Periods (300–250 *mya*). Instead, it likely started in the Ordovician: (1) P. Sheehan and D.R.J. Schiefelbein, "The trace fossil *Thalassinoides* from the Upper Ordovician of the eastern Great Basin: deep burrowing in the Early Paleozoic," *Journal of Paleontology* 58 (1984): 440–447; (2) P. J. Orr, "Ecospace utilization in early Phanerozoic deep-marine environments: deep bioturbation in the Blakely Sandstone (Middle Ordovician), Arkansas, USA," *Lethaia* 36 (2003): 97–106.

p. 212   "During the Silurian Period . . . invertebrates began mixing oceanic sediments on a massive scale . . ." L. G. Tarhan, *et al.*, "Protracted development of bioturbation through the early Palaeozoic Era," *Nature Geoscience* 8 (2015): 865–896.

p. 212   "Once arthropods and other invertebrates began burrowing in marginal-marine environments . . . during the Ordovician . . . they were later joined by insects and vertebrates . . . in the Devonian . . ." (1) Buatois *et al.* (2005); (2) J. S. Bridge, *Rivers and Floodplains: Forms, Processes, and Sedimentary Record* (New York: John Wiley & Sons, 2009).

p. 213   ". . . during the Early Triassic (252–247 *mya*), burrows were clearly reduced in both number and depth . . ." (1) S. Pruss and D. J. Bottjer, "Early Triassic trace fossils of the western United States and their implications for prolonged environmental stress from the end-Permian mass extinction," *Palaios* 19 (2004): 551–564; (2) M. L. Fraiser and D. J. Bottjer, "Opportunistic behaviour of invertebrate marine tracemakers during the Early Triassic aftermath of the end-Permian mass extinction," *Australian Journal of Earth Sciences* 56 (2009): 841–857.

p. 213   ". . . microbial communities were free to grow and spread, covering shallow-marine and lake bottoms once more . . ." (1) S. Pruss *et al.*, "Proliferation of Early Triassic wrinkle structures: Implications for environmental stress following the end-Permian mass extinction," *Geology* 32 (2004): 461–464; (2) D. Chu *et al.*, "Early Triassic wrinkle structures on land: stressed environments and oases for life," *Scientific Reports* 5 (2015): 10109; doi: 10.1038/srep10109.

p. 213   ". . . Middle Triassic rocks (247–235 *mya*) have much more evidence of burrowing . . ." Pruss and Bottjer (2004).

CHAPTER 8: RULERS OF THE UNDERWORLD

p. 215   "When we behold a wide, turf-covered expanse, we should remember that its smoothness . . . is mainly due to all the inequalities having been slowly levelled by worms . . ." C. Darwin,

The Formation of Vegetable Mould through the Action of Worms, with Observations on their Habits (London: John Murray, 1881).

p. 216   Much of this story about my trace-identification ineptitude at Wormsloe in October 2014 was told previously on my blog, *Life Traces of the Georgia Coast*, but has been edited for this book and then merged with content from two other blog posts I wrote about a May 2012 visit to Darwin's home at Down House. If you would like to read the original posted story, it is "Slow Worms at Wormsloe" (posted October 21, 2014) and includes many photos of the worm traces seen that morning: http://www.georgialifetraces. com/2014/10/21/slow-worms-at-wormsloe/. The two blog posts I wrote about my visit to Down House and how this relates to burrowing earthworms and moles are "Of Darwin, Earthworms, and Backyard Science" (posted May 29, 2012) and "Darwin, Worm Grunters, and Menacing Moles" (posted June 18, 2012), with the following respective links: (1) http://www.georgialifetraces. com/2012/05/29/of-darwin-earthworms-and-backyard- science/ (2) http://www.georgialifetraces.com/2012/06/18/ darwin-worm-grunters-and-menacing-moles/

p. 216   "Wormsloe was originally a plantation dating back to the eighteenth century . . ." D. A. Swanson, *Remaking Wormsloe Plantation: The Environmental History of a Lowcountry Landscape* (Athens: University of Georgia Press, 2012).

p. 216   "As mentioned previously, these gopher-tortoise driven communities were far more widespread . . ." Finch *et al.* (2012).

p. 219   "Earthworms, after all, are on strict low-salt diets, and object strenuously to saline water filling their burrows . . ." (1) O. J. Owojori *et al.*, "Effects of salinity on partitioning, uptake and toxicity of zinc in the earthworm *Eisenia fetida*," *Soil Biology and Biochemistry* 40 (2008): 2385–2393; (2) O. J. Owojori and A. J. Reinecke, "Avoidance behaviour of two eco-physiologically different earthworms (*Eisenia fetida* and *Aporrectodea caliginosa*) in natural and artificial saline soils," *Chemosphere* 75 (2009): 279–283.

p. 219   "Consisting of about 7,000 known species (but possibly as many as 30,000), earthworms are broadly classified as annelids . . ." in A. Orgiazzi *et al.* (eds.), *Global Soil Biodiversity Atlas* (Luxembourg: European Commission, Publications Office of the European Union, 2016).

p. 219   "Other annelids include leeches (Hirudinea) . . . and bristleworms (Polychaeta) . . ." L. Margulis and M. J. Chapman, *Kingdoms and Domains: An Illustrated Guide to the Phyla of Life on Earth* (Cambridge, Mass.: Academic Press, 2009).

p. 219   "Oligochaetes . . . have short, stiff projections outside of their bodies called *chaetae* . . ." J. H. Thorp, *Ecology and Classification of North American Freshwater Invertebrates*, (Cambridge, Mass.: Academic Press, 2010).

p. 220 "Darwin's last book (and one of his most popular) was *The Formation of Vegetable Mould through the Actions of Worms* . . ." (1) Darwin (1881); (2) D. Porter and P. Graham, *Darwin's Sciences* (New York: John Wiley & Sons, 2015).

p. 220 "Darwin was a homebody after his formative years of voyaging on the *Beagle* . . ." D. Quammen, *The Reluctant Mr. Darwin: An Intimate Portrait of Charles Darwin and the Making of His Theory of Evolution* (New York: W.W. Norton, 2007).

p. 220 "Darwin never ventured more than 50 kilometers (31 miles) from home and stayed there until his death in 1882 . . ." Quammen (2007).

p. 221 "Darwin began thinking intensively about earthworms in 1837, when his uncle showed him how layers of lime and cinders . . ." (1) S. G. Pemberton, and R. W. Frey, "Darwin on worms: the advent of experimental neoichnology," *Ichnos* 1 (1990): 65–71; (2) Porter and Graham (2015).

p. 221 ". . . one of these stones behind the house, a talisman of a simple but elegant test for assessing the effects of burrowing effects of earthworms . . ." (1) Pemberton and Frey (1990); (2) Porter and Graham (2015).

p. 221 ". . . he used a precise measuring device . . . and flat, circular rocks set out in the fields behind Down House, nicknamed 'wormstones' . . ." (1) Pemberton and Frey (1990); (2) Porter and Graham (2015).

p. 221 "In essence, worms would bury the stones from below . . ." (1) Pemberton and Frey (1990); (2) Porter and Graham (2015).

p. 222 "Based on his measurements taken over nearly twenty years, he calculated an approximate 'sinking' rate of 2.2 millimeters/year . . ." Pemberton and Frey (1990).

p. 222 "As a geologist (and a fine one at that) . . ." S. Herbert, *Darwin, Geologist* (Ithaca, N.Y.: Cornell University Press, 2005).

p. 222 "Or, to put it in Darwin's own words when he responded to (and totally owned) a critic who claimed earthworms were too insignificant . . ." Darwin (1881), in response to a "Mr. Fish" (no first name given) who disagreed with Darwin's original observations about earthworms related in an 1837 paper delivered to the Royal Society of London.

p. 223 ". . . annelids likely originated just after the Ediacaran Period, then modern-style earthworms evolved before the end of the Cretaceous . . ." (1) M. F. Glaessner, "Lower Cambrian Crustacea and annelid worms from Kangaroo Island, South Australia," *Alcheringa* 3 (1979): 21–31; (2) S. C. Morris and J. S. Peel, "The earliest annelids: Lower Cambrian polychaetes from the

Sirius Passet Lagerstätte, Peary Land, North Greenland," *Acta Palaeontologica Polonica* 53 (2008): 137–148; (3) J. Domínguez *et al.*, "Underground evolution: new roots for the old tree of lumbricid earthworms," *Molecular Phylogenetics and Evolution* 83 (2015): 7–19.

p. 223   "In the Paleozoic Era, oligochaetes descended from . . . marine polychaetes, and body fossils in Permian rocks (about 260 *mya*) give us a minimum time . . ." C. S. Morris *et al.*, "A possible annelid from the Trenton Limestone (Ordovician) of Quebec, with a review of fossil oligochaetes and other annulate worms," *Canadian Journal of Earth Science* 19 (1982): 2150–2157.

p. 223   ". . . Mariano Verde and several other South American paleontologists described fossil examples of fecal pellets and aestivation chambers . . ." M. Verde *et al.*, "A new earthworm trace fossil from paleosols: aestivation chambers from the Late Pleistocene Sopas Formation of Uruguay," *Palaeogeography, Palaeoclimatology, Palaeoecology* 243 (2007): 339–347.

p. 223   "In 2013, ichnologist Karen Chin and a few colleagues documented fossil earthworm burrows and feces from just above the end-Cretaceous boundary . . ." K. Chin *et al.*, "Fossil worm burrows reveal very early terrestrial animal activity and shed light on trophic resources after the end-Cretaceous mass extinction," *PLoS One* 8 (2013): e70920; doi: 10.1371/journal.pone.0070920.

p. 224   "In a 2016 study by Anne Zangerlé and her colleagues, they proposed that earthworm burrows and feces were responsible . . ." A. Zangerlé *et al.*, "The *surales*, self-organized earth-mound landscapes made by earthworms in a seasonal tropical wetland," *PLoS One* 11 (2016): e0154269; doi: 10.1371/journal.pone.0154269

p. 224   "Individual worms are as long as a meter (3.3 feet) and can make 5-centimeter (2-inch) wide fecal casts . . ." Zangerlé, A. *et al.* (2016).

p. 225   ". . . This perspective . . . reflects an ichnological worldview appreciated and well articulated by Darwin . . ." Darwin (1881).

p. 225   "In the episode 'Deep Space Homer' (1994) . . ." *The Simpsons*, Season 5, Episode 15, originally broadcast on February 24, 1994.

p. 225   ". . . their biomass at any given moment on Earth is equal to that of our species, with more than one million ants per person . . ." (1) B. Hölldobler and E. O. Wilson, *The Ants* (Cambridge, Mass.: Harvard University Press, 1990); (2) T. R. Schultz, "In search of ant ancestors," *Proceedings of the National Academy of Sciences* 97 (2000): 14028–14029.

p. 226   "Represented by more than 12,000 known species, with 20,000 species likely, ants are diverse . . ." Ohio State University maintains a species counter for ants (classified as the clade Formicidae) at the following link. As of October 9, 2016, the count given there

was 14,096: http://atbi.biosci.ohio-state.edu/hymenoptera/tsa.
sppcount?the_taxon=Formicidae

p. 226   "Thus if an entomologist states that, say, ants are '. . . arguably
the greatest success story in the history of terrestrial metazoans
[animals] . . .'" Schultz (2000).

p. 226   "In a study published in 2009, entomologists reported that massive
Argentine-ant colonies in California . . . Europe, and Japan . . ."
E. Sunamura et al., "Intercontinental union of Argentine ants:
behavioral relationships among introduced populations in Europe,
North America, and Asia," Insectes Sociaux 56 (2009): 143–147.

p. 227   "The European colony alone spans more than 6,000 kilometers
(3,700 miles) along the Mediterranean coast . . ." (1) Sunamura
et al. (2009); (2) E. V. Wilgenburg et al., "The global expansion
of a single ant supercolony," Evolutionary Applications 3 (2010):
136–143.

p. 227   "Originally written and published by Carl Stephenson in 1938, this
story [Leiningen Versus the Ants] . . ." German writer Carl Stephenson
(1893–1960) published this story in the December 1938 issue of
Esquire.

p. 227   ". . . when the TV series Star Trek: The Next Generation (1987–1994)
introduced a new species to its universe: the Borg . . ." The Borg
were officially revealed in "Q Who?," Season 2, Episode 16, of Star
Trek: The Next Generation, first aired on May 5, 1989.

p. 228   "A species is considered as eusocial when it fulfills three main
criteria . . ." Hölldobler and Wilson (1990).

p. 228   ". . . castes often look quite dissimilar, meaning eusocial ant species
are polymorphic . . ." E. O. Wilson, Sociobiology: The New Synthesis
(Cambridge, Mass.: Harvard University Press, 2000).

p. 228   "How they make room for their logic-defying nests is by
removing one grain of soil after another, creating voids . . ."
C. A. Schmidt, Morphological and Functional Diversity of Ant
Mandibles, Tree of Life Web Project (2004): http://tolweb.org/
treehouses/?treehouse_id=2482

p. 229   "In some ant nests, vertical shafts also help to circulate air through
the colony . . ." C. Kleineidam et al., "Wind-induced ventilation
of the giant nests of the leaf-cutting ant Atta vollenweideri,"
Naturwissenschaften 88 (2001): 301–305.

p. 229   ". . . some ants can grow their own food, such as leaf-cutter ants,
consisting of species of Atta and Acromyrmex . . ." Hölldobler and
Wilson (1990).

p. 229   ". . . a 2004 study, in which several Brazilian entomologists
investigated leaf-cutter ant (Atta laevigata) nests by pouring cement
down into them . . ." A. A. Moreira et al., "Nest Architecture of Atta

laevigata (F. Smith, 1858) (Hymenoptera: Formicidae)," *Studies on Neotropical Fauna and Environment* 39 (2004): 109–116.

p. 229   "The nest covered more than 50 square meters (540 square feet) and went more than 7 meters (23 feet) deep . . ." A. A. Moreira *et al.* (2004).

p. 230   "A three-minute-fifteen-second excerpt from this film, later posted on YouTube in 2010 . . ." The excerpt is from a documentary titled *Ants! Nature's Secret Power* (2006), produced by Animal Planet, and the clip is titled "Giant Ant Hill [*sic*] Excavated": https://www.youtube.com/watch?v=lFg21x2sj-M

p. 230   "Ant biologist Walter Tschinkel perfected this technique by using either melted zinc or aluminum for casting ant nests . . ." (1) W. R. Tschinkel, "Nest architecture of the Florida harvester ant, *Pogonomyrmex badius*," *Journal of Insect Science* 4 (2005a): 1–19; (2) W. R. Tschinkel, "The nest architecture of the ant, *Camponotus socius*. *Journal of Insect Science* 5 (2005b): 1–18. An 11:28 minute video shows Tschinkel and an assistant picking a nest, melting the metal in the field, pouring, and excavating an ant nest: https://vimeo.com/68954580

p. 230   "Any one of these beauties could be slipped into a modern art museum and readily accepted . . . as the latest creative endeavor of a Chihuly-influenced metalworker . . ." A photo gallery of these beautiful nest casts and some commentary are provided by Rain Noe in an online article titled "Walter Tschinkel's Aluminum Casts of Ant Colonies Reveals Insect Architecture" (October 9, 2012): http://www.core77.com/posts/23607/walter-tschinkels-aluminum-casts-of-ant-colonies-reveals-insect-architecture-23607

p. 231   "In some ant species, the shafts were straight-down tubes leading to chambers . . ." (1) W. R. Tschinkel, "Subterranean ant nests: trace fossils past and future?" *Palaeogeography, Palaeoclimatology, Palaeoecology* 192 (2003): 321–333; (2) W. R. Tschinkel, "The architecture of subterranean ant nests: beauty and mystery underfoot," *Journal of Bioeconomics* 17 (2015): 271–291.

p. 231   "For example, nests of Florida harvester ants (*Pogonomyrmex badius*) can reach depths of 4 meters (13 feet) . . ." Tschinkel (2005).

p. 231   "Based on molecular clocks and ant body fossils, some of which are beautifully preserved in Cretaceous amber . . ." (1) Schultz (2000), and references therein; (2) V. Perrichot *et al.*, "Extreme morphogenesis and ecological specialization among Cretaceous basal ants," *Current Biology* 26 (2016): 1468–1472.

p. 231   "Trace fossils of subsurface insect colonies from Late Cretaceous rocks of Utah provide an intriguing clue . . ." E. M. Roberts and L. Tapanila, "A new social insect nest from the Upper Cretaceous

Kaiparowits Formation of southern Utah," *Journal of Paleontology* 80 (2006): 768–774.

p. 231   ". . . fossil ant nests from the Paleogene Period (66–23 *mya*), which followed the Cretaceous, closely resemble those of modern ants . . ." E. M. Roberts *et al.*, "Oligocene termite nests with *in situ* fungus gardens from the Rukwa Rift Basin, Tanzania, support a Paleogene African origin for insect agriculture," *PLoS One* 11 (2016): e0156847. doi: 10.1371/journal.pone.0156847.

p. 232   ". . . the advent of eusociality in insects, whether it happened first in termites or ants early in the Mesozoic Era . . ." (1) S. G. Brady *et al.*, "Evaluating alternative hypotheses for the early evolution and diversification of ants," *Proceedings of the National Academy of Sciences* 103 (2006): 18172–18177; (2) M. S. Engel *et al.*, "Termites (Isoptera): their phylogeny, classification, and rise to ecological dominance," *American Museum Novitates* 3650 (2009): 1–27.

p. 232   "A few ant species have even managed to nest in intertidal zones, such as mangrove swamps . . ." M. G. Nielsen, "Nesting biology of the mangrove mud-nesting ant *Polyrhachis sokolova* Forel (Hymenoptera, Formicidae) in northern Australia," *Insectes Sociaux* 44 (1997): 15–21.

p. 232   "Indeed, some evolutionary biologists view ant nests as an excellent example of an "extended phenotype" . . ." N. J. Minter *et al.*, "Morphogenesis of an extended phenotype: four-dimensional ant nest architecture," *Journal of the Royal Society Interface* 9 (2014): 586–595.

p. 232   ". . . an ant nest not only reflects the genes of an ant species, but also shapes its environment . . ." (1) S. T. Meyer *et al.*, "Ecosystem engineering by leaf-cutting ants: nests of *Atta cephalotes* drastically alter forest structure and microclimate," *Ecological Entomology* 36 (2011): 14–24; (2) W. R. Tschinkel and J. N. Seal, "Bioturbation by the fungus-gardening ant, Trachymyrmex septentrionalis," *PLoS One* 11(7) (2016): e0158920; doi: 10.1371/journal.pone.0158920.

p. 233   "Crayfish . . . share an ancestry with lobsters, having diverged . . . either late in the Permian Period or early in the Triassic Period . . ." (1) J. Breinholt *et al.*, "The timing of the diversification of the fresh-water crayfish." In J. W. Martin *et al.* (eds.), *Decapod Crustacean Phylogenetics* (Boca Raton, Fla.: CRC Press, 2009, 343–356); (2) H. Karasawa *et al.*, "Phylogeny and systematics of extant and extinct lobsters," *Journal of Crustacean Biology* 33 (2013): 78–123.

p. 233   "Crayfish today consist of more than 600 species on widely separated landmasses . . ." In T. Kawai, *et al.* (eds.), *Freshwater Crayfish: A Global Overview* (Boca Raton, Fla.: CRC Press, 2015).

p. 234   "Burrowing crayfish are further divided into three categories: Primary, secondary, and tertiary . . ." P.H.J. Horowitz and A.M.M. Richardson, "An ecological classification of the burrows of

Australian freshwater crayfish," *Australian Journal of Freshwater Research* 37 (1986): 237–242.

p. 234   "Modern crayfish burrows typically have at least one vertical shaft connecting a circular exit (or entrance) to the surface . . ." (1) A. J. Martin *et al.*, "Fossil evidence in Australia for oldest known freshwater crayfish of Gondwana," *Gondwana Research* 14 (2008): 287–296; (2) Martin (2013).

p. 234   ". . . burrows of the prairie crayfish (*Procambarus hagenianus*) are as deep as 4 meters (16 feet) . . ." J. F. Fitzpatrick, Jr., "The taxonomy and biology of the prairie crayfishes, *Procambarus hagenianus* (Faxon) and its allies." In "Freshwater Crayfish." In J. W. Avault, (ed.), *Papers from the Second International Symposium on Freshwater Crayfish* (Baton Rouge: Louisiana State University, Division of Continuing Education, 1975, 381–389).

p. 235   ". . . it comes as no surprise that burrows attributed to crayfish . . . show this shelter-building ability was selected early in their evolutionary history . . ." (1) S. T. Hasiotis and C. E. Mitchell, "A comparison of crayfish burrow morphologies: Triassic and Holocene fossil, paleo- and neo-ichnological evidence, and the identification of their burrowing signatures," *Ichnos* 2 (1993): 291–314. (2) E. Bedatou *et al.* "Crayfish burrows from Late Jurassic–Late Cretaceous continental deposits of Patagonia, Argentina: their palaeoecological, palaeoclimatic and palaeobiogeographical significance," *Palaeogeography, Palaeoclimatology, Palaeoecology* 257 (2008): 169–184; (3) Martin *et al.* (2008).

p. 235   ". . . Camp Shelby burrowing crayfish (*Fallicambarus gordoni*) of Mississippi move an estimated 80 metric tons per hectare (180 tons/acre) . . ." S. M. Welch *et al.*, "Seasonal variation and ecological effects of Camp Shelby burrowing crayfish (*Fallicambarus gordoni*) burrows," *American Midland Naturalist* 159 (2008): 378–384.

p. 235   "In a 1991 study done in east Texas, crayfish towers numbered more than 63,000/hectare (157,500/acre) . . ." H. H. Hobbs, Jr., and M. Whiteman, "Notes on the burrows, behavior, and color of the crayfish Fallicambarus *(F.) devastator* (Decapoda: Cambaridae)," *Southwestern Naturalist* 36 (1991): 127–135.

p. 235   ". . . red swamp crayfish in Lake Naivasha (Kenya) expanded their range by taking advantage of hippopotamus tracks . . ." J. Foster and D. Harper, "The alien Louisianan red swamp crayfish *Procambarus clarkii* Girard in Lake Naivasha, Kenya 1999–2003," *Freshwater Crayfish* 15 (2006): 195–202.

p. 236   "These decapods, which belong to the clade Nephropodidae, consist of more than fifty species . . ." T.-Y. Chan, "Annotated checklist of the world's marine lobsters (Crustacea, Decapoda:

Asticidea, Glypheidea, Achelata, Polychelida)," *The Raffles Bulletin of Zoology* 2010 [Supplement to 23] (2010): 153–181.

p. 236 ". . . lobsters are likely the weightiest [of crustaceans], with the record . . . a 20-kilogram (44-pound) specimen of American lobster (*Homarus americanus*) . . ." M. Carwardine *et al., Animal Records* (New York: Sterling, 2008).

p. 236 "Lobsters burrows tend to be roomy, with some nearly twice as wide as their makers . . ." R. A. Cooper and J. R. Uzmann, "Ecology of juvenile and adult *Homarus.*" In S. J. Cobb and B. F. Phillips, (eds.), *The Biology and Management of Lobsters* (Amsterdam: Elsevier, 2012). Also check out the primary references related to lobster burrows and burrowing cited in this chapter.

p. 237 ". . . lobster burrows vary with each species . . . but can be as simple as J-shaped (one entrance) or U-shaped (two entrances) . . ." (1) A. L. Rice and C. J. Chapman, "Observations on the burrows and burrowing behaviour of two mud-dwelling decapod crustaceans, *Nephrops norvegicus* and *Goneplax rhomboides,*" *Marine Biology* 10 (1971): 330–342. (2) R. G. Bromley, *Trace Fossils: Biology, Taphonomy, and Applications* (London: Chapman and Hall, 1996).

p. 237 ". . . their [American lobster] tunnels reach 30–40 centimeters (12–16 inches) deep and more than 70 centimeters (28 inches) long . . ." Cooper and Uzmann (2012).

p. 237 "American lobsters carve out depressions . . . as deep as 1.5 meters (5 feet) and 5 meters (16 feet) wide . . ." Cooper and Uzmann (2012).

p. 238 ". . . species of *Homarus* . . . seal off their burrows and become inactive for weeks or months during the winter . . ." (1) M.L.H. Thomas, "Overwintering of American Lobsters, *Homarus americanus,* in burrows in Bideford River, Prince Edward Island," *Journal of the Fisheries Research Board of Canada* 25 (1972): 2725–2727. (2) Cooper and Uzmann (2012).

p. 238 "Like crayfish, lobsters also have been digging burrows for a long time, minimally since the Early Jurassic . . ." R. G. Bromley and U. Asgaard, "The burrows and microcoprolites of *Glyphaea rosenkrantzi,* a Lower Jurassic palinuran crustacean from Jameson Land, East Greenland," *Grønlands Geologiske Undersøgelse Rapp* 49 (1972): 15–21.

p. 238 "Early Cretaceous (125 *mya*) lobster burrows in Portugal actually contain fossil lobsters . . ." (1) C. N. de Carvalho *et al.,* "*Thalassinoides* and its producer: populations of Mecochirus buried within their burrow systems, Boca do Chapim Formation (Lower Cretaceous), Portugal," *Palaios* 22 (2010): 104–109; (2) C. N. de Carvalho, "The massive death of lobsters smothered within their *Thalassinoides* burrows: The example of the lower Barremian from

Lusitanian Basin (Portugal)," *Comunicações Geológicas* 103 (2016): 143–152.

p. 239 ". . . crabs as a clade (Brachyura) are quite diverse, consisting of nearly 7,000 species . . ." L. M. Tsang *et al.*, "Evolutionary history of true crabs (Crustacea: Decapoda: Brachyura) and the origin of freshwater crabs," *Molecular Biology and Evolution* 31 (2014): 1173–1187.

p. 239 "Fossil burrows attributed to crabs are in Middle Jurassic rocks, dating at about 170 *mya* . . ." C. N. de Carvalho *et al.*, "Patterns of occurrence and distribution of crustacean ichnofossils in the Lower Jurasic-Upper Cretaceous of Atlantic occidental margin basins, Portugal," *Acta Geologica Polonica* 60 (2010): 19–28.

p. 239 "Represented by more than a hundred species under the genus *Uca* . . ." M. S. Rosenberg, "Contextual Cross-Referencing of Species Names for Fiddler Crabs (Genus *Uca*): An Experiment in Cyber-Taxonomy," *PLoS One* 9(7) (2014): e101704; doi: 10.1371/journal. pone.0101704. Connected to this article is a list of 103 species of *Uca* with links to each individual species, online for your reading pleasure: http://www.fiddlercrab.info/uca_species.html

p. 239 ". . . fiddler crabs are often ecosystem engineers, cycling nutrients and aerating sediments that normally would be anoxic . . ." (1) P. Daleo, *et al.*, "Ecosystem engineers activate mycorrhizal mutualism in salt marshes," *Ecology Letters* 10 (2007): 902–908; (2) Martin (2013), and references therein.

p. 239 " Burrows . . . have circular cross-sections, are steeply inclined, and can plunge as deep as 75 centimeters (30 inches) . . ." Martin (2013).

p. 240 ". . . I have even seen where sand fiddler crabs (*Uca pugilator*) burrowed into claw marks of fresh alligator tracks . . ." Although I did not publish a peer-reviewed article about this surprising observation, I wrote a blog post and included lots of photos, one of which includes a fiddler crab in a burrow modified from an alligator claw impression: "Erasing the Tracks of a Monster," by Anthony J. Martin (August 29, 2013), Life Traces of the Georgia Coast (blog): http://www.georgialifetraces.com/2013/08/29/ erasing-the-tracks-of-a-monster/

p. 241 "In a 2016 study, Christina Painting . . . documented how Australian banana fiddlers (*Uca mjoebergi*) use burrows as 'mating traps' . . ." C. J. Painting *et al.*, "Ladies first: coerced mating in a fiddler crab," *PLoS One* 11 (2016): e0155707. doi: 10.1371/journal. pone.0155707.

p. 241 "Most ghost crabs are represented by the genus *Ocypode*, which has 21 species living on temperate to tropical beaches throughout the world . . ." K. Sakai and M. Türkay, "Revision of the genus Ocypode with the description of a new genus, Hoplocypode (Crustacea: Decapoda: Brachyura)," *Memoirs of the Queensland Museum, Nature* 56 (2013): 665–793.

p. 242   "Those made by the largest crabs are often more than a meter (3.3 feet) long and steeply inclined, plunging at 30–50° angles . . ." (1) R. W. Frey *et al.*, "Tracemaking activities of crabs and their environmental significance: the ichnogenus *Psilonichnus*," *Journal of Paleontology* 58 (1984): 333–350; (2) G. A. Duncan, "Burrows of *Ocypode quadrata* (Fabricus) as related to slopes of substrate surfaces," *Journal of Paleontology* 60 (1986): 384–389.

p. 242   "Burrows have either just one opening . . . or two openings next to one another . . ." (1) Frey *et al.* (1984); (2) Duncan (1986); (3) Martin (2013).

p. 242   "Bigger [ghost crab] burrows are farther from the surf, whereas smaller ones are closer . . ." (1) G. W. Hill and R. E. Hunter, "Burrows of the ghost crab *Ocypode quadrata* (Fabricus) on the barrier islands, south-central Texas coast," *Journal of Sedimentary Research* 43 (1973): 24–30.

p. 242   ". . . crabs use burrows to overwinter by plugging the top and staying below for as long as six months . . ." S. Lucrezi and T. A. Schlacher, "The ecology of ghost crabs," *Oceanography and Marine Biology: An Annual Review* 52 (2014): 201–256.

p. 243   "The majority of land crabs belong to the clade Gecarcinidae, which includes species under at least six genera . . ." P. Davie, "Gecarcinidae." *World Register of Marine Species*: http://www.marinespecies.org/aphia.php?p=taxdetails&id=196152

p. 243   "The latter is also intriguingly called the 'zombie crab . . .'" I could not find any information on the origin of this crab species's nickname. However, because *G. ruricola* lives in the Caribbean and the Bahamas, I assume it was related somehow to zombie folklore of that region, which in turn was related to slavery in Haiti. For this historical background, read "The Tragic, Forgotten Story of Zombies," by Mike Mariani, published in *The Atlantic*, October 28, 2015: http://www.theatlantic.com/entertainment/archive/2015/10/how-america-erased-the-tragic-history-of-the-zombie/412264/

p. 243   ". . . blue land crab burrows can be nearly 20 centimeters (8 inches) wide and reach 3–4 meters (10–14 feet) long . . ." (1) C. F. Herreid, II, and G. A. Gifford, "The burrow habitat of the land crab, *Cardisoma guanhumi* (Latreille)," *Ecology* 44 (1963): 773–775; (2) A. W. Pinder and A. W. Smits, "The burrow microhabitat of the land crab *Cardisoma guanhumi*: respiratory/Ionic conditions and physiological responses of crabs to hypercapnia," *Physiological Zoology* 66 (1993): 216–236.

p. 243 ". . . some land-crab burrows are more than 300 meters (1,000 feet) above sea level . . ." R. G. Hartnoll *et al.*, "Population biology of the black land crab, *Gecarcinus ruricola*, in the San Andres archipelago, Western Caribbean," *Journal of Crustacean Biology* 26 (2006): 316–325.

p. 244 "The island [Christmas Island] has an estimated crab [*Gecarcoidea natalis*] population in the tens of millions . . ." (1) A. M. Adamczewska, and S. Morris, "Ecology and behavior of *Gecarcoidea natalis*, the Christmas Island red crab, during the annual breeding migration," *The Biological Bulletin* 200 (2001): 305–320; (2) H. Dingle, *Migration: The Biology of Life on the Move* (Oxford, England: Oxford University Press, 2014).

p. 244 "The coconut crab . . . is not a true crab, but a hermit crab . . ." I. W. Brown, and D. R. Fielder, "Project overview and literature survey." In I. W. Brown and D. R. Fielder, (eds.), *The Coconut Crab: Aspects of the Biology and Ecology of Birgus Zatro in the Republic of Vanuatu*, Australian Centre for International Agriculture Monograph, 8 (1991): 1–11.

p. 244 "Burrows are also good places for bringing and storing food . . ." Other crabs' burrows also serve as places for coconut crabs to find food, as they hunt *Gecarcoidea natalis* by either waiting outside or digging them out of their burrows: J. Krieger *et al.*, "Notes on the foraging strategies of the giant robber crab *Birgus latro* (Anomala) on Christmas Island: evidence for active predation on red crabs *Gecarcoidea natalis* (Brachyura)," *Zoological Studies* 55 (2016): 1–6.

p. 245 "A coconut crab that needs to shed its skin excavates a meter-long and half-meter (1.6-foot) deep burrow . . ." W. J. Fletcher *et al.*, "Moulting and growth characteristics." In I. W. Brown and D. R. Fielder, (eds.), *The Coconut Crab: Aspects of the Biology and Ecology of Birgus Zatro in the Republic of Vanuatu*, Australian Centre for International Agriculture Monograph, 8 (1991): 35–60.

p. 245 "Burrows are also used for mating, and female crabs afterward dig burrows near the ocean . . ." (1) T. Sato and K. Yosheda, "Egg extrusion site of coconut crab *Birgus latro*: direct observation of terrestrial egg extrusion," *Marine Biodiversity Records* 2 (2009): e37: http://dx.doi.org/10.1017/S1755267209000426; (2) S. Harzsch *et al.*, "A review of the biology and ecology of the Robber Crab, *Birgus latro* (Linnaeus, 1767) (Anomura: Coenobitidae)," *Zoologischer Anzeiger: A Journal of Comparative Zoology* 249 (2010): 45–67.

p. 245 "Subsequent search-and-rescue operations failed to find them, and in 1939 . . . the pair was declared officially dead . . ." The sizeable

number of biographies on Amelia Earhart attests to her importance
as an aviator and explorer, but her unexplained disappearance
during an around-the-world flight added an enduring mystique.
(Let's just say a few of the explanations—ranging from her being a
spy captured by the Japanese military to alien abduction—stretch
credulity.) Hence I can't recommend just one book describing
Earhart's last flight and the subsequent search-and-rescue
operation. Here is a brief article by Tony Long describing it, with
some links to further information: "July 2, 1937: Earhart Vanishes
Over the Pacific," in *Wired*, July 2, 2008: https://www.wired.
com/2008/07/dayintech-0702/

p. 245 "Gerald Gallagher, reported that colonists on Nikumaroro
(Kiribati) found part of a woman's skeleton and . . . a sextant and
a woman's shoe . . ." Christopher Joyce, "Bones, Shoes Might Have
Been Amelia Earhart's," *All Things Considered* (NPR), first broadcast
on December 2, 1998: http://www.npr.org/1998/12/02/1032135/
bones-shoes-may-have-been-amelia-earharts

p. 246 "However, the people associated with those expeditions
suggested that the absence of Earhart's and Noonan's remains
was because of coconut crabs . . ." Ric Gillespie proposed this
idea, which is not really a hypothesis because it would be
extremely difficult to test or disprove. Gillespie directs the
International Group for Historic Aircraft Recovery (TIGHAR)
and has organized many expeditions to Nikumaroro, claiming
to have found artifacts linked to Earthart's plane. Other
explorers, in turn, have disputed these findings. Anyway,
the "Earhart's bones got put in a coconut crab burrow" idea
is summarized well in two news articles: (1) Rachel Nuwer,
"Coconut Crabs Eat Everything from Kittens to, Maybe,
Amelia Earhart," in *Smithsonian Magazine*, December 26, 2013,
http://www.smithsonianmag.com/smart-news/coconut-
crabs-eat-everything-from-kittens-to-maybe-amelia-earhart-
180948206/#zGtFqd4b4cHcet6y.99 (2) Lauren Davis, "Did
Coconut Crabs Really Hide Amelia Earhart's Remains?", in *i09*,
May 5, 2014: http://io9.gizmodo.com/did-coconut-crabs-really-
hide-amelia-earharts-remains-1571944416

p. 246 "Like burrowing crayfish and crabs, thalassinidean shrimp
are ecosystem engineers . . ." D. Pillary and G. M. Branch,
"Bioengineering effects of burrowing thalassinidean shrimps on
marine soft-bottom ecosystems," *Oceanography and Marine Biology*
49 (2011):137–192.

p. 246 "Based on fossil burrows in Brazil . . . thalassinidean shrimp
have been affecting marine environments since at least the

Early Permian Period (about 280 *mya*) . . ." (1) R. G. Netto *et al.*, "*Gyrolithes* as a multipurpose burrow: an ethologic approach," *Revista Brasileira de Paleontologia* 10 (2007): 157–160; (2) R. Gandini and R. G. Netto, "*Ophiomorpha* from Lower Permian sandstones of southern Brazil," *Ichnia 2012, The 3rd International Conress on Ichnology, Abstract Book* (St. John's, Newfoundland, Canada: Memorial University, 2012, 36).

p. 247   "The upper portion of its burrow is shaped more like a wine bottle . . ." Martin (2013).

p. 247   "Pleopods then flutter in the opposite direction to reverse the flow and eject feces . . ." (1) Pryor (1975); C. M. Astall *et al.*, "Behavioural and physiological implications of a burrow-dwelling lifestyle for two species of upogebiid mud-shrimp (Crustacea: Thalassinidea)," *Estuarine, Coastal, and Shelf Science* 44 (1997): 155–168.

p. 248   ". . . these openings can exceed 400 per square meter (40 per square foot), all made by just two species, the Carolina ghost shrimp . . . and Georgia ghost shrimp . . ." (1) G. A. Bishop and E. C. Bishop, "Distribution of ghost shrimp, North Beach, St. Catherines Island," *American Museum Novitates* 3042 (1992): 1–17; (2) G. A. Bishop and N. A. Brannen, "Ecology and paleoecology of Georgia ghost shrimp." In K. M. Farrell *et al.* (eds.), *Geomorphology and Facies Relationships of Quaternary Barrier Island Complexes Near St. Mary's Georgia, Georgia Geological Society Guidebook* 13 (1993): 19–29.

p. 248   ". . . the main part of a ghost-shrimp burrow can descend to 3–4 meters (10–13 feet); fossil examples are as long as 5–6 meters (16–20 feet) . . ." (1) Bromley (1996); Martin (2013).

p. 248   ". . . many species fortify burrow exteriors with rolled up balls of muddy sand cemented with all-organic and all-natural shrimp spit . . ." R. W. Frey *et al.*, "*Ophiomorpha*: its morphological, taxonomic, and environmental significance," *Palaeogeography, Palaeoclimatology, Palaeoecology* 23 (1978): 199–229.

p. 249   "At nodes where branches meet, the burrow widens, again providing enough room for a shrimp to turn around . . ." (1) Bromley (1996); (2) Astall *et al.* (1997).

p. 249   "Ichnologists began investigating ghost shrimp burrows and those of other coastal invertebrates in earnest during the 1960s . . ." Martin (2013).

p. 249   "Rather than zinc or aluminum, though, these ichnologists used far less risky epoxy resin to make the casts . . ." R. W. Frey *et al.*, "Techniques for sampling salt marsh benthos and burrows," *American Midland Naturalist* 89 (1973): 228–234.

p. 250   ". . . shrimp that ingest sediment (deposit feeders) are more likely to make lots of galleries . . ." R. B. Griffis and T. H. Suchanek, "A model of burrow architecture and trophic modes in thalassinidean shrimp (Decapoda: Thalassinidea)," *Marine Ecology Progress Series* 79 (1991): 171–183.

p. 250 ". . . shrimp storing these grass clippings likely encourage microbes to grow and nutritionally enrich these . . ." D. Abed-Navandi *et al.*, "Nutritional ecology of thalassinidean shrimps constructing burrows with debris chambers: the distribution and use of macronutrients and micronutrients," *Marine Biology Research* 1 (2007): 202–215.

p. 250 "Probably the most useful architectural classification scheme for thalassinidean burrows (that is, the simplest) came from Richard Bromley . . ." Bromley (1996).

p. 251 ". . . ichnologist Al Curran estimated that if given eight years of time to burrow, one *Glypturus* could potentially mix a cubic meter (more than 35 cubic feet) of sediment . . ." I can't recall the exact time and place where Al Curran mentioned this figure to me, but it was in reference to experimental work he did on San Salvador Island (the Bahamas) a few years after I helped him set up sediment traps in the lagoon there (Pigeon Creek) and just offshore. Still, to get a sense of how much these shrimp burrow, read our paper: H. A. Curran and A. J. Martin, "Complex decapod burrows and ecological relationships in modern and Pleistocene intertidal carbonate environments, San Salvador Island, Bahamas," *Palaeogeography, Palaeoclimatology, Palaeoecology* 192 (2003): 229–245.

p. 251 ". . . many different species of crustaceans, polychaete worms, small fish, and other animals are often found living in thalassinidean burrows . . ." (1) Bromley (1996), and references therein; (2) D. Kneer *et al.*, "Seagrass as the main food source of *Neaxius acanthus* (Thalassinidea: Sthralaxiidae), its burrow associates, and of *Corallianassa coutierei* (Thalassinidea: Callianassidae)," *Estuarine Coastal Shelf Science* 79 (2008): 620–630.

p. 251 "Thalassinidean burrows even affect microbial communities in oceanic sediments, with greater bacterial diversity within burrows . . ." B. Laverock *et al.*, "Bioturbating shrimp alter the structure and diversity of bacterial communities in coastal marine sediments," *ISME Journal* 4 (2010): 1531–1544.

p. 252 "This [feces] is how most mud in coastal environments gets deposited . . ." (1) W. A. Pryor, "Biogenic sedimentation and alteration of argillaceous sediments in shallow marine environments," *Geological Society of America Bulletin* 86 (1975): 1244–1254; (2) J. M. Smith and R. W. Frey, "Biodeposition by the ribbed mussel *Geukensia demissa* in a salt marsh, Sapelo Island, Georgia," *Journal of Sedimentary Research* 55 (1985): 817–825.

p. 252   "In 1987, Edward O. Wilson wrote a succinct essay with a provocative title . . ." E. O. Wilson, "The little things that run the world (the importance of invertebrates in conservation)," *Conservation Biology* 1 (1987): 344–346.

CHAPTER 9: VIVA LA EVOLUCIÓN: CHANGE COMES FROM WITHIN

p. 255   ". . . they [problems] could be traced to the Jurassic Period (about 150 *mya*), when the offshore Juan de Fuca Plate collided . . ." The U.S. Geological Survey provides a synopsis of the geologic history, including helpful illustrations of the plate-tectonic setting for the Cascade Mountains at *Pacific–Cascade Volcanic Province* (last modified October 2, 2014): http://geomaps.wr.usgs.gov/parks/province/cascade2.html

p. 256   "The eruption began at 8:32 A.M. (U.S. Pacific time) . . ." An excellent summary of the geological events before and during the 1980 eruption of Mount St. Helens is also supplied by the U.S. Geological Survey at *1980 Cataclysmic Eruption* (last modified August 27, 2015): https://volcanoes.usgs.gov/volcanoes/st_helens/st_helens_geo_hist_99.html

p. 256   "The breech explosion and collapse of the volcano top dropped the height of Mount St. Helens by more than 400 meters (1,300 feet) . . ." U.S. Geological Survey, *1980 Cataclysmic Eruption*.

p. 257   "This *nuée ardente* ("glowing cloud") extinguished all surface life in a 600-square-kilometer (230-square-mile) area . . ." (1) U.S. Geological Survey, *1980 Cataclysmic Eruption*; (2) R.M.C. Lopes, *The Volcano Adventure Guide* (Cambridge, England: Cambridge University Press, 2005).

p. 257   ". . . as well as thick, fast-moving mudflows called lahars . . ." R. B. Wait, Jr., *et al.*, "Eruption-triggered avalanche, flood, and lahar at Mount St. Helens: effects of winter snowpack," *Science* 221 (1983): 1394–1397.

p. 257   "Consisting of 2.5–3 cubic kilometers (0.6–0.7 cubic miles) of earth material moving at 175–250 kilometers per hour (110–155 miles per hour) . . ." U.S. Geological Survey, *1980 Cataclysmic Eruption*.

p. 258   "'Vancouver [Washington], Vancouver, this is it!' he shouted . . ." David Johnston (1949–1980) was a geologist who gave up his life for his science, and he wanted to help others through his study of volcanic hazards. For instance, because of his on-site observations just before the 1980 eruption, he saved many lives by urging evacuation of the area and keeping others out as the situation became more dangerous. A short biography of Johnston is at the U.S. Geological Survey site *The Legacy of David Johnston*

(last modified August 14, 2013): https://volcanoes.usgs.gov/
observatories/cvo/david_johnston.html

p. 258   "... including this one, whom we will call Loowit ..." Loowit is a
shortened version of a Native American name for Mount St. Helens
from the Puyallup peoples, meaning "Lady of Fire." It is based on
a legend of Loowit—a Native American woman who took care
of their fire source—who later became the mountain. The U.S.
National Park Service also named one of the trails around Mount
St. Helens the Loowit Trail. To read more about this legend, go to
this Oregon State University site: http://volcano.oregonstate.edu/
oldroot/education/livingwmsh/hr/hrho/nam.html

p. 258   "It had a main tunnel 5–8 centimeters (2–3 inches) wide and about
30 centimeters (12 inches) below the soil surface ..." The burrows
of a closely related species of pocket gopher, *Thomomys bottae*, is
described in detail here: D. Vleck, "Burrow structure and foraging
costs in the fossorial rodent, Thomomys bottae," *Oecologia* 49
(1981): 391–396.

p. 260   "... once close to the surface, she encountered a thin, rigid crust ..."
Much of the post-eruption area had a stiff upper crust afterward,
which the pocket gophers broke once they emerged from their
burrows: D. C. Andersen and J. A. MacMahon, "Plant succession
following the Mount St. Helens volcanic eruption: facilitation by a
burrowing rodent, *Thomomys talpoides*," *American Midland Naturalist*
114 (1985): 62–69.

p. 260   "... which took her right into the danger zone" K. Loggins (1986),
"Danger Zone."

p. 261   "... stuffed the 'pockets' in his cheeks with seeds and other
vegetative bits ..." This cheeky trait is why these rodents are called
"pocket gophers."

p. 262   "... they were doing what they liked, and doing it naturally ..."
K. Loggins (1980), "I'm Alright." And yes, the movie *Caddyshack*
(1980) did indeed feature a burrowing gopher as the arch-nemesis
of Bill Murray's character, Carl Spackler.

p. 262   "... other gophers began punching holes through the gray terrain
..." Although the Loowit story is fictional, researchers who flew
over the devastated area reported seeing pocket-gopher burrow
mounds just days after the eruption and by September 1980 were
studying the burrow distributions: D. C. Andersen, "Observations
on *Thomomys talpoides* in the region affected by the eruption of
Mount St. Helens," *Journal of Mammalogy* 63 (1982): 652–655.

p. 262   "... she birthed five pups, three female and two male ..." Litter sizes
of the northern pocket gopher vary considerably, but one researcher
reported sizes of five to seven: G. Proulx, "Reproductive characteristics

of northern pocket gophers, *Thomomys talpoides*, in Alberta alfalfa fields," *The Canadian Field Naturalist* 116 (2002): 319–321.

p. 262   "Each individual pocket gopher was capable of overturning more than a ton of soil each year . . ." (1) N. Huntly, and R. Inouye, "Pocket gophers in ecosystems: patterns and mechanisms, pocket gophers profoundly affect microtopography, soils, plants, and other animals," *Bioscience* 38 (1988): 786–793; (2) O. J. Reichman and E. W. Seabloom, "The role of pocket gophers as subterranean ecosystem engineers," *Trends in Ecology and Evolution* 17 (2002): 44–49.

p. 263   ". . . they pulled up darker, organics-rich soils from below the volcanic deposits and mixed the old with the new . . ." Andersen and MacMahon (1985).

p. 263   "Tunnel casts made under snow during the winters, as well as burrow mounds outside of plugged entrances, also contained seeds . . ." (1) Andersen and MacMahon (1985); (2) J. Knight, "Infilled pocket gopher tunnels: seasonal features of high alpine plateaux," *Earth Surface Processes and Landforms* 34 (2008): 590–595.

p. 263   "Most plants have a form of symbiosis with fungi that wrap around their roots . . ." S. E. Smith and D. J. Read, *Mycorrhizal Symbiosis*, 3rd Edition (London: Academic Press, 2008).

p. 263   ". . . pocket gophers introduced fungal spores from soil below to burrow mounds above . . ." M. F. Allen *et al.*, "Re-formation of mycorrhizal symbioses on Mount St. Helens, 1980–1990: interactions of rodents and mycorrhizal fungi," *Mycology Research* 96 (1992): 447–453.

p. 263   ". . . Their numerous holes . . . caused surface water . . . to flow down into burrow systems . . ." Small-mammal burrows are especially important for downward flow and rentention of water in more arid environments: J. W. Laundre, "Effects of small mammal burrows on water infiltration in a cool desert environment," *Oecologia* 94 (1993): 43–48.

p. 263   ". . . these depressions turned into more than a hundred new ponds . . ." J. J. Major *et al.*, "After the disaster: the hydrogeomorphic, ecological, and biological responses to the eruption of Mount St. Helens," *Geological Society of America Field Guides* 15 (2009): 111–134.

p. 264   "Of the 55 mammal species in the area of Mount St. Helens in May 1980, only 14 survived . . ." (1) D. C. Andersen and J. A. MacMahon, "The effects of catastrophic ecosystem disturbance: the residual mammals at Mount St. Helens," *Journal of Mammalogy* 66 (1985): 581–589. (2) C. M. Crisafulli *et al.*, "Small mammal survival and colonization on the Mount St. Helens volcano: 1980–2002." In V. H. Dale *et al.* (eds.), *Ecological Responses to the 1980 Eruption of Mount St. Helens* (New York: Springer, 2005, 199–220).

p. 264 "One of the few non-rodent survivors was the tiny Trowbridge's shrew (*Sorex trowbridgii*) . . ." S. B. George, "*Sorex trowbridgii.*" *Mammalian Species* 337 (1989): 1–5.

p. 264 "The . . . timing of this disaster at the transition between winter and spring . . . greatly enhanced the chances . . ." Crisafulli *et al.* (2005).

p. 264 ". . . 12 of 15 native species of amphibians . . . also showed up after the explosion . . ." C. M. Crisafulli *et al.*, "Amphibian responses to the 1980 eruption of Mount St. Helens." In V. H. Dale *et al.* (eds.), *Ecological Responses to the 1980 Eruption of Mount St. Helens* (New York: Springer, 2005, 183–197).

p. 264 "Yet the northwestern salamander is also a well-known squatter of small-mammal burrows . . ." J. A. MacMahon *et al.*, "Small mammal recolonization on the Mount St. Helens volcano: 1980–1987," *American Midland Naturalist* 122 (1989): 365–387.

p. 264 ". . . thanks to pocket gopher burrows, which in the absence of trees provided plenty of shady, moist refuges . . ." Crisafulli *et al.* (2005).

p. 265 "Elk lent a helping hoof by walking across shallow tunnels of older gopher burrows . . ." Although this interaction between elk and gopher burrows has not been quantified, it is mentioned in a 2000 U.S. Forest Service publication, *Science Update: Mount St. Helens 30 Years Later: A Landscape Reconfigured*: http://www.fs.fed.us/pnw/pubs/science-update-19.pdf

p. 265 "Traveling elk herds made their own contributions to the plant population by depositing seed-laden feces . . ." M. P. Fleming, "Scat happens: Local influences of elk on primary succession, Mount St. Helens, WA," *94th Meeting of Ecological Society of America* (2009), [abstract]: http://esa.org/meetings_archive/2009/Paper17710.html

p. 265 "Survival rates [of ants] were apparently determined by ash thickness . . ." (1) J. S. Edwards, "Arthropods as pioneers: recolonization of the blast zone on Mt. St. Helens," *Northwest Environmental Journal* 2 (1986): 63–73; (2) J. J. Rango, "A survey of ant species in three habitats at Mount St. Helens National Volcanic Monument," *Psyche* (2012), Article 415183: 1–9.

p. 265 "From 1981 through 1987, researchers sampled other insects and arachnids in the blast zone . . ." (1) J. S. Edwards and P. M. Suggs, "Arthropods as the pioneers in the regeneration of life on the pyroclastic-flow deposits of Mount St. Helens." In V. H. Dale *et al.* (eds.), *Ecological Responses to the 1980 Eruption of Mount St. Helens* (New York: Springer, 2005, 127–138); (2) R. R. Parmenter *et al.* "Posteruption arthropod succession on the Mount St. Helens volcano: the ground-dwelling beetle fauna (Coleoptera)." In V. H. Dale *et al.*

(eds.), *Ecological Responses to the 1980 Eruption of Mount St. Helens* (New York: Springer, 2005, 139–150).

p. 266   "Among these arthropods were aquatic species, such as the signal crayfish (*Pacifastacus leniusculus*) . . ." The signal crayfish is the only native species of crayfish in the Pacific Northwest; all others are invasive, and the signal crayfish is an invasive species in Europe. Oddly enough, some references say it does not burrow, whereas other references say it does, and especially in places it has invaded. So I will default to "burrowing if needed," but whether or not this species used burrows to survive the Mount St. Helens eruption is yet undetermined.

p. 266   "By 2008, at least ten species of ants were thriving in ecosystems around Mount St. Helens . . ." Rango (2012).

p. 266   ". . . it was co-led by an ecologist (Charlie Crisafulli) and a geologist (Jon Major) . . ." Major *et al.* (2009).

p. 266   ". . . Johnston Ridge Observatory, named in honor of David Johnston . . ." U.S. Geological Survey, *Mount St. Helens National Monument* (last modified January 18, 2013): http://volcanoes.usgs.gov/volcanoes/st_helens/st_helens_geo_hist_106.html

p. 266   "From a geological perspective, their [pocket gophers'] effects were astoundingly quick . . ." MacMahon *et al.* (1989).

p. 267   ". . . four out of any ten species [of mammals] are rodents . . ." D. E. Wilson and D. M. Reeder, *Mammal Species of the World: A Taxonomic and Geographic Reference*, 3rd Edition (Baltimore: Johns Hopkins University Press, 2005).

p. 267   "The fossil record of rodents goes back to the Paleocene Epoch . . . around 58 *mya* . . . but molecular clocks hint at 62 *mya* . . ." S. Wu *et al.*, "Molecular and paleontological evidence for a post-Cretaceous origin of rodents," *PLoS One* 7 (2012): e46445. doi: 10.1371/journal.pone.0046445.

p. 268   "Yes, common rats . . . are excellent burrowers . . ." R. G. Pisano and T. I. Storer, "Burrows and feeding of the Norway rat," *Journal of Mammalogy* 29 (1948): 374–383; (2) L. Nieder *et al.*, "Burrowing and feeding behaviour in the rat," *Animal Behavior* 30 (1982): 837–844.

p. 268   "In a fascinating 2013 study, researchers linked burrow architecture with DNA in oldfield mice . . . and deer mice . . ." J. N. Weber *et al.*, "Discrete genetic modules are responsible for complex burrow evolution in *Peromyscus* mice," *Nature* 493 (2013): 402–405.

p. 268   "Other burrowing rodents include . . ." The number of references citing burrowing rodents would be far too long to list here. Instead, take a look at Wilson and Reeder's (2005) comprehensive source on mammals to look up each of burrowing rodent species I have mentioned.

p. 269   "... paleontologists have found beaver skeletons in deep,
          corkscrew-shaped burrows from the Oligocene Epoch ..." L. D.
          Martin *et al.*, "The burrows of the Miocene beaver Palaeocastor,
          Western Nebraska, U.S.A.," *Palaeogeography, Palaeoclimatology,
          Palaeoecology* 22 (1977): 173–193.

p. 269   "Owing to the abundance and diversity of Asian rodents, biologists
          currently think rodents originated there ..." Wu *et al.* (2012).

p. 269   "... humans later aided in the spread of mice and rats ... also from
          Polynesians seafaring to islands throughout the Pacific Ocean ..."
          (1) R. N. Holdaway, "The arrival of rats in New Zealand," *Nature*
          384 (1996): 225–226; (2) G. A. Harper and B. Bunbury, "Invasive
          rats on tropical islands: their population biology and impacts on
          native species," *Global Ecology and Conservation* 3 (2015): 607–627.

p. 269   "They are one of only two species of eusocial mammals; the other
          is the ... Damaraland mole-rat ..." H. G. Thomas *et al.*, "Burrow
          architecture of the Damaraland mole-rat (*Fukomys damarensis*) from
          South Africa," *African Zoology* 51 (2016): 29–36.

p. 270   "They are also the longest-lived rodents, with some in captivity
          exceeding thirty years ..." (1) P. W. Sherman and J.U.M. Jarvis,
          "Extraordinary life spans of naked mole-rats (*Heterocephalus
          glaber*)," *Journal of Zoology* 258 (2002): 307–311; (2) E. N. Kim *et al.*,
          "Genome sequencing reveals insights into physiology and longevity
          of the naked mole rat," *Nature* 479 (2011): 223–227.

p. 270   "Naked mole-rats live in tropical savannas and grasslands at
          relatively high elevations ..." J.U.M. Jarvis and P. W. Sherman,
          "*Heterocephalus glaber*," *Mammalian Species* 706 (2002): 1–9.

p. 270   "A social network for naked mole-rats thus mimics that of ants or ter-
          mites by having a queen ..." J.U.M. Jarvis *et al.*, "Mammalian eusoci-
          ality: a family affair," *Trends in Ecology and Evolution* 9 (1994): 47–51.

p. 270   "To assist in her matriarchy, a breeding queen picks one to four
          subservient males, and ... she becomes pregnant for about 70
          days ..." P. M. Sherman *et al.*, *The Biology of the Naked Mole-Rat*
          (Princeton, N.J.: Princeton University Press, 1991).

p. 270   "Amazingly, queen mole-rats do not go through menopause ..."
          Y. H. Edrey *et al.*, "Endocrine function and neurobiology of the
          longest-living rodent, the naked mole-rat," *Experimental Gerontology*
          46 (2011): 116–123.

p. 270   "... she [the queen] is assisted by a bevy of handmaidens, nannies,
          butlers, and other servants in the colony ..." Sherman *et al.* (1991).

p. 271   "If a rebel female wins and becomes a new queen ... her vertebrae
          loosen and she elongates ..." F. M. Clarke and C. G. Faulkes,
          "Dominance and queen succession in captive colonies of the eusocial
          naked mole-rat, *Heterocephalus glaber*," *Proceedings of the Royal Society*

of London, B 264 (1997): 993–1000; (2) E. C. Henry *et al.*, "Growing out of a caste: reproduction and the making of the queen mole-rat," *Journal of Experimental Biology* 210 (2007): 261–268.

p. 271   "Their teeth are positioned in front of their lips to avoid swallowing dirt . . ." Jarvis and Sherman (2002).

p. 271   ". . . colony members employ a variety of calls, such as soft chirps, to communicate throughout the burrow system . . ." S. Yoshida and K. I. Kobayasi, "Antiphonal vocalization of a subterranean rodent, the naked mole-rat (*Heterocephalus glaber*)," *Ethology* 113 (2007): 703–710.

p. 271   "Digging is likewise collaborative, as workers form conga lines to excavate tunnels . . ." J.U.M. Jarvis and J. B. Sale, "Burrowing and burrow patterns of East African mole-rats *Tachyoryctes, Heliophobius* and *Heterocephalus*," *Journal of Zoology* 163 (1971): 451–479.

p. 272   "Most burrow systems are relatively deep for such small rodents, about 2 meters (6.6 feet) . . ." (1) Jarvis and Sale (1971); (2) Sherman *et al.* (1991).

p. 272   "A single colony may displace more than 4 metric tons (4.4 tons) of soil . . ." (1) N. Hageneh and N. C. Bennett, "Mole rats act as ecosystem engineers within a biodiversity hotspot, the Cape Fynbos," *Journal of Zoology* 289 (2012): 19–26.

p. 272   ". . . living and foraging tunnels in a mole-rat system can add up to 3–5 kilometers (2–3 miles) in total length . . ." J.U.M. Jarvis and J. B. Sale, "Burrowing and burrow patterns of East African molerats *Tachyoryctes, Heliophobius* and *Heterocephalus*," *Journal of Zoology* 163 (1971): 451–479.

p. 272   ". . . how the animals making these burrow systems are normally only 10–15 centimeters (4–6 inches) long . . ." Jarvis and Sherman (2002).

p. 272   "One of these interlopers is their sworn mortal enemy, the rufous-beaked snake (*Rhamphiophis oxyrhynchus*) . . ." J. Kingdon *et al.*, *Mammals of Africa, Volume III: Rodents, Hares and Rabbits* (London: Bloomsbury, 2013).

p. 273   ". . . they can flip back and forth between poikilothermic ("cold-blooded") and homeothermic ("warm-blooded") states . . ." T. Daly *et al.*, "Catecholaminergic innervation of interscapular brown adipose tissue in the naked mole-rat (*Heterocephalus glaber*)," *Journal of Anatomy* 190 (1997): 321–326.

p. 273   "To elevate body temperatures, they either get higher up, or they pile on top of one another in large spaces . . ." S. Yahav and R. Buffenstein, "Huddling behavior facilitates homeothermy in the naked mole rat *Heterocephalus glaber*," *Physiological Zoology* 64 (1991): 871–884.

p. 273   ". . . poo-eating (more elegantly known as coprophagy) . . ." "Poo Turns Naked Mole Rats into Better Babysitters," by Sara Reardon, *Nature News,* October 20, 2015: doi: 10.1038/nature.2015.18606.

p. 273   ". . . pocket gopher burrows provide a succinct summary of what one might expect in a typical rodent burrow . . ." (1) D. Vleck, "Burrow structure and foraging costs in the fossorial rodent, *Thomomys bottae,*" *Oecologia* 49 (1981): 391–396; (2) F. K. Holtmeier, *Animals' Influence on the Landscape and Ecological Importance: Natives, Newcomers, Homecomers* (New York: Springer, 2014).

p. 274   "Marsupial moles (*Notoryctes*) . . . represent a remarkable example of convergent evolution . . ." Marsupial moles also make distinctive burrows that help to identify their presence in places where people might never see one above ground: J. Bensehmesh, "Backfilled tunnels provide a novel and efficient method of revealing an elusive Australian burrowing mammal," *Journal of Mammalogy* 95 (2007): 1054–1063.

p. 274   "Monotremes include the platypus of Australia, as well as four species of spiny, ant-eating echidnas . . ." M. S. Springer and C. W. Krajewski, "Monotremes (Prototheria)." In S. B. Hedges and S. Kumar (eds.), *The Timetree of Life* (Oxford, England: Oxford University Press, 2009, 462–465).

p. 274   ". . . female platypuses can hollow out nesting burrows as long as 20 meters (66 feet) . . ." B. K. Hall, "The paradoxical playpus," *BioScience* 49 (1999): 211–218; (2) T. Grant, *Platypus* (Clayton, Victoria, Australia: CSIRO Publishing, 2007).

p. 274   "Mother echidnas also carve out dens for their young . . ." M. D. Opiang, "Home ranges, movement, and den use in long-beaked echidnas, *Zaglossus bartoni*, from Papua New Guinea," *Journal of Mammalogy* 90 (2009): 340–346.

p. 275   ". . . trace fossils of complex burrow networks from Triassic rocks of Argentina, South Africa, Poland, and elsewhere point to synapsid makers . . ." (1) V. Krapovickas *et al.*, "Large tetrapod burrows from the Middle Triassic of Argentina: a behavioural adaptation to seasonal semi-arid climate?" *Lethaia* 46 (2013): 154–169; (2) Groenewald (1991); (3) S. Voight *et al.*, "Complex tetrapod burrows from Middle Triassic red beds of the Argana Basin (Western High Atlas, Morocco)," *Palaios* 26 (2011): 555–566; (4) M. Tałanda *et al.* "Vertebrate burrow systems from the Upper Triassic of Poland," *Palaios* 26 (2011): 99–105.

p. 275   "Jurassic fossil burrows match sizes and architectures of modern-day small-mammal burrows . . ." (1) S. T. Haisotis *et al.*, "Vertebrate burrows from Triassic and Jurassic continental deposits of North America and Antarctica: their paleoenvironmental and

paleoecological significance," *Ichnos* 11 (2004): 103–124; (2)
D. J. Riese *et al.*, "Synapsid burrows and associated trace fossils in
the Lower Jurassic Navajo Sandstone, southeastern Utah, U.S.A.,
indicates a diverse community living in a wet desert ecosystem,"
Journal of Sedimentary Research 81 (2011): 299–325.

p. 275   "Middle and Late Jurassic fossil mammals from China and Portugal
are also interpreted as burrowers . . ." (1) Q. Ji *et al.* "A swimming
mammaliaform from the Middle Jurassic and ecomorphological
diversification of early mammals," *Science* 311 (2006): 1123–1127;
(2) T. Martin, "Postcranial anatomy of *Haldanodon exspectatus*
(Mammalia, Docodonta) from the Late Jurassic (Kimmeridgian)
of Portugal and its bearing for mammalian evolution," *Zoological
Journal of the Linnean Society* 145 (2005): 219–248; (3) Also, there is
a Late Jurassic mammal from Colorado interpreted as a burrower:
Z.-X. Luo and J. R. Wible, "A Late Jurassic digging mammal and
early mammalian diversification," *Science* 308 (2005): 103–107.

p. 276   ". . . paleontologists have described Late Cretaceous dinosaur
trace fossils . . . made by clawed theropods digging into mammal
burrows . . ." E. L. Simpson *et al.* "Predatory digging behavior by
dinosaurs," *Geology* 38 (2010): 699–702.

p. 276   "In a 2016 analysis of genes linked to eyesight in mammals,
geneticists concluded that low-light vision was an original ancestral
trait . . ." J.-W. Kim *et al.*, "Recruitment of rod photoreceptors from
short-wavelength-sensitive cones during the evolution of nocturnal
vision in mammals," *Cell* 37 (2016): 520–532.

p. 276   "For instance, *Morganucodon*, a cynodont from the Late Triassic (205
mya), retained two jawbones . . . that later evolved into middle-ear
bones . . ." K. D. Rose, *The Beginning of the Age of Mammals*
(Baltimore: Johns Hopkins University Press, 2006).

p. 276   "This assumption is based on its species' descent from burrowing
cynodonts, as well as how nearly all small modern mammals are
nocturnal . . ." A. W. Crompton *et al.*, "Evolution of the mammalian
nose." In K. P. Dial, *et al.* (eds.), *Great Transformations in Vertebrate
History* (Chicago: University of Chicago Press, 2015, 189–204).

p. 277   ". . . we not only have an evolutionary history represented as 'your
inner fish' . . ." Shubin (2009).

p. 277   ". . . mammals expanded into every terrestrial environment and
ecological niche owned previously by dinosaurs . . ." D. R. Prothero,
After the Dinosaurs: The Age of Mammals. (Bloomington: Indiana
University Press, 2006).

p. 277   ". . . rhinoceros relative *Paraceratherium* from the Oligocene Epoch
(34–23 *mya*), which may have weighed as much as 18 metric tons
(20 tons) . . ." Prothero (2006).

p. 278   ". . . As climate continues to change too quickly for most animals to adapt, small burrowing mammals . . . are the ones most likely to make it . . ." Pike and Mitchell (2013).

p. 278   ". . . the movie *Tremors* (1990) was the best ichnology-horror-comedy film of all time . . ." Kevin Bacon reportedly said *Tremors* was a "low point" in his career and is quoted as saying, "I can't believe I'm doing a movie about underground worms!": "23 Fun Facts about 'Tremors'" by Mark Mancini, in *Mental Floss*, January 19, 2015: http://mentalfloss.com/article/61181/23-fun-facts-about-tremors

p. 279   ". . . a Bacon Number of 1 on the Six Degrees of Kevin Bacon scale . . ." The Six Degrees: Kevin Bacon at TEDxMidwest: https://www.youtube.com/watch?v=n9u-TITxwoM&ab_channel=TEDxTalks

p. 279   ". . . a glorious tradition since the Cambrian Period of predatory worms using burrows for ambush predation . . ." Recent research points out that "burrows for predation" behavior may have actually extended back into the Ediacaran and partially contributed to the demise of Ediacaran animals: Darroch *et al.* (2016).

p. 280   "Geologists in Argentina and Brazil noted the massive structures in the 1920s and 1930s as features cutting into a variety of . . . bedrock" (1) S. F. Vizcaíno *et al.*, "Pleistocene burrows in the Mar del Plata area (Argentina) and their probable builders," *Acta Palaeontologica Acta* 46 (2001): 289–301; (2) H. T. Frank *et al.*, "Cenozoic vertebrate tunnels in southern Brazil," *Ichnology of South America, SBP Monographs*, 2 (2012): 141–157; (3) H. T. Frank *et al.*, "Underground chamber systems excavated by Cenozoic ground sloths in the state of Rio Grande do Sul, Brazil," *Revista Brasileira de Paleontologia* 18 (2015): 273–274.

p. 280   "In the 1970s and 1980s, archaeologists studied some of the tunnels . . ." Frank *et al.* (2012).

p. 280   "The smallest, however, are only about 60 centimeters (24 inches) wide, 50 centimeters (20 inches) tall . . ." Frank *et al.* (2012).

p. 280   ". . . the largest tunnels are much more voluminous, with some as wide as 4 meters (13 feet), 2 meters (6.6 feet) tall . . ." Frank *et al.* (2015).

p. 280   ". . . archaeologists realized the tunnels were not of human origin . . ." Frank *et al.* (2012).

p. 281   "paleontologists and geologists took over investigating the tunnels, with intense research in early twenty-first century . . ." (1) Vizcaíno *et al.* (2001); Frank *et al.* (2012); (3) H. T. Frank *et al.*, "Description and interpretation of Cenozoic vertebrate ichnofossils in Rio Grande du Sol State, Brazil," *Revista Brasileira de Paleontologia* 16 (2013): 83–96.

p. 281   ". . . paleontologists concluded that giant ground sloths were the likely makers of the largest burrows, and giant armadillos were credited with the smaller ones . . ." (1) H. T. Frank *et al.*, "Karstic features generated from large palaeovertebrate tunnels in southern Brazil," *Espeleo-Tema* 22 (2011): 139–153; (2) Frank *et al.* (2012); Frank *et al.* (2015).

p. 281   "Adults of the South American ground sloth *Scelidotherium* from the Late Pleistocene were . . ." R. A. Fariña *et al.*, *Megafauna: Giant Beasts of Pleistocene South America* (Bloomington: Indiana University Press, 2013).

p. 282   ". . . *Glossotherium*, was even bigger . . ." Fariña *et al.* (2013).

p. 282   "First, parallel grooves on burrow walls acted as 'fingerprints' for identifying their perpetrators . . ." Vizcaíno *et al.* (2001); (2) Frank *et al.* (2012).

p. 282   "Such spacing probably reflects digging cycles, in which a ground sloth dug a chamber just short of its body length . . ." Frank *et al.* (2012).

p. 282   ". . . paleontologists conjectured that the tunnels and chambers might have been constructed and occupied over hundreds of years . . ." Frank *et al.* (2015).

p. 282   "Giant ground sloth anatomies . . . confirm these mammals were well equipped to scratch out the gigantic burrows . . ." Vizcaíno *et al.* (2001), and references therein.

p. 282   ". . . calculated forces and stresses generated by these forelimbs equaled or exceeded those of galloping mammals . . ." M. S. Bargo *et al.*, "Limb bone proportions, strength, and digging in some Lujanian (Late Pleistocene-Early Holocene) mylodontid ground sloths (Mammalia, Xenarthra)," *Journal of Vertebrate Paleontology* 20 (2000): 601–610.

p. 283   "South America is where armadillos evolved originally, and is still home to twenty species . . ." K. D. Rose, "Xenarthra and Pholidota." In K. D. Rose and J. D. Archibald (eds.), *The Rise of Placental Mammals: Origins and Relationships of the Major Extant Clades* (Baltimore, Maryland: Johns Hopkins University Press, 2005, 106–126).

p. 283   "Giant armadillos are typically 1–1.5 meters (3.3–5 feet) long and 20–30 kilograms (44–66 pounds (1) R. M. Nowak, *Walker's Mammals of the World, Volume I* (Baltimore: Johns Hopkins University Press, 1999); (2) Carwardine *et al.* (2008).

p. 283   "For instance, two genera of Pleistocene armadillos . . . weighed at least twice as much as the modern giant armadillo . . ." Fariña *et al.* (2013).

p. 283   "The Pleistocene armadillos *Eutatus* and *Propraopus*, however, were closer in size to *Priodontes* . . ." Vizcaíno *et al.* (2001).

p. 283   "Present-day giant armadillos dig dens that are slightly taller (45 centimeters/18 inches) than wide (35 centimeters/14 inches) . . ."

(1) N. Ceresoli and E. Fernandez-Duque, "Size and orientation of giant armadillo burrow entrances (*Priodontes maximus*) in Western Formosa Province, Argentina," *Edentata* 13 (2012): 66–68; (2) A.L.J. Desbiez and D. Kluyber, "The role of giant armadillos (*Priodontes maximus*) as physical ecosystem engineers," *Biotropica* 45 (2013): 537–540.

p. 283   "These threats included saber-toothed cats (*Smilodon*) and short-faced bears (*Arctotherium*) . . ." Frank *et al.* (2012).

p. 284   ". . . Armadillos and ground sloths dug tunnels and chambers to better cope with climates much drier than today . . ." Frank *et al.* (2012).

p. 284   "Lacking cave-forming limestones, local bedrock geology also gave incentives for these mammals to burrow . . ." Frank *et al.* (2015) also discuss in detail how caves dug by large mammals differ from those formed by other natural means.

p. 284   ". . . zoologists assumed body sizes restricted burrowing, preventing big animals from living below the surface . . ." White (2006).

p. 285   "Surprisingly, it is the grizzly bear (*Ursos arctos*), with the Kodiak bear (*Ursos arctos middendorfii*) as its largest subspecies . . ." G. A. Feldhamer *et al.*, *Wild Mammals of North America: Biology, Management, and Conservation* (Baltimore: Johns Hopkins University Press, 2003).

p. 285   "Grizzlies dig dens with their robust claws and powerful forelimbs, sometimes displacing more than a metric ton (1.1 tons) of soil . . ." Butler (1995) once figured that the "average" grizzly bear den displaced 4.3 cubic meters of soil, the weight of which depends on the density of the soil, including moisture. Based on that measure, these bears probably move much more than a metric ton per den.

p. 285   "Once ensconced, grizzly bears can hibernate for as long as seven months . . ." (1) R. A. Nelson *et al.*, "Behavior, biochemistry and hibernation in black, grizzly, and polar bears," *Bears: Their Biology and Management* 5 (1983): 284–290; (2) S. Manchi, and J. E. Swenson, "Denning behaviour of Scandinavian brown bears *Ursus arctos*," *Wildlife Biology* 11 (2005): 123–132.

p. 285   "Female polar bears (*Ursus maritimus*), which are about half the size of males, can still approach 350 kilograms (770 pounds) . . ." Feldhamer *et al.* (2003).

p. 285   "They first make a narrow tunnel, which leads to a main chamber about 1.5–2 meters (5–6.6 feet) wide and a meter high . . ." Descriptions and diagrams of polar bear dens in both snow and soil are in: C. J. Jonkel *et al.*, "Further notes on polar bear denning habits," *Bears: Their Biology and Management* 2, IUCN Publications New Series No. 23 (1972): 142–158.

p. 286   "This situation is forcing polar bears to swim longer distances when hunting for prey . . ." E. V. Regehr *et al.*, "Survival and breeding of polar bears in the southern Beaufort Sea in relation to sea ice," *Journal of Animal Ecology* 79 (2009): 117–127.

p. 286   "In a long-term study of polar-bear behaviors published by Todd Atwood, Elizabeth Peacock, Melissa McKinney, and other researchers in 2016 . . ." T. C. Atwood *et al.*, "Rapid environmental change drives increased land use by an Arctic marine predator," *PLoS One* 11 (2016): e0155932; doi: 10.1371/journal. pone.0155932.

p. 286   "'The Sixth Extinction,' following the big five of the end-Ordovician, Devonian, Permian, Triassic, and Cretaceous, began about 50,000 years ago in Australia . . ." (1) D. A. Burney and T. F. Flannery, "Fifty millennia of catastrophic extinctions after human contact," *Trends in Ecology and Evolution* 20 (2005): 395–401.

p. 286   "This scenario was repeated toward the end of the Pleistocene in broad swaths of Europe, North America, and South America . . ." Burney and Flannery (2005).

p. 287   "Large islands, such as Madagascar and New Zealand, also experienced rapid extinctions of their big land-dwelling animals . . ." S. Turvey, *Holocene Extinctions* (Oxford, England: Oxford University Press, 2009).

p. 287   "Upright bipeds . . . dealt a lethal wild card to the hand of the megafauna . . . also via habitat alteration . . ." (1) Burney and Flannery; (2) P. S. Martin, *Twilight of the Mammoths: Ice Age Extinctions and the Rewilding of America* (Berkeley: University of California Press, 2005).

p. 287   In a 2016 study, paleontologists Nicholas Longrich and others examined end-Cretaceous mammal extinctions in North America . . ." N. R. Longrich *et al.*, "Severe extinction and rapid recovery of mammals across the Cretaceous–Palaeogene boundary, and the effects of rarity on patterns of extinction and recovery," *Journal of Evolutionary Biology* 29 (2016): 1495–1512.

p. 288   ". . . Douglas Robertson and his coauthors published a speculative-science article in 2004 titled 'Survival in the First Hours of the Cenozoic' . . ." Robertson *et al.* (2004). Another article acknowledging how burrowing animals in shallow marine environments were among the survivors of the end-Cretaceous is: L. A. Weist *et al.*, "Ichnological evidence for endobenthic response to the K-PG event, New Jersey, USA," *Palaios* 31 (2016): 231–241.

p. 288   ". . . in a recent (2016) study focusing on the South American megafaunal extinctions from about 12,000 years ago . . ." J. L. Metcalf *et al.*, "Synergistic roles of climate warming and human occupation in Patagonian megafaunal extinctions during the last deglaciation," *Science Advances* 2 (2016): e1501682; doi: 10.1126/sciadv.1501682.

p. 289   "Geologists are even considering (and debating) naming this new time the Anthropocene Epoch . . ." S. C. Finney and L. E. Edwards, "The 'Anthropocene' epoch: Scientific decision or political statement?" *GSA Today* 26(3–4) (2016): 4–10.

p. 289   ". . . citizens of the island nation of Kiribati in the Pacific Ocean are making plans to abandon their islands . . ." "A Remote Island Nation, Threatened by Rising Seas," by Mike Ives, *New York Times*, July 2, 2016: http://www.nytimes.com/2016/07/03/world/asia/climate-change-kiribati.html?_r=0

p. 289   "In a 2013 research article, ecologists David Pike and John Mitchell proposed that burrowing ecosystem engineers . . . provide 'thermal refugia' for other species . . ." D. A. Pike and J. C. Mitchell, "Burrow-dwelling ecosystem engineers provide thermal refugia throughout the landscape," *Animal Conservation* 16 (2013): 694–703.

p. 290   "This in turn allowed geologists to more easily explore for economically valuable deposits of fossil fuels, such as oil and natural gas . . ." S. G. Pemberton, *Applications of Ichnology to Petroleum Exploration: A Core Workshop* (Calgary, Alberta: SEPM Core Workshop No. 17, 1992).

p. 291   "Curiosity-driven research about the lasting effects of burrows on the properties of rocks bearing fossil fuels . . ." M. K. Gingras *et al.*, "Porosity and permeability in bioturbated sediments," *Developments in Sedimentology* 64 (2012): 837–868.

p. 291   ". . . curiosity-driven research about changes in fossil-burrow assemblages aided geologists in more accurately interpreting when sea level went up or down . . ." (1) C. R. Fielding, "Cyclicity in the nearshore marine to coastal, Lower Permian, Pebbley Beach Formation, southern Sydney Basin, Australia: a record of relative sea-level fluctuations at the close of the Late Palaeozoic Gondwanan ice age," *Sedimentology* 53 (2006): 435–463.
(2) A. J. Martin, "Applications of trace fossils to interpreting paleoenvironments and sequence stratigraphy." In M. S. Duncan and R. L. Kath (eds.), *Fall Line Geology of East Georgia: With a Special Emphasis on the Upper Eocene*, Georgia Geological Society Guidebook 29 (2009): 35–42.

p. 291   ". . . 2015 was the warmest year on record, with eight of the previous ten years' global temperatures also among the top 10 warmest . . ." NASA, NOAA Analyses Reveal Record-Shattering Global Warm Temperatures in 2015," Press Release, January 20, 2016: http://www.nasa.gov/press-release/

nasa-noaa-analyses-reveal-record-shattering-global-warm-
temperatures-in-2015

p. 291 "The latest DOD report on this topic in 2015 said . . ." "DoD
Releases Report on Security Implications of Climate Change," July
29, 2015, with link to Report on National Security Implications
of Climate-Related Risks and a Changing Climate: http://www.
defense.gov/News/Article/Article/612710. These aren't granola-
crunching hippies and ivory-tower scientists saying this, folks.
Either do something about it or start digging.

Acknowledgments

Most animals on our planet are hidden from view, living below us as we go about our daily lives. Moreover, these and other fauna invented and reinvented ecosystems with their burrowing, while also sharing a common ancestry that originated with the security and stability of the underground world.

These ideas and a lifetime fascination with burrowing animals led to my writing this book, which began with its summary title, *The Evolution Underground*. The subtitle, however, was another matter, as I vacillated between *Better Surviving through Burrows* and *How Burrows Changed the World*. Which sounded more pretentious, I thought? Fortunately my editor provided a caption that better reflected the book's content, avoiding the need to defend such over-the-top statements. Still, when we put on our time-machine glasses and peer back into the geologic past, our deep connection (pun intended) to subsurface living pokes its head out of the seafloor, smiles, and winks. So my first acknowledgment is to all of the ancient and modern makers of burrows that shaped the world we live in today, from the Ediacaran to the present and from aardvarks to zorillas: Thanks for making our world.

As for humans, I thank the first ones I ever knew, my parents Richard and Veronica. They gave me food, shelter, clothing, a public-school education, and much love as I grew up an odd and sickly little kid in Terre Haute, Indiana. (I'm healthy now, albeit still elfin and quite happily odd.) Neither my father nor mother went to college, but they knew the value of a formal education and did everything they could to ensure I obtained one. They also supported my early predilection for getting outside, which started in our backyard and progressed from there. Many days of my father taking me to hunt and fish in southern Indiana cultivated a life-long appreciation for natural history and a passion for exploration. Unfortunately, his only travel outside of the U.S. was as a U.S. Army soldier, having served in the Pacific theater during World War II. Like many of the "Greatest Generation," the experience of war damaged him, but he found emotional solace in woods, rivers, and forests, and passed on that good medicine to me. His death in 1985 only a few months before I began my PhD study in Georgia meant he never saw how his devotion helped make me the field-oriented scientist I am today. Hey Dad, it worked. You did good. Thank you.

My mother fortunately lived to witness me write a few books on natural history, books she proudly displayed at the same Terre Haute home where I grew up. Although she never had much money nor saw an ocean, she made sure her marine-obsessed child got swim lessons at the local YMCA. Years later, while scuba diving at the Great Barrier Reef, I wished she could have seen how her investment paid off, even though I know she would have warned me to be careful with all of those sharks, jellyfish, and giant clams. When she died in 2014, this book had not yet been conceived, but it exists because of her. Hey Mom, it worked. You did good. Thank you. This book is dedicated to you and all of the women who raise sons and daughters in everyday life, whether they are your own or those of others.

The most inspiring person in my current (and future) life is my wife, Ruth Schowalter. In rapid succession, she supported and encouraged my writing *Life Traces of the Georgia Coast* (2013), *Dinosaurs Without Bones* (2014), and this book, making me love her all the more while also questioning our collective sanity. As our lives often intertwine with field trips, academic conferences, book readings, and otherwise, it is no wonder she makes so many appearances in this book. Her simple daily inquiry, "What were you writing today?" always provoked spirited discussion, and her feedback and advice often influenced my storytelling. She is also a visual artist and applied her considerable talents to create a lovely and evocative depiction of the burrowing *Lystrosaurus* heroines of Chapter 5. Thank you, Ruth. And I promise to not to write another book any time soon.

My literary agent, Laura Wood (FinePrint Literary Management), aided in making my previous book *Dinosaurs Without Bones* a critical and commercial success, which led to this one. When I first pitched *The Evolution Underground* as a book idea to her, though, she was a bit skeptical. After several lengthy phone calls, e-mail exchanges, and written outlines, it began resembling its present form, as she ably kept the book from branching off in more directions than a leaf-cutter ant nest. Laura also made the brilliant suggestion of reversing the chronology so that the first few chapters started in the present and went progressively farther back in geologic time: the deeper the time, the shallower the burrows. Thank you, Laura. (P.S.: I promised Ruth I wouldn't write another book any time soon, but you might be able to talk her out of it.)

Working with Pegasus Books again was a pleasure, returning to their familiar and less-than-painful procedures for writing, revising, and resubmitting a book manuscript, as well as the myriad of details needed to produce a book that's only an abstract concept until I'm holding a hardbound copy. Jessica Case especially went above and beyond the call of duty as my editor: While I was birthing this book, she birthed a new human. Her chapter-by-chapter feedback and helpful suggestions for improvement—while enduring my penchant for near-fatal doses of corny humor and pop-culture asides—turned a not-bad book into a much better one. Maria Fernandez (Pegasus Books) is also commended for her fine detail-oriented interior design of the book, and the evocative book cover is by Tim Green (FaceOut Studio).

Peer reviewers of the book manuscript supplied me with more protection than a U.S. government–approved bunker and greatly increased the odds that my science was sound and my writing was clear. Drs. Patricia (Tricia) Kelley (University of North Carolina-Wilmington), Andrew (Andy) K. Rindsberg (University of West Alabama), and Sally Walker (University of Georgia) were each outstanding in their own ways with their meticulous reading, fact-checking, and valuable suggestions for improving the original manuscript. Considering the vast expanse of time and sheer variety of burrowing animals mentioned in this book, their breadth of knowledge and educational expertise was necessary and valued. Rest assured, any errors remaining in the book are mine and mine alone: You won't hear any complaints from me.

Chapter 6 of this book revisits my graduate-school days at the University of Georgia (Athens), when I learned from one of the best ichnologists in the world, Robert (Bob) Frey. Bob died in 1992, but I fortuitously reconnected with his daughter Valerie and son Eric, both of whom live in Athens. Eric loaned me an extraordinary cast of a crustacean burrow made by one of Frey's graduate students on the Georgia coast in the 1970s, which I am holding in the photo on the jacket sleeve of this book. Thus I extend much appreciation to Bob Frey for his molding me into the unrepentant ichnologist I am today, and to Valerie and Eric as living (and cheerful) reminders of my academic legacy. Who took the photo of me with the burrow cast? That was by Lisa Streib, who I appreciate for making me look almost as interesting as the burrow.

Much of this book discusses burrowing animals of Georgia, which is not a coincidence because it is both my home and a place blessed with a rich natural history. Various people in Georgia and throughout the southeastern U.S. who contributed in some way to the field work mentioned in the book include (in alphabetical order): Craig Barrow, Gale Bishop, Tim Chowns, John Crawford, Jon Garbisch, Royce Hayes, Stephen Henderson, Jim Henry, Jenifer Hilburn, Jessica McGuire, Michael Page, Sarah Ross, Sheldon Skaggs, Lora Smith, Robert Kelley Vance, Gracie Townsend, and Lula Walker. Ann and Andrew Hartzell, both of whom have become among my most ardent "book groupies," hosted me at their lovely home in Savannah, Georgia while I wrote part of the book. Y'all have been far too kind.

Parts of the book take place in Australia, where I have been lucky enough to do research on and off for the past ten years. My childhood dream of traveling to the Land Down Under was made possible by Patricia (Pat) Vickers-Rich (Monash University), who graciously extended an invitation for me to visit and work with her there. Later, I worked with Thomas (Tom) Rich (Museum Victoria), and became friends with Mike Cleeland, Michael and Naomi Hall, Lesley and Gerry Kool, Dave Pickering, Peter Trusler, and Mary Waters. Looking forward to shouting a coldie (or two) for all of you next time I'm there.

A list of all of the ichnologists and paleontologists who have taught me so much about burrows and burrowing animals are too long to list, but here are a few who come to mind: Richard Bromley, Luis Buatois, Karen Chin, Carlos de Carvahlo, Al Curran, Tony Ekdale, Jorge Genise, Jordi de Gilbert, Murray Gregory, Stephen Hasiotis, Dan Hembree, Liam Herringshaw, Gabriela Mángano, Jack Matthews, Duncan McIlroy, Radek Mikuláš, Renata Guimarães Netto, George Pemberton, Roy Plotnick, Dolf Seilacher, Alfred Uchman, Trish Weaver, Lothar Vallon, Mariano Verde, Andreas Wetzel. My special thanks go to Dolf Seilacher for his inspiration as a consummate ichnologist and a fierce intellectual who encouraged engagement and discouraged complacency. Dolf died just before I started writing this book, which is really too bad because he would have vehemently disagreed with much of it, making for a terrific argument.

Lastly, I need to thank the generations of students I have taught at Emory University (Atlanta, Georgia), whether in classrooms bounded by walls or the ones I prefer outdoors. Teaching is a gift that gives back, as I often learn from my students while trying to pass on observations and lessons about the world around us and below. Whether we are peering into alligator dens, measuring ghost crab burrows, or walking together in dinosaur tracks, we acquire knowledge together, and once in a great while this knowledge results in wisdom. We could all use a little more of both.

Index

INDEX

Burrows, 1–14
Burton, Tim, 205
Butler, David, 68–69

C

Caecilians, 174–175
"Cambrian Explosion," 207
Cambrian Period, 152, 158, 182, 186, 190, 198, 202–212, 279
Camel crickets (*Ceuthophilus latibuli*), 64
Canada, 35–37, 183–193
Cappadocia, Turkey, 16–28, 32, 39–41, 249, 280
Carboniferous Period, 12, 13, 112, 128, 133, 163, 175
Carolina ghost shrimp (*Callichirus major*), 248
Carroll, Lewis, 204
Caves, 15–45
Cenozoic Era, 287–288
Chelonia, 74
Chemical fossils, 197–198
Cheney, Dick, 27
Chin, Karen, 223
Chinchillas (*Chinchilla*), 268
Chinese alligator (*Alligator sinensis*), 10
Chipmunks, 264, 268
Christmas Island, 244
Clepsydrops, 128
Climate change, 132–133, 136, 143, 253, 286–292
Cloudina, 203–204, 209
Coconut crab (*Bigurus latro*), 244–245, 289
Cold-blooded creatures, 9–10, 75–76, 170
Commensals, 58
Congo eels, 173–175
Conservation Biology, 252
Continental drift, 127
Conure, 105
Coober Pedy, Australia, 34–35, 173, 274
Copris beetle (*Copris gopheris*), 61
Coraciiformes, 111
Coypus (*Myocastor coypus*), 268
Crabs, 149, 159–161, 166, 173, 217, 236, 239–247, 289
Crayfish, 173–174, 233–238, 265
Cretaceous Period, 5, 12, 34–35, 72–78, 87, 93–96, 110–113, 138–144, 166–178, 181, 213, 223–238, 267, 275–277, 286–287
Crickets, 64
Crisafulli, Charlie, 266
Crocodiles, 4–5, 9–11
Crustaceans, 39–42, 91, 212, 233, 233–239, 251, 256
Curran, Al, 251
Cut Cay, Bahamas, 176–178
Cynodontia, 128–129
Cynodonts, 129–131, 138, 276
Cynognathus, 127

D

Damaraland mole-rat (*Fukomys damarensis*), 269–270
Darwin, Charles, 156, 213, 215, 220–225
Darwin, Emma, 220
Darwin, Horace, 221, 224
Decapods, 233–239, 242–250
Deer mice (*Peromyscus maniculatus*), 264, 268
Denning behavior, 92–93, 286, 289
Derinkuyu, Turkey, 21–24, 39–40
Desert rain frog (*Breviceps macrops*), 172
Desmatochelys padillai, 78
Devonian Period, 13, 29, 152, 162–166, 175, 180–181
Diamictites, 197
Diamondback rattlesnakes (*Crotalus adamanteus*), 58–59
Diapsids, 127, 178
Dickinsonia, 200
Dicynodontia, 128–129
Dicynodonts, 129–131, 138–139
Diictodon, 131
Dimetrodon, 128
Dinilysia patagonica, 181
Dinosaurs, 5, 12, 72–79, 81, 86–98, 108–113, 127–130
Dropstones, 196–197
Dung beetle, 61–62
Dzik, Jerzy, 206

E

Earhart, Amelia, 245–246
Earthworms, 151, 174, 215–225
Eastern diamondback rattlesnakes (*Crotalus adamanteus*), 58–59
Eastern indigo snakes (*Drymarchon couperi*), 58–59
Echinerpeton, 128
Ecological collapse, 71, 133, 136, 139, 241
Ectotherms, 9–10, 170
Ediacara Hills, 185–186
Ediacaran Period, 182–190, 194–198, 203, 209–213, 223, 253, 290
Ellesmere Island, 164
Emperor penguins (*Aptenodytes forsteri*), 82
Endotherms, 10, 130
Enigmatic fossils, 198
Eocene Epoch, 111
Estivation, 76–77, 124, 167, 171, 174, 222–223. *See also* Hibernation
Eunotosaurus, 74–75
European badger (*Meles meles*), 43
European bee-eaters (*Merops apiaster*), 104
Eusocial behavior, 228–232, 269, 273
Eutatus, 283
Exoskeletons, 158–159, 204–207, 237, 244–245
Extinction selectivity, 144
Eyes, development of, 148–149, 207–208

INDEX

INDEX

INDEX

INDEX